Advanced Studies in Theoretical and Applied Econometrics

Volume 47

For further volumes:
http://www.springer.com/series/5667

Władysław Welfe

Macroeconometric Models

Springer

Władysław Welfe
Institute of Econometrics
University of Łódź
Łódź, Poland

ISSN 1570-5811 Advanced Studies in Theoretical and Applied Econometrics
ISBN 978-3-642-44014-4 ISBN 978-3-642-34468-8 (eBook)
DOI 10.1007/978-3-642-34468-8
Springer Heidelberg New York Dordrecht London

Printed on acid-free paper

Springer is part of Springer Science+Business Media (www.springer.com)

To my son—Aleksander

Foreword by Lawrence R. Klein

The growth of econometric model building is known and used for understanding what has been taken throughout the academic and business activities the worldover. Many of the earliest studies grew as the subject of economics developed at academic centres and have become established in many countries; the earliest activities are now to be taken for granted and are still expanding.

In the remarkable expansion of the underlying subject is now covered over the world every day, in small investigations and have grown to many of the largest investigations of the meaning of economics, and the whole economic profession is thankful that Professor Władysław Welfe of University of Łódź, Poland has served the world's Academy with the beautiful subject for all readers of this book to digest.

Lawrence R. Klein

Foreword by Carlo D'Adda

My knowledge of Władysław Welfe goes back many years and is indirectly connected with our participation in the Project Link, an ambitious enterprise started at the beginning of the Nineteen Seventies and aiming at constructing a network of national econometric models, with the ultimate purpose to permit the simulation of the whole world economy. The project was guided by Lawrence Klein. When he was awarded the Nobel prize the Project Link was explicitly mentioned. I remember that Władysław Welfe joined a yearly meeting of Project Link organized in Perugia (Italy) and hosted by the Bank of Italy to give the participants, coming from many countries, the chance to exchange their results and the occasion to discuss problems of mutual interest. Professor Welfe was already a recognized master in his discipline, but coming from Poland in the days of Solidarnosc he was also the bearer of fresh news from the Eastern Europe that everybody was eager to know. There was much attention around him. Scientifically many of the participants had the opportunity of realizing that econometrics was not only interesting for the western researchers, but for their eastern colleagues as well. Some piece of the wall was already falling.

In the course of time Władysław told me much of his life. He spoke to me about the years spent in the obstinate and unsuccessful search for some news about the end of his father (a medical officer of the Polish military) during the second World War. At the same time he was pushed by the purpose of learning statistical methods and techniques applied to economic problems. Later on, when it became possible, Władysław went to the UK and became familiar with econometric research as it was practiced and thought in the western countries. In Oxford he came in touch with Lawrence Klein starting a dialogue and a friendship that are still going on. During his professional career Władysław has had the opportunity, and also the pleasure, of visiting centres for economic and econometric research in many countries of the world. Notably he played a relevant role in promoting the progress of his discipline in several East European countries, including Czech, the Slovak Republic and Bulgaria. In Europe his Institute of Econometrics and Statistics in Łódź has become a well known scientific centre to which many researchers look with interest. Awarding him the title of Dr.hc. the University of Łódź wanted to recognize his merits.

Introducing the present book, Macroeconometric Models, is an unexpected honour for me and also an intellectual task that would require an experience comparable to the one of the Author. This fruit of Władysław Welfe's last labour is a monumental work in the sense that he reviews, classifies and analyses an incredible number of contributions and publications. The reader of this book who is not an economist specialized in the field of macromodels may even not imagine that there are countries, in particular industrial countries, for which dozens and or hundreds of models of the national economy have been constructed over the years.

The prerequisite to build an econometric model of the whole national economy is the availability of adequate statistics. It must be born in mind that the existence of a sophisticated system of national economic statistics such as the one to which we are now used is the result of relatively recent efforts. When Keynes's *General Theory of Employment Interest and Money* was firstly published (1936) a satisfactory statistical data basis wasn't available and the model that appeared so appealing to understand the possibility for an economic system to be the prisoner of a depression was not susceptible of statistical estimation and test. Yet that very situation signalled the existence of a gap between demand for and supply of statistical macrodata. Over about a quarter of a century that gap was filled and since then much further progress has been done in the area of data collection.

But data were not the only problem. Almost in coincidence with the appearance of the Keynesian ideas the problem of an appropriate scientific methodology to test of economic theories (in analogy to what was happening since a century in the field of hard sciences) started to be needed. In 1933 a group of distinguished economists including R. Frish, J.A. Schumpeter and Keynes himself founded the Econometric Society "for the advancement of economic theory in its relation to statistics and mathematics". *Econometrics*, a new discipline with scientific location between economics and statistics, had come into existence.

A question comes natural at this point: do econometric models, in particular macro models, help discriminating among economic theories enabling us to separate false from true theories? The question is by far more complicated than a naïf reader of Welfe's book may imagine. As a matter of fact the Author of this book is careful in distinguishing the parenthood of the models he reviews and classifies but doesn't want to enter the philosophical problem of false and true theories. In the field of social sciences theories are always partial representations of reality and it happens that "if you make assumptions enough" most theories have a right to exist.

Along the way that has led to the biggest and more sophisticated models of either the national economies or of the whole world economy (often including hundreds of equations) and after the pioneering contributions of the fathers of modern econometrics, T. Haavelmo and J. Tinbergen, a fundamental step ahead is represented by the well-known Klein-Goldberger 1955 model (*An Econometric Model of the United States 1929–1952*). A phase of assimilation was necessary but after that, especially since the end of the Nineteen Sixties, the Klein-Goldberger model was recognized as the vanguard of a blooming of national and international macro models. Nonetheless it would be a mistake thinking that this fortunate research stream has developed without theoretical doubts and contradictions. The most relevant theoretical disputes in economics of the last decades of the nineteenth century, such as the monetary controversy (Friedman versus Keynesian school), as well as the upsurge of the supply side versus demand theories, let a clear imprinting on the stream of econometric macro models that progressively came to existence. But even more dangerously some theoretical attacks seemed so radical at their onset to possibly compromise the future of macro models. I am thinking of the Lucas critique (1976) focused on the connection between expectations and the set of possible economic policy rules, with the consequence to make prediction theoretically impossible. Have applied econometricians proceeded with their large scale models remaining dumb to the Lucas' critique? This may sometimes have been the impression, but it is not necessarily so. Practitioners like to say that the use of their large scale models is an "art" meaning that the way a model is run may allow for corrections to merely mechanical results. In their view (in the wait of the necessary theoretical improvements) intuition may to some extent remedy the shortcomings of the existing models. Think for example of the practice of presenting scenarios to which subjective probabilities may be attached. Twenty years after the Lucas' critique (1996) C. Sims, even if raising new doubts about the traditional (probabilistic) approach to identification, has considerably downsized the weight of the Lucas' critique. So large scale macro models survive and seem to enjoy still a prosperous season.

The reader of this book must be warned that in spite of the great number of contributions taken into account by the Author, he has been forced to do some choices as regards the macro models to consider with major emphasis. His choice has been an interesting one: priority goes to models constructed by institutions rather than by individual researchers. I think that this is a good decision because institutions build their models pushed by the need to solve their problems, such as analysing the likely effects of alternative policy decisions or advising governments about the size required by policy actions to be implemented. This is an important perspective from where to look at macro models: they serve decision making.

Carlo D'Adda

Contents

Part I The History in Outline

1 Introduction 3
References 7

2 The Origins of Macroeconometric Models 9
2.1 First Attempts of the Quantitative Description of National Economies 9
2.2 General Equilibrium Models 9
2.3 Models of Business Cycles 10
2.4 J.M. Keynes' Macroeconomic Theory 12
References 13

3 Macroeconometric Models of the United States and Canada 15
3.1 First Macromodels by L.R. Klein 15
3.2 The Klein-Goldberger Model 16
3.3 The Quarterly Macroeconometric Models. The Brookings Model . 18
3.4 Models Developed at Public and Commercial Institutions. Models Built by Individual Scholars or Groups of Scholars 21
3.4.1 The Wharton Models 21
3.4.2 The DRI Model 22
3.4.3 The Michigan Model 23
3.4.4 The BEA Model 23
3.4.5 The MPS Model 24
3.4.6 The Hickman-Coen Model 25
3.4.7 The Fair Model 25
3.4.8 The St. Louis Model 26
3.4.9 The Liu-Hwa Model 26
3.5 Models of the US Economy in the 1990s and Later 27
3.5.1 The DRI Model 28
3.5.2 The Quarterly WEFA Model 28

3.5.3 The Quarterly DRI-WEFA Model—The GLOBAL INSIGHT Model 29
3.5.4 The Quarterly Michigan Econometric Model and S. Hymans RSQE Model 30
3.5.5 The Current Quarterly Model 30
3.5.6 The FRB/US Model 31
3.5.7 The Fair MC Model 32
3.6 The Macroeconometric Models of Canada 33
3.6.1 The Models of the Government of Canada 34
3.6.2 The Rhomberg Model 34
3.6.3 The Bank of Canada Models 35
3.6.4 The University of Toronto Models 35
3.6.5 The CANDIDE Model 41
3.6.6 Other Models 41
3.6.7 Models Constructed in the 1990s and Later 42
References 42

4 Modelling Activities in Europe. Macroeconometric Models of the United Kingdom 47
4.1 Introduction 47
4.2 First Macroeconometric Models of the UK Economy 48
4.3 The Annual Multisectoral Model of the UK Economy 48
4.4 The Quarterly Models 49
4.4.1 The London Business School (LBS) Model 50
4.4.2 The National Institute of Economic and Social Research (NIESR) Model 50
4.4.3 The H.M. Treasury Model 50
4.4.4 The Bank of England Model 51
4.4.5 The Liverpool University (LPL) Model 51
4.5 From Keynesian to Neoclassical Specification 52
4.5.1 Supply Side and Production Functions 52
4.5.2 Modelling Labour Markets 52
4.5.3 Rational Expectations 53
4.6 Developments in the Last Twenty Years 54
4.6.1 The Oxford Economic Forecasting Model 54
4.6.2 Reconstruction of the LBS Model 54
4.6.3 The COMPACT Model 55
4.6.4 The Bank of England Models 55
4.6.5 ESRC Macroeconomic Modelling Bureau 59
4.6.6 The Long-Run Structural Macroeconometric Model 60
References 60

5 Models of the French Economy 63
5.1 Annual Models and Indicative Planning 63
5.1.1 FIFI Model 63
5.1.2 The Annual Forecasting INSEE Models 64

5.2 Annual Models of the Second Generation 64
5.2.1 The Annual DMS Model 65
5.2.2 The Annual GAMA Models 65
5.2.3 Other Annual Models 69
5.3 Quarterly Models of the French Economy 69
5.3.1 The METRIC Model 69
5.3.2 The PROTEE Model 70
5.4 Disequilibrium Models 70
5.5 Models Built in the Last Twenty Years 70
5.5.1 The MESANGE Model 71
5.5.2 The MANEGE Model 71
References 71

6 The Dutch Macroeconometric Models 75
6.1 Introduction. J. Tinbergen's Contributions 75
6.2 Annual Models of the Central Planning Bureau (CPB) 75
6.2.1 The Verdoorn-Koyck Model 75
6.2.2 The CS, VINTAF and FREIA Models 76
6.3 The Quarterly CPB Models 76
6.3.1 The Driehuis Model 76
6.3.2 The KOMPAS, FREIA-KOMPAS and FKSEC Models 76
6.4 The CPB Models in the 1990s and Later 77
6.4.1 The JADE Model 77
6.4.2 The ATHENA and MIMIC Models 77
6.4.3 The Model SAFE 82
6.5 The MORKMON Model of the Central Bank 82
6.6 The Models of the Academic Institutions 82
6.6.1 The SECMON Model 83
6.6.2 The Groningen University Models 83
References 83

7 The Models of the Nordic and Baltic Countries 85
7.1 Introduction 85
7.2 The Models of the Danish Economy 85
7.2.1 The ADAM Model 85
7.2.2 The SMEC and CLEO Models 86
7.2.3 The MONA Model 87
7.3 The Models of the Estonian Economy 87
7.3.1 The ESTMOD Model 87
7.4 The Models of the Finnish Economy 87
7.4.1 The KESSU Model 87
7.4.2 The BOF and QMED Models 88
7.5 The Models of the Lithuanian Economy 93
7.5.1 The LITMOD Model 93
7.6 The Models of the Norwegian Economy 94
7.6.1 The MODIS and MODAG Models 94

7.6.2 The Quarterly KVARTS Model 95
7.6.3 The Quarterly RIMINI Model 95
7.6.4 Modelling Inflation in an Open Economy 95
7.7 Models of the Swedish Economy 95
7.7.1 The KOSMOS Model 95
7.7.2 The STEP Model 97
7.7.3 The SNEP Model 97
7.7.4 The GUESS Model 97
7.7.5 The Riks Bank RIXMOD and SOE Models 97
References 98

8 Models of the Central and South European Countries 101
8.1 Introduction 101
8.2 The Models of the Austrian National Economy 101
8.2.1 The LIMA Model 101
8.3 The Models of the Belgian Economy 102
8.3.1 The MIRABEL Models 102
8.3.2 The Belgian National Bank Models 102
8.3.3 The NONAME Model 103
8.4 The Models of the German National Economy 103
8.4.1 The BONN Models 104
8.4.2 The SYSIFO Model 105
8.4.3 The RWI Models of Business Cycles 105
8.4.4 The Freiburger Model 106
8.4.5 The Bundes Bank Quarterly Model 106
8.4.6 Disequilibrium Models 106
8.4.7 New Quarterly Models 107
8.5 The Models of the Greek Economy 108
8.6 The Models of the Irish National Economy 108
8.6.1 The Bank of Ireland Model 108
8.6.2 The HERMES and HERMIN Models 109
8.7 The Models of the Italian National Economy 109
8.7.1 The PROMETEIA Model 109
8.7.2 The Bank of Italy Model 109
8.8 The Models of the Portuguese National Economy 116
8.8.1 The HERMIN Model 116
8.9 The Models of the Spanish National Economy 116
8.9.1 The Academic Centres' Models 116
8.9.2 The Bank of Spain Models 117
8.10 The Models of the Swiss National Economy 117
8.10.1 The CREA Model 117
8.10.2 The Quarterly KOF Model 117
8.10.3 The Quarterly Model of the University of St. Gall 118
8.11 The Models of the Turkish National Economy 118
References 118

9 The Models of the Middle and East European Countries 123
9.1 The Models of the Centrally Planned Economies 123
9.1.1 Introduction 123
9.1.2 The Models of the Czechoslovak National Economy 125
9.1.3 The Models of the German Democratic Republic 126
9.1.4 The Models of the Hungarian National Economy 127
9.1.5 The Models of the Polish Economy 128
9.1.6 The Models of the United Soviet Socialist Republics (USSR) 136
9.1.7 The Models of the Yugoslav National Economy 137
9.1.8 The Models of the Council of Mutual Economic Assistance (CMEA) Countries 137
9.1.9 The Models of the Centrally Planned Countries. Recapitulation 137
9.2 The Models of the Transition Economies 147
9.2.1 Introduction 147
9.2.2 The Models of the Bulgarian National Economy 148
9.2.3 The Models of the Czechoslovak National Economy 148
9.2.4 The Models of the National Economy of the Czech Republic 149
9.2.5 Models of the Hungarian National Economy 150
9.2.6 The Model of the Montenegrin National Economy 150
9.2.7 The Macroeconometric Models of the Polish National Economy 151
9.2.8 The Macroeconometric Models of Romania 155
9.2.9 The Macroeconometric Models of the National Economy of the Russian Federation 171
9.2.10 The Macroeconometric Models of the Slovak National Economy 174
9.2.11 The Macroeconometric Models of the Slovenian National Economy 176
9.2.12 The Macroeconometric Models of the Ukrainian National Economy 177
9.2.13 The Multicountry Models of the Middle and East European Countries 178
9.2.14 Final Remarks 180
References 181

10 The Macroeconometric Models of the European Union and the EURO Area 191
10.1 The Models of the European Union 191
10.1.1 Introduction 191
10.1.2 The EUROMON Model 192
10.1.3 The MAC SIM Model 192
10.1.4 The HERMIN Model 194
10.1.5 The GEM-E3 Model 194

10.1.6 The COMPACT Model 194
10.1.7 The Aggregated Model for the European Union 195
10.2 The Models of the EURO Area 195
10.2.1 Introduction 195
10.2.2 The Models of the European Central Bank. The AWM Model 196
10.2.3 The MZE Model 197
10.2.4 The MORANA Model 200
10.2.5 A Structural Model for the EURO Economy 200
References 200

11 The Macroeconometric Models of the Rest of the World Countries . 203
11.1 Introduction 203
11.2 The Macroeconometric Models of the Japanese National Economy 204
11.2.1 The Early Models 204
11.2.2 The Economic Planning Agency Models 204
11.2.3 The Bank of Japan Quarterly Models 205
11.2.4 The Macroeconometric Models of Academic Institutions . 205
11.2.5 The Macroeconometric Models of Commercial Institutions 206
11.3 Macroeconometric Models for the Far East Asian Countries 206
11.3.1 The Systems of Models for Far East Asian Countries 206
11.3.2 The Macroeconometric Models of the Indonesian National Economy 207
11.3.3 The Macroeconometric Models of the Philippine National Economy 208
11.3.4 The Macroeconometric Models of the South Korean National Economy 209
11.3.5 The Macroeconometric Models of the Economy of Taiwan Province 210
11.4 The Macroeconometric Models of the Australian and New Zealand Economies 211
11.5 The Macroeconometric Models of the Chinese National Economy . 212
11.5.1 The Early Models 212
11.5.2 The Models of the State Economic Information Centre . . . 212
11.5.3 The Demand-Oriented Models 213
11.6 Macroeconometric Models of India, Pakistan and Sri Lanka 213
11.6.1 The Early Macroeconometric Models for India 214
11.6.2 The Macroeconometric Models of the Indian Economy in the Late 1960s and 1970s 214
11.6.3 Later Macroeconometric Models of India 215
11.6.4 The Macroeconometric Model of Pakistan 225
11.6.5 Macroeconometric Models of the National Economy of Sri Lanka 225
11.7 The Macroeconometric Models of the North African and West Asian Countries 226

11.7.1 The Structure of the Models for the Oil Extracting Countries . . . 227
11.7.2 The Macroeconometric Models of the Iraqi National Economy . . . 227
11.7.3 The Macroeconometric Model of the Kuwaiti National Economy . . . 228
11.7.4 The Macroeconometric Models of the Egyptian and Iranian National Economies . . . 229
11.7.5 Modelling the Maghreb Countries . . . 230
11.8 Macroeconometric Models of the African Countries . . . 230
11.8.1 The Macroeconometric Models of the Ghanaian National Economy . . . 230
11.8.2 The Macroeconometric Models of the Kenyan National Economy . . . 231
11.8.3 The Macroeconometric Model of the Malawian National Economy . . . 232
11.8.4 The Macroeconometric Models of the Nigerian National Economy . . . 233
11.8.5 The Macroeconometric Models of the South African National Economy . . . 234
11.8.6 The Macromodel of the Togolese National Economy . . . 234
11.9 The Macroeconometric Models of the Latin American Countries . 235
11.9.1 The Macroeconometric Models of the Mexican National Economy . . . 235
11.9.2 The Macroeconometric Models of the Venezuelan National Economy . . . 236
References . . . 236

12 Macroeconometric Multicountry Models . . . 241
12.1 Introduction . . . 241
12.2 Project LINK . . . 242
12.3 The World Trade Models . . . 244
12.4 The OECD-INTERLINK Model . . . 245
12.5 The Models of the World Bank . . . 247
12.6 The Models of the International Monetary Fund (IMF) . . . 248
12.6.1 The MULTIMOD Models . . . 248
12.6.2 The GEM Model . . . 250
12.7 The Models of European Community. The QUEST Models . . . 250
12.8 The Models of the US Federal Reserve Bank . . . 261
12.8.1 The MCM Quarterly Models . . . 261
12.8.2 The FRB/WORLD Model . . . 261
12.9 The MEMMOD Model of the Deutsche Bundesbank . . . 262
12.10 The Models of the National Institute of Economic and Social Research (NIESR): GEM and NIGEM . . . 262
12.10.1 The GEM Model . . . 262
12.10.2 The NIGEM Model . . . 263

12.11 The Multicountry Model MIMOSA . . . 264
12.12 The Model FUGI . . . 265
12.13 The Model of Oxford Economic Forecasting . . . 267
12.14 The Multicountry Model PRIAMO . . . 267
12.15 The Multicountry MC Model by R.C. Fair . . . 268
12.16 Other Multicountry Models . . . 269
References . . . 270

13 The Epilogue . . . 275
References . . . 278

Part II The Structure of Macroeconometric Models

14 Introduction . . . 281
References . . . 283

15 Macroeconometric Models—The Classification . . . 285
15.1 Introduction . . . 285
15.2 Classifications of the National Economy in Macromodels . . . 285
15.3 Macroeconometric Models. Their Major Types . . . 288
15.4 A Stylized Structure of the National Economy Models . . . 291
15.5 Model Dynamics. Rational Expectations. Long- and Short-Term Relationships . . . 293
15.6 The Microeconomic Foundations of Models' Specification . . . 295
15.7 The Modelling of Supply . . . 296
15.8 The Stylized Version of the Macroeconometric Model Structure in the Last Twenty Years . . . 298
15.9 An Outline of Estimation Methods and Computational Techniques . . . 300
References . . . 301

16 Modelling the Behaviour of Households . . . 303
16.1 The Consumer Demand Function . . . 303
16.2 Equations Explaining Household Investment Demand . . . 310
16.3 Equations Explaining Labour Force Supply . . . 311
References . . . 312

17 The Models of the Enterprise Sector . . . 315
17.1 Introduction . . . 315
17.2 Determination of Output . . . 316
17.3 Equations Explaining the Demand for Production Factors and the Production Function . . . 317
17.4 The Demand Functions for Fixed Capital and Business Investment . . . 320
17.5 The Demand Functions for Employment and Working Time . . . 326
17.6 The Production Supply and Production Function . . . 330
17.7 Total Factor Productivity . . . 333
17.8 Equations Explaining Producer Prices . . . 337

17.9 Equations Explaining Average Wages 340
References 343

18 Modelling Disequilibria and Equilibria 347
18.1 Introduction 347
18.2 Modelling the Global Disequilibria 348
18.3 Modelling Demand of Public Institutions (Social Consumption) . . 350
18.4 Modelling Inventories 350
18.5 Modelling Adjustments in the Production Sector 352
18.6 Modelling Unemployment 353
18.7 Modelling Foreign Trade 356
18.7.1 Introduction 356
18.7.2 Equations Explaining the Demand for Exported Goods . . . 357
18.7.3 The Exports Supply Functions 358
18.7.4 The Imports Functions 359
18.7.5 The Equations Explaining the Supply of Imported Commodities 361
18.8 Price Systems 362
18.8.1 Introduction 362
18.8.2 Equations Explaining the Prices of Domestic Final Goods . 362
18.8.3 Equations Explaining Export and Import Prices 363
18.9 The Economic Mechanisms in the Real Sphere 364
References 368

19 The Modelling of Financial Flows 371
19.1 Introduction 371
19.2 The Models of Households' Incomes, Expenditures and Wealth Changes 371
19.3 The Modelling of Financial Flows of Enterprises 374
19.4 The Modelling of Public Finances 377
19.5 Financial Flows—The Interim and Outside Links 379
References 380

20 Modelling the Money and Capital Markets 383
20.1 Introduction 383
20.2 The Modelling of Credit Activities 383
20.3 Equations Explaining the Demand for Money and Financial Assets 385
20.4 Interest Rates 386
20.5 The Balance of Payments 387
20.6 The Exchange Rate Functions 388
20.7 Equations Systems 392
References 393

21 Prospects of Macroeconometric Modelling 395

Author Index 397

Subject Index 409

Abbreviations

3 SLS	three-stage least squares
FIML	full information maximum likelihood
GDP	gross domestic product
i.a.	inter alia
i.e.	it is
IIV	iterative instrumental variable
LIML	limited information maximum likelihood
ML	maximum likelihood
MPS	material product system
NA	national accounts
OLS	ordinary least squares
PPP	purchasing power parity
UIP	uncovered interest parity
TSLS	two-stage least squares

List of Figures

Fig. 15.1 Consumer multiplier and accelerator 294
Fig. 15.2 Inflationary loop 295
Fig. 18.1 Complete demand determined models. Real flows. Source: Klein et al. (1999, p. 113) 365
Fig. 18.2 Inflation. Source: Klein et al. (1999, p. 124) 366
Fig. 18.3 Complete supply determined models. Real flows. Source: Klein et al. (1999, p. 112) 367

List of Tables

Table 3.1 The models of Canada 36
Table 4.1 Models of the United Kingdom 56
Table 5.1 Models of the French economy 66
Table 6.1 Models of the Dutch economy 78
Table 7.1 Models of Nordic and Baltic countries 89
Table 8.1 Models of the Central and South Europe 110
Table 9.1 Models of Central and Eastern Europe in the period of centrally planned economies 138
Table 9.2 Models of the Central and Eastern Europe in the period of transition to the market economies 156
Table 10.1 Models of the European Union 193
Table 10.2 Models for the EURO area countries 198
Table 11.1 Models of the rest of the world countries 216
Table 12.1 Multicountry models 252
Table 16.1 Household consumption elasticities 305
Table 16.2 Long-term elasticities of personal consumption w.r.t. the Y, W, and R variables 306
Table 16.3 Parameter estimates of the consumption function for developing countries 307
Table 17.1 The parameter estimates of the investment demand functions . . 323
Table 17.2 The elasticity estimates of the investment demand functions. The 1974–1997 sample 324
Table 17.3 Estimates of the parameters of the employment demand functions. The 1974–1997 sample 328
Table 17.4 Output elasticities with respect to the average number of schooling years 336
Table 17.5 Parameter estimates of the producer prices equations 339
Table 17.6 The elasticities of average wages. The 1965–1992 sample 342

Part I
The History in Outline

Chapter 1
Introduction

The modelling of national economies has a long-standing and remarkable tradition. Filling the gap between macroeconomic theory and empirical descriptions of the functioning and growth of national economies, it has become the primary tool in the empirical testing of macroeconomic postulates that has allowed researchers to expand their knowledge of economic mechanisms, their stability and universality. Econometric modelling is based on systems of equations whose parameters are typically estimated using time series. As well as becoming the major instrument used for economic forecasting and policy analyses, thus generated models have provided public institutions, banks and large corporations with a decision-making basis.

The long history of macroeconometric model building can be split into several subperiods, depending on the changing goals, macroeconomic concepts, econometric and statistical methodologies and institutional conditions. The first attempt at describing the initial efforts of model builders can be found in the paper by Nerlove (1966). An excellent description of the first fifty years of model development can be found in the monograph "A History of Macroeconometric Model-Building" edited by Klein and his collaborators Bodkin and Marwah (1991). This publication may be regarded as a convincing proof that during that period macroeconometric modelling became an independent discipline among economic sciences, closely related to macroeconomics. The next major publication on the subject was the monograph "A Course in Macroeconomic Modelling and Forecasting" by Whitley (1994), which mainly laid emphasis on specification issues concerning macromodels' equations. Several monographs presented the surveys of the development of macroeconometric models for particular regions, e.g. Scandinavia. The catalogues of macromodels were collected and shown in the books by Uebe and Fischer (1992) and Uebe (1995). As regards the monographs on macromodelling methodology, two volumes of "Macroeconometric Modelling" edited by Wallis (1994) must be mentioned.

The description of the last nearly three decades of macroeconometric modelling as provided in this book has been mostly based on the papers and reports from modelling institutions. An invaluable source of information has been the journal "Economic Modelling" and its library of models. Being aware that this description

W. Welfe, *Macroeconometric Models*,
Advanced Studies in Theoretical and Applied Econometrics 47,
DOI 10.1007/978-3-642-34468-8_1,

frequently misses certain pieces of information, the author believes that incomplete information is better than none.

This part has been structured following the concept of the "A History..." editors. The history of macroeconometric modelling in particular countries will be described starting with the USA, then the European and other countries' achievements will be discussed, and finally the multicountry models will be presented. The emphasis will be laid on economic justifications underlying models' structures, equation specifications and their changes. The evolution of the associated econometric methodology will only be mentioned.

It is evident that trying to describe all macroeconometric models that have been built so far in particular countries would be futile. For instance, before 1989 more than 200 models were built for Western Germany (Heilemann and Wolters 1997), and Beltran del Rio (cf. Bodkin et al. 1991) made a list of 182 models built for Latin America between 1965 and 1985. This abundance made us select macromodels for presentation in this book. The key criterion for accepting a model was whether its history of continuous utilization was sufficiently long and whether it had a serious impact on the model building activities.

The way selected to characterise macroeconometric modelling activities that were developed in particular countries has an obvious disadvantage—it makes difficult for the reader to grasp the main macroeconometric modelling tendencies and their international propagation. Hence, a summary of these trends will be provided below.

The first models of the US economy were constructed by J. Tinbergen, and after World War II more models were built and inspired by L.R. Klein. The models were initially annual and drew on the Klein-Goldberger model (1955), but the quarterly forecasting models were to follow soon: the most elaborated of them was the Brookings model (Duessenberry et al. 1965). Because all the models were used in regular forecasting and policy simulations, the number of their equations increased to several hundreds. The subsequent disaggregation of activities most frequently linked the models to the input-output submodels.

These initially demand-oriented models remained neo-Keynesian over time. Only the production functions retained the neoclassical origin, but their main use was to generate employment. The general profile of the models was macroeconomic: they contained final demand (consumption, investment), demand for labour, as well as prices, wages and financial flows (Klein et al. 1999). The model builders followed the "Cowles Commission" methodology, using in the estimation process not only the OLS, but also the TSLS, IIV and FIML methods.

These "mainstream" models were constructed in the 1960s and 1970s in Western Europe and Japan and then in the other parts of the world. Particularly in France and Japan they were initially treated as auxiliary instruments supporting the planning of national economies. The centrally-planned economies (CPE) and developing economies (DVE) going through industrialization gave them a similar role. Later on they were primarily used in forecasting and policy simulations.

Their structure gradually evolved, following changes in the economic conditions and economic policy goals. In response to the oil shocks of the 1970s and 1980s, the models' supply sectors were enlarged, mainly by adding energy consumption as an additional variable in the production functions. The financial sector was more deeply elaborated and linked to the endogenization of activities conducted by central banks and administration (Shapiro and Halabuk 1976).

The CPE and DVE models were generally supply-oriented. The barriers to commodity supplies that emerged as a result of production factors' becoming less available because of price rigidities generated chronic excess demand. The processes were reflected in the disequilibrium models that were built based on the production functions, with production allocated according to special, somewhat arbitrary rules (Welfe 1992).

The first tendencies heralding a departure from the mainstream macroeconometric models appeared in the late 1970s. They developed because of the Lucas critique, who insisted that the economic agents could anticipate the economic policy measures, so the forecasts could be misleading (Lucas 1976). This induced a search for the possibility of using rational expectations that were treated as a remedy. Many models, especially those built in the USA and the United Kingdom, introduced rational expectations mainly into the equations explaining prices, wages and exchange rates. However, many modelling centres adhered to the concepts of adaptive expectations, as they believed that many economic agents (small firms, households) were not knowledgeable enough to behave rationally.

As a result of developing macroeconomic theory that showed strong tendencies to rest on solid microeconomic foundations, the structure and specifications of the macromodels' equations were deeply revised towards the neoclassical concepts. Firstly, the equation blocks distinguished in the macromodels were made to cover households' behaviour (not only their demand for goods, but also residential investments, labour supply and finance), enterprises (prices, supply, output, employment, wages) and finances, public institutions, money markets and the foreign sector (Fair 1984). Secondly, not only the production functions but also the demand functions were characterised by the neoclassical origin. The links between the real and financial sectors were strengthened to accentuate the transmission mechanisms. Macromodels of this orientation were widely applied, especially in the analyses of the effects of fiscal and monetary policies (Welfe and Welfe 2004).

The changes coincided with the Sims' critique, who insisted that the restrictions imposed on the structural models were "arbitrary" (Sims 1980). This paved the way for the use of the VAR and VAR-related models. What attracted researchers' attention afterwards was that economic theory was basically applied to static, long-run relationships in the macromodels, which were dynamized in an arbitrary manner. In the UK proposals were put forward (the LSE methodology) to use a "from general to specific" specification procedure (Hendry 1995). To estimate the short-term equations' parameters the error correction models (ECM) being a transformation of ADL equations were proposed. The techniques were initially applied within models explaining wages and prices (Sargan 1964) and consumer demand (Davidson et al. 1978). These developments provided a solid background, after new parameter estimation methods for equations containing non-stationary variables were proposed

(Johansen 1988). A two-step procedure based on the cointegration analysis was developed (Engle and Granger 1987). In this procedure, the parameters of the long-term equations and of the short term equations are estimated with the OLS and ECM methods, respectively. It has been commonly used in macroeconometric modelling since the late 1990s (Brooks and Gibbs 1994). The use of the VAR models attracted criticism, because the estimated parameters had no economic interpretation. In response, appropriate restrictions were imposed on the VAR models, which lead to their structural forms (SVAR).

In the last 20 years the construction and use of macroeconometric models has been the domain of public institutions and research institutes. Academic centres have concentrated their research activities on constructing small systems with several equations describing particular economic relationships (such as inflationary processes) (Welfe 2009). Owing to the new estimation methods based on the cointegration techniques it has become possible to estimate the parameters of small systems of equations only. Some suggestions have been recently put forward to start the construction of large macromodels by linking particular submodels piecewise (Bårdsen et al. 2005).

In the 1990s the World Bank' staff made first attempts to build the computable general equilibrium models (CGE) in order to create an instrument with which the financial situation of DVEs could be thoroughly investigated. The CGE models were then constructed in many other countries. Presenting the advantage of deep disaggregation, they suffered from being static.

The developments associated with the construction of the modern business cycle models gave a new impulse to the development of large macroeconometric models (Kydland and Prescott 1982). The empirical models are derived from neoclassical theories like the CGE models, but their equations are stochastic and dynamic, so they are called dynamic, stochastic general equilibrium models (DSGE). The parameters of the long-term equations based on economic theory are frequently calibrated. ECM is used when the short-term equations are to be estimated. The DSGE models may cover the entire national economy, but in practice their scope is limited. Although the DSGE models were mainly built at research centres associated with central banks, many economists maintain that the building and use of the DSGE model has opened a new stage in the development of macroeconometric models, where the results of academic centres' research are linked again with their applications and use by central banks and administration (Pagan 2003).

Acknowledgements This monograph has been based on the lectures that the author delivered to econometrics and statistics students enrolled at the Faculty of Economics and Sociology, University of Łódź, Poland. We wish to thank the students and staff of the Department of Econometric Models and Forecasts—W. Florczak, R. Kelm, and M. Majsterek—who led the conversatoria and made valuable suggestions. Special thanks are going to W. Maciejewski who read the whole manuscript of Part I and to numerous colleagues who reviewed particular chapters: G. Barabas, R.G. Bodkin, R. Courbis, C. D'Adda, S.G. Hall, B. Hickman, M. Olexa, K.F. Wallis, A. Welfe and P. Winker. Nedless to say, I am responsible for any errors and omissions in this book. Thanks are also due to J. Kwitecki, B. Strojecka, I. Szczepaniak for efficient help in the preparation of the manuscript.

References

Bårdsen, G., Eitrheim, Ø., Jansen, E. S., & Nymoen, R. (2005). *The econometrics of macroeconometric modelling*. Oxford: Oxford University Press.

Bodkin, R. G., Klein, L. R., & Marwah, K. (1991). *A history of macroeconometric model-building*. Aldershot: E. Elgar.

Brooks, R., & Gibbs, D. (1994). A model of the New Zealand economy, Reserve Bank Model XII. *Economic Modelling*, *11*, 5–86.

Davidson, J. E. H., Hendry, D. H., Srba, E., & Yeo, S. (1978). Econometric modelling of the aggregate time-series relationship between consumers expenditures and income in the United Kingdom. *Economic Journal*, *88*, 661–692.

Duessenberry, J. G., Fromm, G., Klein, L. R., & Kuh, E. (Eds.) (1965). *The Brookings quarterly econometric model of the United States*. Amsterdam: North-Holland.

Engle, R. F., & Granger, C. N. J. (1987). Co-integration and error correction: representation, estimation and testing. *Econometrica*, *55*, 251–276.

Fair, R. C. (1984). *Specification, estimation and analysis of macroeconometric models*. Cambridge: Harvard University Press.

Heilemann, U., & Wolters, J. (1997). Gesamtwirtschaftliche eokonometrische Modelle in der Bundesrepublik Deutschland: Erfahrungen und Perspektiven. *RWI-Mitteilungen*, *47*, 31–46.

Hendry, D. H. (1995). *Dynamic econometrics*. Oxford: Oxford University Press.

Johansen, S. (1988). Statistical analysis of cointegration vectors. *Journal of Economic Dynamics and Control*, *12*, 231–254.

Klein, L. R., & Goldberger, A. S. (1955). *An econometric model of the United States, 1929–1952*. Amsterdam: North-Holland.

Klein, L. R., Welfe, A., & Welfe, W. (1999). *Principles of macroeconometric modeling*. Amsterdam: North-Holland.

Kydland, F. E., & Prescott, E. C. (1982). Time to build and aggregate fluctuations. *Econometrica*, *50*, 1345–1370.

Lucas, R. J. Jr. (1976). Econometric policy evaluation. A critique. In K. Brumer & A. H. Meltzer (Eds.), *The Phillips curve and labour market* (pp. 19–46). Amsterdam: North-Holland.

Nerlove, M. (1966). A tabular survey of macroeconometric models. *International Economic Review*, *7*, 127–175.

Pagan, A. (2003). Report on modelling and forecasting at the Bank of England. *Bank of England Quarterly Bulletin*, pp. 1–29.

Sargan, J. D. (1964). Wages and prices in the United Kingdom: a study in econometric methodology. In P. E. Hart, G. Mills, & J. K. Whitaker (Eds.), *Econometric analysis for national economic planning* (pp. 22–54). London: Butterworth.

Shapiro, H. T., & Halabuk, L. (1976). *Macro-econometric model building in socialist and non-socialist countries. A comparative study* (Mimeo). University of Michigan, Dept. of Economics.

Sims, C. A. (1980). Macroeconomics and reality. *Econometrica*, *48*, 1–48.

Uebe, G. (1995). *World of economic models, a catalogue of typical specifications of economic models*. Aldershot: Averbury.

Uebe, G., & Fischer, J. (1992). *Macroeconometric models* (2nd ed.). Aldershot: Averbury.

Wallis, K. F. (Ed.) (1994). *Macroeconometric modelling* (Vols. I, II). Aldershot: E. Elgar.

Welfe, A. (2009) *Ekonometria. Metody i ich zastosowanie (Econometrics. Methods and applications)* (4th ed.). Warszawa: PWE.

Welfe, W. (1992). *Ekonometryczne modele gospodarki narodowej Polski (Econometric models of the Polish economy)*. Warszawa: PWE.

Welfe, W., & Welfe, A. (2004). *Ekonometria stosowana (Applied econometrics)* (2nd ed.). Warszawa: PWE.

Whitley, J. (1994). *A course in macroeconomic modelling and forecasting*. New York: Harvester Wheatsheaf.

Chapter 2
The Origins of Macroeconometric Models

2.1 First Attempts of the Quantitative Description of National Economies

The first attempts to provide a quantitative picture of the national economy viewed as a whole date from the turn of the 18 century. The French scientist, F. Quesnay, constructed then "economic tables" (tableau economique) with output, employment and other variables providing the quantitative characterization of the national economy. These schemes preceded in time the structures of empirical macromodels that were built almost 100 years later.

Another forerunner was K. Marx who discussed macroeconomic relationships in his Capital. Marx distinguished two schemes of simple commodity reproduction, extended commodity reproduction and analysed numerically the relationships between two divisions—one being production and the other the use of production—as well as the income distribution between the working class and the capitalists. These relationships were formalized much later, but this had little effect on economic growth theory.

The first attempts to formalize the description of the national economy as a whole took place towards the end of the 19th c. and in the early 20th c. Three trends in the literature could be distinguished then. The first one stemmed from general equilibrium theory developed by L. Walras, the second one rested on the foundations of business cycle theory laid by R. Frisch and M. Kalecki that encouraged J. Tinbergen to construct the first macroeconometric models, and the third one referred to J.M. Keynes' fundamental writings that resulted in the construction of the macroeconometric "mainstream" models by L.R. Klein (cf. Bodkin et al. 1991; Whitley 1994).

2.2 General Equilibrium Models

The concept of general equilibrium was formulated by L. Walras, whose theories were later developed by V. Pareto. According to the concept, a national economy

W. Welfe, *Macroeconometric Models*,
Advanced Studies in Theoretical and Applied Econometrics 47,
DOI 10.1007/978-3-642-34468-8_2,

can be described using a system of equations explaining the behaviour of particular economic agents. An economy is assumed to be purely competitive when in static equilibrium. The price-setting process is determined by equalizing demand and supply. As a result, the equations of the system explain both demand and supply and the (equilibrium) prices obtained are the system's solution. The system would have to contain several million equations, which seems as intellectually challenging as impractical. This property caused that it remained nothing but a theoretical proposal for many years (Friedman 1955).

However, more than 20 years ago a group of American economists led by D. Jorgenson proposed making a special aggregation of economic agents' activities. Particular brands or articles were to be replaced by their groups. An operationally manageable number of equations would explain the demand for and supply of several groups of commodities. Although the distance between theory and empirical applications turned out to be long, because not only the aggregation rules had to be defined, but also the respective data banks had to be established, the empirical computable general equilibrium (CGE) models were ultimately born (Wallis 1994).

2.3 Models of Business Cycles

The models for business cycle analysis were formalized between 1933 and 1935. Their major builders were Ragnar Frisch (Frisch 1933) and Michał Kalecki. M. Kalecki had built a system of equations explaining business cycles some 2 years before the fundamental works of J.M. Keynes became known. The system was published in Polish first and then in "Econometrica" in 1935, i.e. 2 years after it had been built (Kalecki 1935).

In the above models the investment functions were given a major role. In the Kalecki's model investments depended on the initial stock of fixed capital (replacement investment) and profitability represented by the interest rate. The investment function in the R. Frisch model was specified likewise. As a result, the models showed oscillations and cyclical behaviour. Both models were dynamic but remained theoretical, as no attempt was made to feed empirical data into them. The parameter values were calibrated, but not estimated.

The above concepts paved the way for the empirical macromodels built by Jan Tinbergen. In the years 1935–1936 Tinbergen constructed a macroeconometric model for the Netherlands, but it was in the Dutch language, so it took some time before the League of Nations asked him to build a macroeconometric model for the US economy. The advantage of the proposal was that rich databases on this country were available, mostly owing to the activities of S. Kuznets. J. Tinbergen complied and constructed a large model of the US economy, which was published in 1939 (Tinbergen 1939). Its followers were few, but because it was very dynamic it served as a point of departure for many models. The Tinbergen's model can be regarded as an effective attempt at employing the theoretical concepts developed by Frisch, Kalecki and Haberler to describe cyclical fluctuations in the dynamics of the US economy (von Haberler 1939).

The US model Tinbergen built was an annual, medium-sized model based on the 1919–1932 sample. It had 17 identities and 32 stochastic linear equations whose parameters were estimated with OLS. Most equations were recursive, because Tinbergen tried to avoid simultaneous equation biases that became apparent after T. Haavelmo published his writings (1943). Given that the only tool he could use was calculators his achievement is nothing short of impressive.

The model's equations can be grouped into 4 blocks explaining final demand, prices and wages, financial sector and income distribution. The major role in generating cyclical fluctuations was given to the equations explaining final demand. Consumer demand was represented by one equation, with nominal expenditures on the left-hand side and particular income components, i.e. wages and salaries, non-rural property incomes, capital gains and two deflators and a time trend, placed on the right-hand side. The first estimates of the marginal propensities to consume revealed considerable differences between particular items: 0.95 with respect to wages and 0.77 to non-rural incomes, but only 0.28 for gains. This decomposition of disposable incomes was repeated in many subsequent models, as long as the changes in social structures were regarded as essential.

The model had two investment functions to explain enterprise sector's demand for fixed-capital investment and the demand for investments in residential construction, respectively. Their specifications emphasised the availability of investment funds, ignoring the expectations of future growth (accelerator). As for the first equation, the enterprises had to decide what part of their corporate profits they would invest in plant and equipment and to take account of the lagged changes in price differentials between finished goods and investment goods, as well as wage costs and interest rates. The dynamization of this equation generated fluctuations. Developers made their decisions based on their expected profits, but they also considered the trends in prices and interest rates.

The model served as a basis for many analytical exercises. The computed multipliers could be used for monetary policy simulations, but not for fiscal policy, as its representation was scarce.

The Tinbergen's model was strongly criticized by J.M. Keynes in the late 1930s. Keynes claimed that the theoretical background was not sufficient and particularly that macroeconomic theories were impossible to test with econometric techniques (Keynes 1939, cf. Tinbergen's replay 1940). In the 1940s J.M. Keynes changed his mind, though, and acknowledged the importance of Tinbergen's contributions in two essays. J. Tinbergen did not have many followers.

It is worth noting, however, that in the last several years new business cycle theory has been developed, mainly by Kydland and Prescott (1982). These new concepts differ seriously from Tinbergen's. Tinbergen viewed cyclical changes as an immanent property of capitalist economies determined by investment activity changes. The new theories attribute oscillations to stochastic shocks that bring about recurrent changes, for instance as a result of rising oil prices.

2.4 J.M. Keynes' Macroeconomic Theory

The macroeconomic theory developed by J.M. Keynes, especially his "The General Theory of Employment, Interest and Money" (1936), became a cornerstone of the concepts that led to the construction of a class of macroeconometric models mainly based on L.R. Klein's works that predominated in the USA and Europe for over 30 years. Keynes analysed macroeconomic relationships applying to aggregates, but not to individuals. This means that they concerned total consumption, but ignored the consumption of particular households, showed total investments, but omitted investment projects of particular firms. This approach made the construction of macromodels much easier, because the number of interrelated variables could be substantially reduced and their disaggregation was performed only exceptionally.

Since Keynes, the main stress was laid on the analysis of demand, as a crucial element of macro analysis. Keynes stated that the national economy was not in equilibrium, i.e. the production factors were not fully utilized. Quite the contrary, a capitalist economy was characterised by a state of disequilibrium such that the final demand was not enough to ensure the full utilization of production capacity and full employment. Therefore, unemployment was long term and not exclusively frictional, as the classical theorists claimed. Consequently, macroeconomic analyses and economic policy started to stress final demand. As far as consumer demand analysis is concerned, Keynes defined a concept of marginal propensity to consume that became the integral component of the (linear) consumption function. He subordinated household consumer demand to GDP or national income and eventually to other factors. Investment activity was also determined by demand, but decisive were the factors determining its profitability, mainly interest rates in the capital markets. Keynes has strongly accentuated the role played by firms' and households' liquidity preferences that in the macroscale translate into preferences for maintaining global liquidity (Hicks 1937).

These concepts produced very significant and far-reaching effects on macromodelling. If demand plays a central role and no barriers exist to block it because labour force is still available in excess (unemployment) and productive capacity is not fully utilized, then it is justified to assume that the observations (the data) represent the realizations of demand. If so, they can be used to estimate the demand functions' parameters. Moreover, in the macroeconometric models the central role must be given to the final demand functions, i.e. consumer and investment demand functions. Because supply is assumed to follow demand, there is no need to construct any commodity supply equations. Final demand will indirectly determine the demand for production factors, including employment. To meet this goal, the model builders commonly used inverted production functions and compared the demand for employees with the available labour force to estimate the level of unemployment.

It is worth knowing that the economic policy that many European countries and the USA pursued in the late 1930s was intended to stimulate demand, especially by forcing public investments. The multiplier and accelerator mechanisms caused that the entire economy was stimulated. In our view, the most important thing about J.M. Keynes' concepts is that they significantly contributed to the expansion of macromodelling activities after the Second World War (cf. Klein 1951; Pesaran and Smith 1985; Bodkin et al. 1986).

References

Bodkin, R. G., Klein, L. R., & Marwah, K. (1986). Keynes and the origins of macroeconometric modelling. *Eastern Economic Journal, 12*, 442–450.

Bodkin, R. G., Klein, L. R., & Marwah, K. (1991). *A history of macroeconometric model-building*. Aldershot: E. Elgar.

Friedman, M. (1955). Leon Walras and his economic system. *American Economic Review, 45*, 1190–1201.

Frisch, R. (1933). Propagation problems and impulse problems in dynamic economics. In *Economic essays in honour of Gustav Cassel* (pp. 171–205). London: Frank Cass.

Haavelmo, T. (1943). The statistical implications of a system of simultaneous equations. *Econometrica, 11*, 1–12.

Hicks, J. R. (1937). Mr Keynes and the "Classics"; a suggested interpretation. *Econometrica, 5*, 147–159.

Kalecki, M. (1935). A macrodynamic theory of business cycles. *Econometrica, 3*, 327–344.

Keynes, J. M. (1936). *The general theory of employment, interest and money*. New York: Harcourt Brace & Co.

Keynes, J. M. (1939). Professor Tinbergen's method. *Economic Journal, 49*, 558–568.

Klein, L. R. (1951). The life of John Maynard Keynes. *Journal of Political Economy, 59*, 443–451.

Kydland, F. E., & Prescott, E. C. (1982). Time to build and aggregate fluctuations. *Econometrica, 50*, 1345–1970.

Pesaran, M. H., & Smith, R. (1985). Keynes on econometrics. In T. Lawson & M. H. Pesaran (Eds.), *Keynes' economics: methodological issues* (pp. 134–150). London: Croom Helm.

Tinbergen, J. (1939). *Business cycles in the United States of America 1919–1932, Part II of Statistical testing of business-cycle theories*. Geneva: Economic Intelligence Service of the League of Nations.

Tinbergen, J. (1940). On a method of a business-cycle research: a replay. *Economic Journal, 50*, 140–154.

von Haberler, G. (1939). *Prosperity and depression* (revised and enlarged ed.). Geneva: Economic Intelligence Service of the League of Nations.

Wallis, K. F. (Ed.) (1994). *Macroeconometric modelling* (Vols. I, II). Aldershot: E. Elgar.

Whitley, J. (1994). *A course in macroeconomic modelling and forecasting*. New York: Harvester Wheatsheaf.

Chapter 3
Macroeconometric Models of the United States and Canada

3.1 First Macromodels by L.R. Klein

The favourable conditions created within the Cowles Commission for Research in Economics in the late '40s encouraged L.R. Klein to construct macroeconometric models of the interwar US economy. His research took advantage of the newly developed econometric methods. Their results were published in "Economic Fluctuations in the United States 1921–1941" (Klein 1950). Klein made an attempt at exploring the possibilities offered by macroeconometric modelling with a view to describing an economy whose development was well recognized and for which rich data sets were available. As a result, three small annual models (versions) were built.

Model I was comprised of 6 equations: 3 behavioural and 3 identities. Their specifications were similar to those prepared for the Tinbergens' model, but they additionally contained some special elements that were developed in the subsequent models. Consumer demand was a function of wages, salaries and property incomes. Investments were determined by property incomes (current and lagged) and the initial stock of fixed capital only. No links to current or expected output were introduced. The model contained an equation explaining employment that was derived from the inverted production function. This model as well as its versions defined the fundamental macro-identity, i.e. national income, as being equal to the sum of consumption, investment and net exports. The stock of fixed capital at the period end was defined as its initial stock plus subsequent investments.

All versions explicated the major feedbacks that included a consumer multiplier, where consumption depended on national income and was one of the national income components. They also included a special type of accelerator: investment was dependent on the stock of fixed capital that expanded owing to investment activity. Another characteristic feature of Klein's early models was that their author took advantage of the new econometric techniques that were developed after the Haavelmo critique, namely, two-stage least squares (TSLS), full maximum likelihood (ML), and limited-information maximum likelihood (LIML). The last two methods and OLS were used for estimating the above models' parameters. This was the prototype application of effective estimation procedures to the simultaneous equation

W. Welfe, *Macroeconometric Models*,
Advanced Studies in Theoretical and Applied Econometrics 47,
DOI 10.1007/978-3-642-34468-8_3, © Springer-Verlag Berlin Heidelberg 2013

systems. Models II and III were more elaborate than model I. Model III that contained 12 behavioural equations is regarded as the precursor of the Klein-Goldberger model.

3.2 The Klein-Goldberger Model

In the mid-1950s, L.R. Klein and A.S. Goldberger built a new macroeconometric model of the US economy in Michigan. This new model, called the Klein-Goldberger model, was an annual structure. The two researchers shared the work between them: while Klein was responsible for specifying the equations, Goldberger was in charge of the estimation of their parameters. Goldberger tried the newly developed estimation methods for systems of simultaneous equations. As the application of full maximum likelihood was found to be too complicated numerically, limited information maximum likelihood (LIML) was used instead (Goldberger 1964).

The Klein-Goldberger model paved the way for the builders of many other medium-term, annual models of the US economy. Many researchers tested its static and dynamic properties and its sensitivity using multiplier analysis and stochastic simulations. The contemporaneous econometricians are usually familiar with the model's structure and properties, because everyone involved in model building learns about them at some point in the process.

The Klein-Goldberger model covering the pre-war period 1927–1941 and the post-war years 1944–1952 included 20 equations, 15 of which were stochastic. The use of the pre-war data was necessary because only 9 observations were available for the post-war years and parameter estimation with a very small sample seemed rather risky (Klein and Goldberger 1955).

The Klein-Goldberger model had a Keynesian orientation and emphasised the specification of demand components, starting with final demand. Consumer demand was dependent on real disposable income. However, households' incomes were disaggregated by income source. The propensities to consume seemed to differ depending on whether the households mainly relied on earned income, capital gains or incomes from agricultural activities. The estimates confirmed this hypothesis—the short-term marginal propensity to consume was 0.55 for real wages and salaries, but for real after-tax incomes of entrepreneurs and for farmers' incomes it was 0.40 and 0.34, respectively, in the latter case the reason being farmers' inclination to save money and invest in the development of their farms. For the first time consumption was dependent on liquid assets being a proxy for households' financial wealth. No variable was introduced to account for competitive savings that could constrain expenditures (in the future, this role was to be given to interest rates). Because total expenditures were an explained variable—the population size was introduced to serve as an additional explanatory variable. It is noteworthy that the models built in the next years omitted this variable, because increasing real incomes absorbed the effect of population growth. Lagged consumption was also introduced into the model. A low estimate (0.26) of this variable's parameter pointed to much higher estimates of the long-term parameters; for instance, for the real earned incomes the marginal propensity to consume would be 0.74.

The investment function was specified like that in Klein' previous model. Investment demand depended on real financial resources, their major source being firms' after-tax profits and depreciation. The profits were determined by the level of firms' activity. Because it was assumed that the investment processes consumed only a fraction of firms' profits, the rest being spent on the financial assets, the actual shares had to be estimated. Another assumption was that no hard barriers impeding the fund-raising processes existed, because credit lines stayed open to firms.

This specification differs from the specifications where current investments are related to the expected demand for products that could be made owing to new investment projects. The attempts to introduce these determinants proved unsuccessful. The variables representing investment project profitability also turned out to be insignificant. The initial stock of fixed capital was introduced as an additional explanatory variable. A negative value (−0.07) of the variable's parameter estimate indicated, rather surprisingly, that firms having substantial production capacity invested less than others.

The transition from final demand to demand for production factors followed the solution that Klein proposed in his earlier model. Demand for labour was obtained from the inverted production function. The number of hours worked was related to output and the initial stock of fixed capital.

Real earned incomes in the enterprise sector were a function of national after-tax income and employees' payments received from the budget. The Klein-Goldberger model was the first to explain wage rates assuming that their growth depended on the rate of unemployment. This was three years before Phillips formulated this hypothesis in his seminal paper (1958). In the Klein-Goldberger model the first differences of the wage rate logs depend on the rate of unemployment and the lagged first difference of price logs. Hence, in the American literature this specification is credited to Klein or Klein and Phillips.

The import equation in the model was new and its specification has become almost a classical solution. Imports were dependent on national income, lagged imports and relative prices. The latter were not treated as an individual explanatory variable, but introduced together with national income: the real national income was multiplied by the ratio between national income deflators and imports. Hence, elasticity with respect to such defined variable represented the joint income and substitution effects.

The Klein-Goldberger model included also 4 equations for the monetary sector explaining households' and enterprises' liquidity preferences, as well as the short-term interest rates.

The Klein Goldberger model emphasises the role of the nominal incomes. Households' incomes that depend on the level of economic activity are deflated and affect consumer demand that codetermines real national income representing the levels of economic activities. This defines a special multiplier. Similarly, a specially defined accelerator mechanism functions, which takes account of incomes, but omits commodity flows. National income determines firms' incomes that, after appropriate deflators are applied, turn into real investments that enter into the volume of national income.

There is also a feedback concerning imports in the model. Imports depend on the level of economic activities and given exports they negatively affect the level of national income. If exports grow less than imports, the growth rate of national income decelerates. Therefore, the Klein-Goldberger model takes account of all types of internal feedbacks characteristic of macromodels, even though they often run through different variables.

The Klein-Goldberger model was analysed from all possible angles. The list of researchers who participated in these exercises is much too long to be quoted *in extenso*. Let us mention, however, Adelman and Adelman (1959) who tested the dynamic properties of the model with stochastic simulations, introducing random shocks. Their study showed that the model had the desirable dynamic properties—it was not explosive but had stabilizing tendencies as regards the rates of growth. They concluded that the model could be the prototype solution for the future model builders.

Indeed, many new annual macroeconometric models of the US economy were constructed in the subsequent years, such as the long-term model (1869–1953) built by Valavanis (1955). Then Duessenberry, Eckstein and Fromm constructed their model (Duessenberry et al. 1960), which was used for analysing recessions. Suits (1962) developed an annual Michigan model comprising 33 equations (16 stochastic), whose parameters were estimated with OLS using first differences. The model was mainly used for forecasting purposes. Successive models were built in almost all developed countries.

The Klein-Goldberger model was generally utilised in the numerous studies on the pre-war and post-war properties of the U.S. economy that extensively made use of the multiplier analysis. Its applications to forecasting the annual time series were less frequent and attracted criticism. The critics maintained that the short-term forecasts and analyses would be of greater practical value for the administration and large corporations, as their having access only to the annual forecasts greatly reduced the range of reactions and interventions they could undertake. These demands obviously called for the construction of the quarterly, if not monthly, macromodels (Bodkin et al. 1991).

3.3 The Quarterly Macroeconometric Models. The Brookings Model

In response to this criticism, L.R. Klein began to work on a quarterly model covering the post-war years. As a result, the first short-term macroeconometric model of the U.S. economy "A Post-war Quarterly Model (1948–1958)" was built in the early 1960s. It became a stepping stone for the construction of the subsequent quarterly models of the U.S. economy. The model was initially small and contained 37 equations (29 were behavioural). The sample started in 1948 and covered seasonally adjusted quarterly data. The parameters of the model equations were estimated with TSLS and LIML (Klein 1964).

Because of its structure, the quarterly model resembled its annual predecessor. There were some differences, though; unlike the annual model where many equations were expressed in terms of nominal values, the quarterly model mostly used variables given in constant prices. Besides, its equations explaining consumer demand and investment demand were extended. Consumer demand was broken down into demand for durables (automobiles), non-durables and services. The changing demand for durable goods induced changes in the activities in the business cycle. Klein made an attempt to include variables representing consumers' and investors' expectations and extended equations explaining inventory changes.

Additionally, Klein introduced the concept of potential output and the rate of its utilization. This helped explain the impacts of market tensions, especially in the equations that determined prices that in the former models depended on wage costs only. The rate of capacity utilization was obtained as a ratio between actual and potential output. Potential output was determined by linking outputs generated in the business cycle peak quarters that represented the points close to full capacity utilization. This ratio has become known as the Wharton School Index of Capacity Utilization. It was used for many years before it was replaced by an official indicator.

The model became operational in 1961 and started producing forecasts. In the next years, when Klein engaged himself in the building of a new, large quarterly model, it was transferred to the Bureau of Economic Analyses at the US Department of Commerce. Frequently modified and substantially enlarged, it was used in regular forecasting service for a long time. It still functions under a new name—as the BEA model being a large quarterly model extended to several hundred equations. Its structure is largely the same as its predecessor's, but an extensive block of equations has been added to cover the public and monetary sectors.

At the same time, T.C. Liu constructed an experimental quarterly model (Liu 1963). It was fitted to the 1947–1951 data and had 36 equations, 19 of which were stochastic. The parameters were estimated with the OLS and TSLS methods. Its structure was similar to the Klein's model, but employment was not covered and foreign trade was exogenous. It was used in forecasting, but mainly for generating simulations analysing the effects of fiscal and monetary policy.

In the early 1960s, L.R. Klein and J.S. Duessenberry initiated a new, major project in the Brookings Institution. The quarterly model which was built in the next years was given the name of the Brookings Model. It was the very first case in the history of econometric macromodelling that a model was constructed not by a single person or a group, but by a team of outstanding scholars from different universities who came together to carry out a project. The project's chairmen were L.R. Klein representing the University of Philadelphia and J.S. Duessenberry from the Harvard University. The project attracted many contributions on equation specifications, estimation methods and model use. The research work and the numerous discussions it prompted led to the publication of three volumes of the project proceedings (Duessenberry et al. 1965, 1969; Fromm and Klein 1975). The striking originality and depth makes the contributions relevant to date.

The theoretical foundations of macromodelling that were formulated in the course of the project have remained outstanding in many respects. An excellent case in point is the investment demand function proposed by Jorgenson (1965). Investment was to be determined not only by the available financial sources, but also by the user's investment costs. The 'user cost of investment' variable he constructed was a ratio between the expected return on investment and investment costs, the interest rate being the most important item among the latter. It was for the investors to decide whether they wanted to borrow paying the current interest rate or rather exchange their assets for bonds and stocks. The construction of the user cost indicator was a complex task, as the indicator had to account for the differences between the prices of the products being sold and of the necessary equipment. To deal with the requirement, the ratio between investment deflators, including the interest rate, and the deflator of producer sales was adopted to represent user costs. This concept has remained useful to date.

The Brookings model had to be large, because its mainly sectoral disaggregation was meant to satisfy the needs of various customers. It originally comprised more than 200 equations, nearly 150 of which were stochastic. With more data becoming available, the model was enlarged to more than 400 equations. The data spanned the years 1949–1960 and were seasonally adjusted.

Final demand was deeply disaggregated. Consumption was subdivided into 5 groups—cars and other durables, non-durables, foodstuffs and beverages, and others. The specifications of the consumer demand functions were similar to those used in the earlier Klein's model, but consumer demand was determined by aggregated real disposable income, relative prices and lagged consumption. Special variables were introduced into some equations; for instance, the equation explaining cars and other durables was given interest rates and credit availability. In addition to real incomes (mainly profits), the investment function also contained user investment costs, including interest rates.

The fairly large sample allowed dynamizing the model's equations. The lags were partly estimated, but, if there were a large number of them, for instance 8–10, special lag distributions were introduced, such as the hypergeometric distribution, assuming descending weights. The polynomial distributions, especially the Almon distribution, became popular in the later periods (Schmidt and Waud 1973).

The model did not explicitly define the production function. After inversion, it was used to generate demand for working time. The function explaining working time contained the level of output and its increase, the initial stock of fixed capital and a lagged endogenous variable. Labour supply was determined by the level of economic activity and its increase, and by demographic variables. The equations explaining wages were built following the Klein-Phillips rule, i.e. prices, labour productivity and unemployment rate were used as the wage determinants. The price equations took into account the profits and unit costs, the latter being mainly composed of labour costs. The model contained fairly elaborated monetary and public sectors and made an attempt at endogenizing the current expenditures of the federal budget.

Owing to the project-related research, two special methodological problems could be solved. Firstly, the strategy of estimating the parameters of large simultaneous equation systems was formulated. According to the strategy, the equations had to be linearized to allow the application of the known numerical procedures and then rearranged to make them block-recursive. The equation parameters within the block could be estimated separately afterwards, assuming that the relationships between the blocks were negligible (Fisher 1965). The theoretical underpinnings belong to Wold (1960), who maintained that in fact all economic activities were recursive and only their aggregation in time made them simultaneous. The above procedures were of historical significance, because the use of the Gauss-Seidel iterative simulation and the Newton procedures for solving large systems became common in the not-so-remote future; this new approach did not require model linearization.[1]

The second problem that was solved was the integration of the input-output relationships into the model. The project participants formulated formal relationships between final output and gross output, on one hand, and between producer prices and value added, on the other, thus making a permanent contribution to the I-O methodology.

The Brookings model was frequently used in policy simulations, not so often in forecasting (Fromm and Klein 1975). Its construction and development significantly influenced further evolution of the macroeconometric models built in the US modelling centres and the contributions it made to macromodelling found broad applications. The project was discontinued in 1972.

3.4 Models Developed at Public and Commercial Institutions. Models Built by Individual Scholars or Groups of Scholars

By the end of 1960s and at the beginning of 1980s several macroeconometric models of the US economy were constructed, their group including the MPS model with an extended financial sector for the Federal Reserve, the BEA model, the St. Louis model, models built at commercial institutions, such as the Wharton, DRI and Michigan models, as well as models created by particular scholars, for instance the Hickman-Coen, Fair, and Liu-Hwa models (Kmenta and Ramsey 1981). Their structures and properties will be briefly characterized below.

3.4.1 The Wharton Models

The Wharton models were built in Philadelphia at the Wharton Econometric Forecasting Associates (WEFA), under the direction of L.R. Klein. The Quarterly Wharton Model being partly a continuation of Klein's Post-war Quarterly Model and the Brookings Model (Evans and Klein 1967) became operational in 1963. From that

[1] This concept returned quite recently with the attempts to construct large models composed of separate recursive blocks, with parameters estimated using the cointegration techniques (Bårdsen et al. 2005).

year regular quarterly forecasts of the American economy and the results of numerous policy simulations were published and presented to the community of the WEFA project's members (representing major corporations and public administration). The numerous mutations of the model increased its size and multiplied its uses. For instance, its 1968 version had approximately 80 equations, while Mark III contained 200 (McCarthy 1972). By the early 1980s the number of the equations rose to nearly 1000 and by the end of the 1980s it was close to 750, 280 of which were stochastic.

The Wharton models had a Keynesian structure. Final demand determined output and, indirectly, employment. The investment function used Jorgenson' neoclassical concepts, though. An increase in inventory was introduced to balance demand and supply. The models generated potential output by means of the Cobb-Douglas production function and its use by applying the Wharton Capacity Utilization Index. The price equations were based on the 'mark-up on unit costs' rule. To explain wages, the Klein-Phillips' approach was used (Duggal et al. 1974).

The models were very dynamic and utilised many polynomial lag distributions. Its Anticipation Version (Adams and Duggal 1974) included consumer and investor expectations, the first type of expectations being represented by the Michigan Index of Consumer Sentiment and the second by the BEA Survey of Investment Intentions. The models were chiefly used for forecasting purposes, but also in simulation analyses, i.a. for computing the multipliers showing the likely effects of fiscal policy. They become members of Project LINK (Waelbroeck 1976).

Almost 10 years were needed for a large, medium-term, annual macroeconometric model containing an I-O subsystem to be built (Preston 1972, 1975). It was called the Wharton Annual and Industry Forecasting Model and initially consisted of nearly 400 equations, but in the 1980s it was expanded to more than 3000. The model covered all areas of economic activity, which were grouped into 8 blocks of equations. It had two submodels of I-O equations. One of them determined the inter-industry relationships between 50 or 63 industries and the other defined price relations. Several attempts were made to endogenize the parameters of the relevant matrices; in the last version of the model they depended on the relative prices. The model was used in medium-term forecasting (up to 10 years) and in numerous multiplier exercises.

The Wharton models were constantly operated until 2001, when they were merged with the DRI model being part of the forecasting and simulation system called Global Insight.

3.4.2 The DRI Model

Built in 1969, the DRI model was a quarterly model constituting the centrepiece of a large information system operated by Data Resources, Inc. that assembled the time series on the US economy. It was re-estimated every year under the direction of

Eckstein (1983). The model followed from an earlier model that Duessenberry, Eckstein and Fromm had built (1960) to study recession, and referred to the Brookings model.

The model had approximately 700 equations, around 400 of which were stochastic, and distinguished 7 blocks that covered the entire economy. The largest of them was the industry block (representing more than half of the equations). Its equations were specified differently than in its predecessors. The Cobb-Douglas production function was extended by including energy use and R&D expenditures. This helped to explain the recession of the 1970s. The flow of funds sector was considerably enlarged. Thus the transmission from the financial to the real sector could be carried out through interest rates, share and bond prices as the channels.

The model was systematically used in short-term forecasting, with increasing accuracy. It also participated in numerous policy simulations, especially those dealing with the oil shock impacts. The counterfactual analyses of the Great Recession that were made with the model earned it high reputation (Eckstein 1978).

3.4.3 The Michigan Model

The Michigan Quarterly Econometric Model (MQEM) was built in the 1970s (Hymans and Shapiro 1974). In 1974 it had 59 equations, 35 of which were stochastic, so it was the smallest among the quarterly models. The MQEM replaced the Suits model and shared some characteristics with the BEA and MPS models built in the 1960s. The model specifically assumed that in the long-run a trade-off takes place between the rate of unemployment and the rate of inflation.

In the 1980s the MQEM was systematically developed and used in research. In the mid-80s it had 111 equations, 61 of which were stochastic, and an elaborated financial sector. The decomposition of the model's final demand was fairly deep. The consumer demand for durables was disaggregated into four groups, the most important of them being the demand for cars. The enterprise sector's demand for investment was split into buildings and machinery and equipment for industry, agriculture and other branches.

Employment was derived from an inverted production function and the rate of unemployment was determined not from an identity but using a stochastic equation.

When the forecasting precision of the MQEM model and the Fair's model were compared it was found that MQEM was the most accurate in predicting prices, wages and the rates of unemployment and the least reliable for investment (Fair and Alexander 1986).

3.4.4 The BEA Model

As it has already been mentioned, the BEA model followed of Klein's "Post-war Quarterly Model". It was the major instrument that the Bureau of Economic Analyses (BEA), an agency of the US Department of Commerce, used to prepare forecasts and economic analyses (Liebenberg et al. 1966). The model was systematically developed—its initial version having 49 equations in 1966 (36 were stochastic) was expanded to 117 equations in 1971, 67 of which were stochastic (Liebenberg et al. 1971). In the next years, their number increased even further (Hirsch et al. 1974).

The model's structure remained basically the same. Seven blocks of equations covering the real and financial sectors were formed. The equations explaining final demand had standard specifications. Separate equations explained the demand for employees and working time, an important element of a business cycle analysis. The producer prices mainly accentuated cost pressures. The model was used for short-term forecasting and in many policy simulations, such as analyses of the multiplier effects of fiscal policy (Hirsch 1972).

3.4.5 The MPS Model

The construction of the MPS model (MIT, PENN and SSRC) was undertaken in 1966 by a team led by A. Ando and F. Modigliani supported by F. de Leeuw representing the Fed (de Leeuw and Gramlich 1968). Because the model was designed for the Fed to carry out its stabilization policy, a strong stress was laid on the specification of the financial sector. The model became operational in 1970 and its original version had 60 stochastic equations (Rasche and Shapiro 1968). The number of the equations increased to reach more than 170 in the mid-70s, 67 of which were stochastic. Their parameters were estimated with OLS and were fitted to the post-war quarterly data (Ando et al. 1972).

The model's structure distinguished 6 blocks of equations. The most sophisticated of them was the financial sector that had links with other sectors. In the short-run, the model had a Keynesian orientation: final demand determined output and employment, wages were specified following the Phillips curve, and prices were given as unit costs plus a mark-up. In the long-run, though, the dynamics of the output and production factors was shaped according to neoclassical theory of growth.

The per capita consumer demand was determined based on the real disposable income per capita (the long-term marginal propensity to consume was 0.61) and the real wealth per capita, allowing for long lag distributions. A distinction was made between consumption and consumer expenditures. The investment demand was determined, i.a. by user costs. Separate blocks were used to explain employment and unemployment and, very carefully, prices and wages.

As its main purpose was to meet the needs of monetary policy analyses, the model contained elaborate blocks of financial sector. A special block of equations explained the generation of taxes and transfers. The federal budget expenditures were maintained as exogenous, but the expenditures made by the states and local authorities were endogenized. The monetary sector was linked to the real sector via three channels: interest rates affecting user investment costs, households' financial wealth and by rationing the loans for residential investments.

The model was constantly improved and modified. To allow for the repercussions of the 1974 oil shock the production function was extended with energy use introduced as an additional explanatory variable. The equations explaining energy and foodstuff prices were specified. In the 1970s the system of equations explaining the balance of payments was extended by an attempt at endogenizing the exchange rate. More improvements were made to the model in the 1980s and in the early 1990s. The equations explaining money demand, including M2, were respecified, the equations explaining changes in the federal reserve were dropped, and the system of equations explaining residential building loans was augmented.

The MPS model was actively used during the numerous analyses of stabilization policy effects. One of its major applications was to verify the hypotheses put forward by the monetarists, which frequently presented the role of the monetary policy instruments in a simplified manner (Ando et al. 1972). It was exploited until 1995, when it was substituted by the FRB/US model of the second generation.

3.4.6 The Hickman-Coen Model

In the mid-70s, B. Hickman and R. Coen constructed an annual model for medium-term analyses and forecasts spanning the horizon of at least 10 years (Hickman and Coen 1976). The parameters of the model equations were estimated with OLS and TSLS based on data from the pre-war and post-war years, i.e. 1926–1940 and 1949–1965, respectively. The model had 170 equations, 50 of which were stochastic. The demand equations were Keynesian, whereas the supply side had a neoclassical orientation.

The model generated two versions of potential output, one assuming full employment and the other full utilization of fixed capital. This allowed calculating the rate of productive capacity utilization. The first version was used more frequently, partly because of more available estimates of full employment. The model used the Cobb-Douglas production function and carefully verified the assumption about constant returns to scale. The model was frequently applied in simulation analyses, such as the scenario analyses of the long-term development of the US economy (Coen and Hickman 1985). Systematically improved in the subsequent years, the model was applied for the last time in 2006, when its authors published the results of their research on the dynamics of potential output, productivity and resource utilization (Coen and Hickman 2006).

3.4.7 The Fair Model

In the early 1970s, R.C. Fair constructed a quarterly model, which was initially intended to support short-term forecasting. With its 19 equations, 4 of which were stochastic, and a limited number of exogenous variables, it was the smallest operational model of the US economy (Fair 1971). Its structure followed the setup of the standard models. Its author's main goal was to improve forecasting accuracy, so he frequently re-estimated the equations, instead of fine-tuning the model. After some time, Fair encouraged the customers to use the model as they deemed appropriate (Fair 1974, 1976).

The Fairs' model was systematically extended and improved. As a result, the number of its equations grew to 128 in 1984, 30 of which were stochastic. Most of them concerned the behaviour shown by households (9) and firms (12). The extended model was used in many policy simulations and for testing economic hypotheses. Most important was the result that expectations are not model consistent in the financial sector too. This result coincided with the view that economic agents are constrained in using the available information, especially the information generated by macroeconometric models (Fair 1984).

At the beginning of the 21st century the Fair's model became the major component of the MC, a multicountry model to be characterized in the last chapter of this monograph (cf. Fair 1994).

3.4.8 The St. Louis Model

In the early 1970s a small monetary model comprising only 9 equations (5 were stochastic) was built at the Federal Reserve Bank of St. Louis (Andersen and Carlson 1970, 1974). The equations' parameters were estimated with OLS. Initially, the 1955–1968 quarterly data were used to this end, but then their range was extended to the year 1973. The model had a special structure. The nominal GNP was dependent on money supply and high-employment expenditures of the federal budget, allowing for lags. Prices were determined with excess demand and anticipated changes in prices, so real GNP was determined from the identity. Potential output was exogenous and served to establish excess demand, as well as being used to calculate the rate of unemployment. Besides, two equations explaining the long- and short-term interest rates were introduced into the model. Although the forecasts it produced were found not to be very accurate, the simulation exercises yielded interesting results regarding the monetary policy issues (Elliot 1985).

3.4.9 The Liu-Hwa Model

Constructed in the mid-70s, the Liu-Hwa model was the first monthly model of the US economy (Liu and Hwa 1974) that partly built on the Liu model (1963). It was a recursive model with many long lags. Its initial version had 33 equations, 16 of which were stochastic, but in the late 1970s their number increased to 131 (51). The sample consisted of the 1954–1971 monthly data that were used for OLS-based parameter estimation. The model was structured following the quarterly Wharton and BEA models. Its most important feature was the extended set of equations explaining inventory changes. A substantial role was given to the equations explaining price fluctuations that depended on labour costs including the wage rate changes. Although it was treated as an experimental model, it was used in many interesting simulation exercises.

3.5 Models of the US Economy in the 1990s and Later

New tendencies in macroeconometric modelling manifested themselves in that the new macromodels were abandoning the traditional structure of the mainstream model. This partly followed from the Lucas criticism, who maintained that the economic policy decisions based on the models' findings could be inaccurate, because the economic agents supposedly anticipated their outcomes (the parameter estimates were biased) (Lucas 1976). Therefore, expectations should be recognised as rational. It was shown, though, that forecast errors were not significant (Eckstein 1983). Another circumstance that makes this hypothesis look somewhat far-fetched is that most economic agents do not have the model-generated information at their disposal (model consistent forecasts) (Fair 2004). It has become a common practice, though, to introduce rational expectations into models, especially when the monetary phenomena were being explained. Pertinent estimation procedures have been elaborated, operable owing to the high speed computers (Taylor 1993). Provided with appropriate fiscal and monetary policy rules, this "new" class of models can be expected to become the major analytical instrument of the administration (Taylor 2001).

Many scholars emphasised that macroeconomic relations must have strong microeconomic foundations. A development of neoclassical concepts followed. It was assumed that the economic agents—households and enterprises—accomplish intertemporal optimisation of their behaviour under monopolistic competition. This resulted in the deep respecifications of the consumer and investment demand equations in many models, one of them being the FRB/US model.

In response to the Sims' critique (1980) new estimation procedures were tried. The reduced forms of equation systems with no restrictions, i.e. VAR and related techniques, were consequently applied.

The recognition that there exist the long-run stable (equilibrium) relationships that are different from short-term adaptations led to the application of the error correction models (ECM). Afterwards, the cointegration relations were tried, but their use was restricted to the small-size submodels (Brayton and Tinsley 1996).

The above directions determined the development of the models constructed at the Federal Reserve Board (FRB/US) and partly of the Fair model. Interestingly, the centre of gravity of macroeconometric modelling shifted from academic centres to commercial and public institutions, especially to the Fed. This change can be exemplified by the forecasting activity and policy simulations run by Global Insight that merged the Wharton and DRI models, as well as the Michigan model.

3.5.1 The DRI Model

Over the many years of its being in use, the DRI model was systematically improved and extended, becoming a renowned forecasting instrument utilised by the Energy Information Administration and other organizations (Documentation 1993).

The DRI model was one of the largest quarterly models of the US economy (Eckstein et al. 1974; Eckstein 1983). Its US 89 version contained nearly 1000 equations, more than 300 of which were stochastic. The reason for the model to have that large number of equations was the deep disaggregation of final demand (into consumption and investment demand, including residential investment and foreign trade). Its extended system of the I-O equations took account of over 100 industries. At the beginning of the 1990s, an aggregated version of the model called a Personal Computer Input-Output Model (PC 10) was built. It served as a point of departure for sectoral analyses, mainly those dealing with production and energy use.

The structure of the DRI model followed the standard mainstream models. Its Keynesian-type equations explaining final demand were highly disaggregated. Final demand was used to determine the demand for industrial output using transfer matrices based on I-O. The model had an extensive block of equations explaining supply. Potential output was generated from the Cobb-Douglas production functions that were extended by introducing additional explanatory variables. Energy use and technological progress represented by total factor productivity were dependent on the cumulated real expenditures on R&D.

Producer prices were determined by the unit costs of labour, materials, energy and import prices, as well as a mark-up dependent on the rate of capacity utilization. The model included an extended sector of financial flows and broadly used expectations that were identified from surveys. They were entered into the equations explaining consumer demand for durables, investment and interest rates.

The model was systematically used in forecasting and simulation analyses of potential energy threats, etc. In 2001, it was merged with the WEFA model.

3.5.2 The Quarterly WEFA Model

During the decade of the 1990s, the quarterly WEFA model that had already been in use for many years was frequently updated and improved. Its restructuring was necessitated by the deep modifications made to the way production and price indices were calculated in the system of national accounts (NIPA). The chain indices the BEA introduced in 1996 changed the GDP's time series and components, as well as their deflators. The respective equations of the Wharton model were respecified and its new version was called MARK 11 (Bachman et al. 1998).

MARK 11 was a very large model, because its builders decided to disaggregate final demand down to a very low level. It had 25 groups of consumer expenditures, 22 investment categories, 9 categories of administration expenditures, 10 export and

11 import commodity groups. Besides, it included an I-O submodel and a large block explaining financial flows.

The model was used in short-term forecasting (up to 2 years) and in economic analyses. It had a Keynesian orientation: consumer demand was determined by permanent real income and expectations, investment demand responded to the accelerator and user costs. Federal expenditures were exogenous in the short-run. Unlike the previous versions, the MARK 11 model also made it possible to conduct the medium-term analyses (5–10 years ahead), mainly focused on the monetary policy issues. An important role was given to the equations explaining wages that depended on the expectations and the ratio between the observed unemployment rate and the (exogenous) natural rate of unemployment. The model was also useful for long-term analyses (15 years or more). Potential output was generated from the Cobb-Douglas production function where technological progress was treated as exogenous. The rate of capacity utilization codetermined prices, thus affecting final demand and consequently modifying the rates of growth.

The equations' parameters were estimated with the ECM model, taking into account the cointegration relationships. The parameters of the long-term (equilibrium) equations were estimated before the parameters of the short-term adjustment equations. Owing to this approach, the use of distributed lags (mainly of the ALMON type) could be given up. The model used information on expectations derived from various surveys.

The model was used in systematic forecasting and during numerous policy simulations, distinguishing itself for its ability to provide many details. In 2001, it was merged with the DRI model.

3.5.3 The Quarterly DRI-WEFA Model—The GLOBAL INSIGHT Model

The quarterly DRI-WEFA model came into being after the DRI and WEFA models were merged in 2001, as mentioned above, so its size and structure resembled its predecessors' (DRI-WEFA 2002). Its more than 1200 equations resulted from deep sectoral disaggregation.

The model had a mixed orientation. In the short-run, the Keynesian specification prevailed—output and employment were determined by final demand. In the medium and long-term, though, it had a neoclassical orientation. The financial sector was subordinated to the monetarist views. Expectations were mainly based on the survey results. The supply sector equations showed that endogenous growth theory was involved in their specification. Potential output (GDP) was generated from the extended Cobb-Douglas production function. The explanatory variables represented the following production factors: fixed capital, full employment and energy. TFP growth depending on the cumulative real R&D expenditures linked to investment stood for the impact of technological progress. The model's characteristics allowed its authors to conclude that it presented a macroeconometric model of dynamic balanced growth.

The model was and is still used to perform short and medium-term forecasts of the US economy; it also supports numerous policy simulations conducted within the Global Insight system.

3.5.4 *The Quarterly Michigan Econometric Model and S. Hymans RSQE Model*

In the 1990s the aforementioned Michigan model (MQEM) was significantly extended. In 1994 it had 213 equations (108 were stochastic) grouped into 7 blocks (Howrey and Hymans 1995). Its structure was eclectic too. Being Keynesian in the short-run, in the long-run it reflected the neoclassical concepts; the specification of the equations representing the money market was treated likewise. It defined a long-term relationship by linking the rate of inflation with the rate of money growth.

The model helped perform numerous policy simulations and quarterly forecasts with a 2–3 year horizon, which were known for their high accuracy. It was actively used until the Soul Hymans RSQE model replaced it very recently.

3.5.5 *The Current Quarterly Model*

Several years ago L.R. Klein entered into a cooperative effort with the Global Insight system in order to build a special quarterly model for short-term forecasting and analyses of the US economy at the University of Pennsylvania.

The model is special in that it uses monthly data on 75 indicators, which cover both real and financial activities of economic agents in the USA. Its equations link the major components of the quarterly production accounts and income national accounts (NIPA) with the selected indicators. The quarterly forecasts generated by the model are updated on a weekly basis after the monthly data become available (Klein et al. 2007).

It is worth emphasising that the GDP forecasting method has been meticulously designed. The forecasts are prepared taking into account (1) the final demand making use of the bridge equations, (2) incomes and (3) the result of GDP regression on the main components of the key indicators.

With the application of the monthly data to short-term forecasting of the current and future quarters, short and ultra-short forecasting started to develop in many other countries, especially in France (cf. Klein 2009).

3.5.6 The FRB/US Model

The short-term quarterly model of the second generation, FRB/US, was built in 1996. Its construction was inspired by the developments within economic theory and econometric methods, among which the applications of the intertemporary optimization of economic agents' behaviour and the introduction of rational expectations should be counted among the most important (Brayton et al. 1997).

The initial version of the model contained more than 250 equations, 40 of which were stochastic, whose parameters were estimated with ECM. Its structure differed from the previous models, because the equation blocks were distinguished not by sectors, but they explained the behaviour of households, enterprises, public institutions and foreign agents.

The model had a mixed orientation. In the short-run, production and employment were determined by effective demand. Prices and wages were assumed to be sticky and adjusting only in the long-run. The long-run economic growth was determined by the supply side factors. Rational expectations had a decisive effect on dynamic adjustments.

The consumption functions were built on the life-cycle hypothesis, meaning that the long-run consumption depended on the expected real disposable income, personal wealth (decomposed by kind) and the rate of capacity utilization, while the short-run consumption was determined by the lagged and expected consumption and current real incomes. For the household block of equations, investment functions for consumer durables and residential building were built.

The enterprise block equations were specified based on profit maximization under imperfect competition. The demand for fixed capital was determined by the level of output in the long run and by a lagged and expected increase in fixed capital in the short run. Appropriate investment functions were constructed, which separately treated machinery and equipment. The long-run investment depended on the level of output and depreciation, negatively on user costs (with calibrated parameters), and on the estimated output increase. For the short-run investment lags and leads were introduced, as well as lagged financial resources of corporations.

The model had an extended block of equations explaining financial flows. The block's equations took account of the arbitrage equilibria. The model builders assumed that adjustments in this sector were immediate, unlike in the real sector, where adjustments were costly and so taking a long time.

The above trends in macroeconometric modelling made some of the modelling centres conclude that the long-term relationships should be specified separately, using mainly the theoretical results of the optimization of economic agents' behaviour, as mentioned above. In this way, the parameters in the long-run equations, for instance those explaining consumer demand and investment, could be calibrated referring to the relevant economic theory, which would facilitate the construction of independent submodels. This approach brought the long-term structure of macroeconometric models closer to that characterizing the CGE or DSGE models. It was characteristic of the structure of the "hybrid" models, such as the Central Bank of Canada and the Bank of England (BEQM) models.

It is noteworthy that a small forecasting Vector Error Correction model was also built at the Federal Reserve Bank of St. Louis. It had 6 equations explaining GDP and deflators and variables representing the money market; 4 cointegrating vectors were distinguished in it. The accuracy of its forecasts did not diverge much from the accuracy of other forecasts (Anderson et al. 2002).

3.5.7 The Fair MC Model

The quarterly model of the US economy constructed by R.C. Fair is a vital component of a multicountry world economy model containing 39 countries (Fair 2004). It is the most recent version of the Fair's models built in 1984 and 1994. High productivity and low inflation that characterised the US economy after 1994 caused that Fair has used the MC model in many simulation analyses.

The structure of the model is founded on several assumptions that Fair formulated in his earlier monographs: (1) macroeconomic relationships should result from microeconomic relations, (2) the prevalence of disequilibria in particular markets must be acknowledged, (3) in the financial flows sector several restrictions must be observed (Fair 2004, p. 6).

The model further assumes that households and firms make decisions to solve the optimization problems. Firms function in a monopolistic competition environment. Prices and wages do not necessarily clear the markets. In the commodity markets disequilibria manifest themselves through inventory changes and in the labour market via changing unemployment rates. The main source of the disequilibria is forecasting errors that develop from the non-rational expectations formed by economic agents. Fair used the model to test the rational expectation hypothesis—the results were negative.

The backbone of the model's financial sector is interest rates. They change because of changing inflation and unemployment rates.

The above assumptions have been reflected in the specification of the model's equations. It has 31 stochastic equations. Their parameters, fitted to a sample beginning in 1954, have been estimated with TSLS. The variables have been assumed to be trend-stationary.

The households' block has the following equations: the consumer demand functions for durables, non-durables and services, the investment functions for apartments and residential buildings, and the labour supply functions split by worker gender and age.

As for consumer demand, it is dependent on net real income, real personal wealth and nominal short-term and long-term interest rates that have proved superior to the real interest rates. Besides, three age distribution characteristics have been used. The equations explaining the demand for durables have been additionally enhanced with the stock of durables and depreciation, under the assumption about gradual adaptation of demand. The equation explaining residential investment has been specified likewise.

To estimate the labour force supply, the coefficients of active population are explained separately for men and women aged 25–54 years and others. The coefficients' fluctuations are explained in terms of variations in personal wealth and unemployment rate (discouraged worker hypothesis). Both impacts have been negative. In the equations explaining the supply of female and other workers real wages have been additionally introduced to show the positive effect of substitution.

In the enterprise sector, organizations' decisions follow intertemporary profit maximization. The demand for domestic production has been determined from the difference between final demand and imports. The resulting volume of sales, adjusted by inventory changes, has been used to determine the volume of output. An increase in output determined the desirable relative increase in fixed capital. The effective increase in fixed capital has been linked to its desired increase, the excess stock of fixed capital and user costs, taking into account appropriate lags and leads. It determines investments in the enterprise sector after allowing for depreciation.

The desired increase in employment depends on output increase, allowing for the estimated employment reserve. It determines the actual size of employment from the equation of adaptive expectations. Separate equations have been constructed to explain the average number of the paid regular and overtime hours worked.

An important part of the enterprise sector is the equation explaining producer prices (deflators). It is of mixed specification. Price levels (expressed in logs) depend on the lagged prices and wages adjusted for potential changes in labour productivity and import prices that represent unit costs. The prices are also determined by the rate of unemployment that has been introduced to represent market pressures; this approach has turned out to be more efficient than using the estimates of the demand gap. At the end, a time trend is included. This specification differs from the standard approach in that the equation explains price levels instead of relative price changes.

The equations explaining average hourly wages have been given a special structure. In the long run, the ratios between hourly wages and potential labour productivity have been linked to current and lagged prices (in logs), lagged endogenous variable and time trend. The impacts of the tensions in the labour market have been found statistically insignificant.

The model has a standard imports equation. It also has many equations explaining the changes in the money market. The equations determine the money demand and interest-rate fluctuations. Most of the equations are dynamic, including the carefully tested lags and leads.

The model has supported regular forecasting activities and numerous policy simulations. For instance, it has been used for analysing the properties of the "new economy" and the possible results of inflationary shocks, and for evaluating the efficiency of different interest rate setting rules and many others (Fair 2004).

3.6 The Macroeconometric Models of Canada

In Canada and the USA macroeconometric modelling developed almost in parallel. In Canada, among the first scholars to engage themselves in this activity were

M. Brown and R. Rhomberg. Models were built in the research departments run by public administration units and the Central Bank of Canada, but also in academic centres, for instance at the University of Toronto (de Bever et al. 1979). Their characteristics have been summarised in Table 3.1.

3.6.1 The Models of the Government of Canada

The first model of the Canadian economy was built under the leadership of T.M. Brown at the Department of Trade and Commerce already in 1947; in the same year it was used for short-term forecasting. It was a small annual model that contained 13 equations, 6 of which were stochastic. Its successive versions were similar. Among the 15 equations of the IIIE model 7 were stochastic. Its parameters were fitted to the data from the years 1927–1940 and 1946–1949 and estimated with OLS and LIML. The model's structure resembled the structures of the early Klein and Klein-Goldberger models. Its special feature was that the consumption function used the concept of habit persistence developed by Brown (1952).

The next versions of the model had more equations. For instance, the IX model (1959) had 49 equations, the XIV model built by May (1966) using the 1927–1941 and 1946–1961 sample included 69 equations (11 were stochastic), and the XVI model built by Kuiper (1970) had 90 equations (20). In the 1980s, the Kuiper's model was replaced with a large quarterly model called a Quarterly Forecasting Model that in 1985 included 668 equations (196 were stochastic). The model incorporated an elaborate financial sector and was used for regular forecasting and in numerous policy simulations.

3.6.2 The Rhomberg Model

The small quarterly model built by Rhomberg (1964) has gained large importance. Its 19 equations (17 were stochastic) were fitted to the 1952–1959 data and their parameters were estimated with LIML. The equations explaining the USD/CanD exchange rates were provided in two variants, distinguishing respectively the floating and non-floating regimes. The model also contained equations explaining US investments in Canada. It was used in numerous policy simulations.

The next quarterly model was constructed by Officer (1968). It had 108 equations (50 were stochastic), whose parameters were estimated with TSLS. The sample covered the years between 1951 and 1962. The model accentuated the openness of the Canadian economy. It was regarded as the predecessor of the RDX models developed at the Bank of Canada.

3.6.3 The Bank of Canada Models

A series of quarterly macroeconometric models RDX (Research Department Experiments) has been built at the Bank of Canada. The RDX 1 model constructed by the team led by J.F. Helliwell (Helliwell et al. 1969) had 110 equations (50 were stochastic). It was fitted to the 1952–1965 data and estimated with OLS and IIV. The equations were tested using dynamic simulations.

The RDX 2 model that was used in the 1970s and 1980s was an important contribution to the development of Canadian macromodelling. Its first version built in the early 1971 had 258 equations (141 were stochastic) and starting from the second half of that year as many as 516. The parameters were estimated using OLS (Maxwell 1978). The model was decomposed into 21 blocks of equations, including production functions. It was used for forecasting purposes and in many policy simulations. Its successor was the somewhat smaller RDXF model designed for short-term forecasting that had 403 equations around 1980, 208 of which were stochastic.

In the mid-1990s a new quarterly macroeconometric model of the Canadian economy QPM (the Quarterly Projection Model) was constructed, which replaced the RDX models (Black et al. 1994; Poloz et al. 1994). Its structure differed from the earlier "mainstream" models in that it distinguished four main groups of economic agents and the associated markets. These were households, enterprises, public institutions and foreign agents. For each group of economic agents the long-term (steady-state) equations were constructed that together formed the SSQPM submodel. The equations were specified based on the relevant theoretical assumptions. Households' behaviour was explained in terms of overlapping generations theory developed by Yaari (1965) and expanded by Blanchard (1985). Enterprises maximized profits under monopolistic competition. After solving the above optimization problems, four main stock variables could be determined: households' financial wealth, enterprises' capital, public debt and net foreign assets. The stock variables determined the flow variables, i.e. consumption, savings, investment, public incomes and expenditures, etc. The parameters of the long-term equations were mostly calibrated; as a result, the structure of the submodel became more similar to the CGE models.

The results of the long-run specifications were used in the QPM model in the ECM procedures that made it possible to introduce short-term adjustments, including the appropriate lags and leads (Coletti et al. 1996). Appreciated by the research centres of many central banks, this special model structure was used as the underpinning of the "hybrid" macromodels constructed in the UK, New Zealand and Sweden (Duguay and Longworth 1998). The model found applications in short-term forecasting and monetary policy simulations.

3.6.4 The University of Toronto Models

The Institute of Policy Analysis at the Toronto University earned an outstanding position among the Canadian model builders. Between 1968 and 1979, the Insti-

Table 3.1 The models of Canada

Model Version	Institution Author	Frequency Sample	Equations total stochastic	Decomposition	Estimation method	Properties	Applications	Multipliers impulse sustained after 10 years
Models of the Government								
I 1947	Trade & Commerce T.M. Brown	Quarterly	12 6	–			Forecasts 1947–1948	
III E 1950		Quarterly 1927–1941 1946–1949	15 7		OLS LIML	Consumption function habit persistence	Short term forecasts, multiplier analysis	1.15
IX 1959		Quarterly	49 12	8 blocks	TSLS			
XIV 1963	Dep. of Finance S.J. May	Quarterly 1927–1941 1946–1961	69 11	Disaggregation of consumption	OLS TSLS	Instruments of financial policy	Forecasts, policy simulations	
XVI 1970	Dep. of Finance J. Kuiper	Quarterly 1927–1941 1946–1968	90 20		OLS			
QFS 1985	GRADY ECONOMIC, Dep. of Finance	Quarterly	668 196			Financial sector	Forecasts, policy simulations	

Table 3.1 (Continued)

Model Version	Institution Author	Frequency Sample	Equations total stochastic	Decomposition	Estimation method	Properties	Applications	Multipliers impulse sustained after 10 years
Rhomberg Model 1964	R.J. Rhomberg	Quarterly 1952–1959	19 12		LIML OLS	Constant and flexible exchange rates	Simulations, multipliers	1.5 (flexible exchange rate)
Officer Model 1968	L.H. Officer	Quarterly 1951–1962	108 50		TSLS		Simulations	1.1–1.7
RDX Models	Bank of Canada							
RDX 1 1969	J.F. Helliwell et al.	Quarterly 1952–1965	110 50	5 blocks	OLS	Verified using dynamic simulations	Simulation analyses, multipliers	
RDX 2 1971	T. Maxwell	Quarterly 1957–1965	258 141	21 blocks	OLS	Production function	Policy simulations, forecasts	1.4
After 1975		1950–1972	516		OLS			
RDXF 1980		Quarterly	403 208		OLS	Forecasting model	Forecasts	
TRACE 1968–1979	Institute for Policy Analysis, University of Toronto J.A. Sawyer	Annual 1947–1971				Neo-Keynesian in 1968–1979, in the Project LINK	Medium-term forecasts 5–10 years, policy simulations	

Table 3.1 (Continued)

Model Version	Institution Author	Frequency Sample	Equations total stochastic	Decomposition	Estimation method	Properties	Applications	Multipliers impulse sustained after 10 years
MARK I 1968 MARK III 1976 MARK IV E			183 44	14 blocks	OLS		Forecasts	1.9 0.5
QFM 1972–1979 Quarterly Forecasting Model		Quarterly	ca. 75				Forecasts up to 8 quarters	1.4 0.8
Focus 1979 (Forecasting and User Simulations)		Quarterly 1965	250 180			Extended financial sector block	Forecasts	1.4 0.8
Focus-CE		Quarterly				Consistent expectations	Forecasts	
CANDIDE 1970	Economic Council of Canada R.G. Bodkin, F.T. Denton, R.S. Preston	Annual 1955–1971	2049 616	8 blocks 2 blocks I-O	OLS	Links with the USA economy	Medium-term forecasts 5–10 years	1.7 0.5
1.0 1977					OLS			
1.1					OLS			
1.2 M					OLS			
2.0 1979					OLS	Extended financial sector	Numerous multiplier analyses	
TiM 1984			ca. 3900		OLS			
3.0 1984–1985			ca. 8000					

Table 3.1 (Continued)

Model Version	Institution Author	Frequency Sample	Equations total stochastic	Decomposition	Estimation method	Properties	Applications	Multipliers impulse sustained after 10 years
TSURUMI 1970 1970–1972	H. Tsurumi	Annual	86 52	4 sectors	TSLS nonlinear LS	Links with the USA economy via prices and wages	Policy simulation	
MOP-CAP 1978	K. Marwah	Quarterly 1954–1974	22 10		TSLS		Policy analyses	
DRI Canada 1977 2001	Data Resources Inc. of Canada R.M. Hyndman	Quarterly 1961–	515 180 700	10 blocks 2 blocks I-O	OLS	Cooperation with DRI of USA	Forecasts and simulation analyses	
AERIC 1976	Conference Board of Canada	Quarterly 1960–1975	151		OLS	Neo-Keynesian strongly dynamic	Short term forecasts up to 8 quarters	
MTFM 1987 2001	Conference Board of Canada	Quarterly 1961–1979 1981–	ca. 800 ca. 300 350	14 blocks and I-O	OLS	Production sector, submodel I-O	Medium-term forecasts up to 20 quarters	
MACE (Macro and Energy) 1985	University of British Columbia J. Helliwell et al.	Annual 1955–1980	60[a] 26[a]	7 block	TSLS and others	Supply oriented, extended energy sector	Simulation analyses	

Table 3.1 (Continued)

Model Version	Institution Author	Frequency Sample	Equations total stochastic	Decomposition	Estimation method	Properties	Applications	Multipliers impulse sustained after 10 years
CEFM Canada Economic and Fiscal Model 1998	Department of Finance	Quarterly 1970	113			Extended public sector	Forecasts, simulations	
LPM Limited Participation Model 2001	Bank of Canada	Quarterly	24			Monetarist CGM, consistent expectations		
QPM Quarterly Projection Model 2001	Bank of Canada	Quarterly 1960				Optimization of economic agents decisions, submodel steady-state	Forecasts, policy simulations	

[a] Equations for energy not included

[b] In 2001, DRI Inc. of Canada was associated with Wharton Economic Forecasting Association, forming DRI-WEFA

tute operated an annual model TRACE containing 183 equations, 44 of which were stochastic. The model was fitted to the 1947–1971 data and the parameters were estimated with OLS. Its structure was Keynesian (Choudhry et al. 1972). The model was used for preparing 5–10 year forecasts and participated in the Project LINK (Sawyer et al. 1976).

In the 1970s, a quarterly model containing around 75 equations, the QFM, was constructed. It was used for making short-term forecasts, up to 8 quarters (Jump 1972). After 1979 it was replaced with a quarterly medium-term model called FOCUS (Forecasting and User Simulation) that in 1985 had 250 equations, approx. 180 of which were stochastic (Institute for Policy Analysis 1977). Their number increased in time to around 300. The short-run dimension of the model was neo-Keynesian, with stress being laid on price and wage rigidity. This means that the quantitative adjustments played an important role. In the long run, production was solely determined by the production factors. The model used a Cobb Douglas production function with diminishing returns to scale. Wages were determined from an extended, nonlinear Phillips curve and prices were shaped by labour costs augmented by a mark-up. The new version of the model, the CE, contained model-consistent expectations (Dungan 1998).

3.6.5 The CANDIDE Model

In the middle of the 1970s, a large, annual model was built within the Economic Council of Canada. It was characterised by a high degree of disaggregation and contained two I-O submodels. Its purpose was to support the medium-term analyses and forecasts (Preston et al. 1979; Waslander 1979).

The model had numerous versions that were given successive numbers. The 1.1 version had 2049 equations, 616 of which were stochastic. Their parameters were estimated with OLS and the sample covered the years 1955–1971. The I-O submodel had 42 identities (Bodkin et al. 1975) and consisted of 8 superblocks and 43 consumption categories. The model was frequently used in policy analyses and for preparing medium-term forecasts ranging over the next 5–10 years. The subsequent versions of the model were even larger; the 2.0 model had approximately 3000 equations and in the 3.0 model the monetarist concepts were introduced (Bodkin 1976).

3.6.6 Other Models

Particular researchers built the macroeconometric models of the Canadian economy almost in parallel. Noteworthy are the annual medium-size model TSURUMI built in 1970, which distinguished 4 sectors (Tsurumi 1973) and the MARWAH model called MOP-CAP (Marwah 1978).

Over time, the large macroeconometric models were constructed. The first of them was the large quarterly model DRI of the Canadian economy that referred to the DRI model of the US economy (Hyndman 1977). The Conference Board of Canada (1976) sponsored the construction of the quarterly model AERIC. It had a Keynesian orientation and was used for short- and medium-term forecasting and policy analyses. In 1982 it was replaced by a large, quarterly model, the MTFM (Medium-Term Forecasting Model), which regularly supported the preparation of short-term forecasts and policy simulations (London and Stokes 1982).

The annual model MACE (Macro and Energy) built by Helliwell et al. (1987) had a special structure. It was supply-oriented and accentuated the energy sector. The parameters of its equations fitted to the 1955–1980 data were mostly estimated with TSLS. The model was frequently used in the analyses of economic impacts generated by the energy sector, especially by the oil-shock in the early 1980s.

3.6.7 Models Constructed in the 1990s and Later

In the late 1990s, the quarterly, medium-sized Canadian Economic and Fiscal Model was built at the Commerce Department (Robidaux and Wong 1998). It had a neoclassical orientation and incorporated a Cobb-Douglas production function. Its extensive public sector was utilized in many simulations analysing the impacts of fiscal policy.

At the beginning of the 21st c. a small monetarist model showing similarities to the CGE models was built at the Bank of Canada (Hendry et al. 2003). It was applied in the studies on the monetary policy effects. Almost at the same time a small quarterly model NAOMI (North American Open Economy Macroeconometric Integrated Model) was built. It had a standard structure (Murchison 2001) and 24 equations, of which 6 were stochastic. Their parameters were estimated with FIML. The properties of the models were compared in an extensive report (Côté et al. 2006).

References

Adams, F. G., & Duggal, V. G. (1974). Anticipation variables in an econometric model: performance of the anticipations version of the Wharton Mark III. *International Economic Review, 15*, 267–284.

Adelman, I., & Adelman, F. (1959). The dynamic properties of the Klein-Goldberger model. *Econometrica, 27*, 596–625.

Andersen, L. C., & Carlson, K. M. (1970). A monetarist model for economic stabilization. *Review of the Federal Reserve Bank of St. Louis, 52*, 7–25.

Andersen, L. C., & Carlson, K. M. (1974). St. Louis model revisited. *International Economic Review, 15*, 305–327.

Anderson, R. G., Hoffman, D. L., & Rasche, R. H. (2002). A vector error-correction forecasting model of the US economy. *Journal of Macroeconomics, 24*, 569–598.

Ando, A., Modigliani, F., & Rasche, R. H. (1972). Appendix to part one: Equations and definitions of variables for the FRB-MIT-PENN econometric model. In B. G. Hickman (Ed.), *Econometric models of cyclical behavior, 1, No. 36*. Conference on research in income and wealth, November, 1969 (pp. 543–598). New York: Columbia University Press.

Bachman, D., Jaquette, P., Karl, K., & Rocco, P. (1998). The WEFA US macro model with chain-weighted GDP. *Journal of Economic and Social Measurement, 24*, 143–155.

Bårdsen, G., Eitrheim, Ø., Jensen, E. S., & Nynoen, R. (2005). *The econometrics of macroeconometric modelling*. Oxford: Oxford University Press.

Black, R., Laxton, D., Rose, D., & Tetlow, R. (1994). *The Bank of Canada's new quarterly projection model, part I, The steady-state model: SSQPM* (Technical Report No. 72). Ottawa: Bank of Canada.

Blanchard, O. J. (1985). Debt, deficits, and finite horizons. *Journal of Political Economy, 93*, 223–247.

Bodkin, R. G. (1976). A large-scale input-output econometric model of the Canadian economy (CANDIDE). In K. R. Polenske & J. V. Skolka (Eds.), *Advances in input-output analysis* (pp. 27–44). Cambridge: Ballinger.

Bodkin, R. G., Tanny, S. M. (Eds.) (1975). *CANDIDE Model 1.1* (CANDIDE Project Paper No. 18, Vol. 1). Ottawa: Information Canada.

Bodkin, R. G., Klein, L. R., & Marwah, K. (1991). *A history of macroeconometric model-building*. Aldershot: E. Elgar.

Brayton, F., & Tinsley, P. (Eds.) (1996). *Finance and economics discussion series. A guide to FRB/US: a macroeconomic model of the United States*. Washington: Board of Governors of the Federal Reserve System.

Brayton, F., Levin, A. T., Tryon, R. W., & Williams, J. C. (1997). The evolution of macro models at the Federal Reserve Board (Mimeo). Washington: Federal Reserve Board.

Brown, T. M. (1952). Habit persistence and lags in consumer behaviour. *Econometrica, 20*, 355–371.

Choudhry, N. K., Kotowitz, Y., Sawyer, J. A., & Winder, J. W. L. (1972). *The TRACE econometric model of the Canadian economy*. Toronto: University of Toronto Press.

Coen, R. M., & Hickman, B. G. (1985). *Social security and macroeconomic activity in 1985–2010: a forecast with the Hickman-Coen model* (Center for Economic Policy Research Publication No. 70). Stanford University.

Coen, R. M., & Hickman, B. G. (2006). An econometric model of potential output, productivity growth and resource utilization. *Journal of Macroeconomics, 28*, 645–664.

Coletti, D., Hunt, B., Rose, D., & Tetlow, R. (1996). *The Bank of Canada's new quarterly projection model, part 3, The dynamic model: QPM* (Technical Report No. 75). Ottawa: Bank of Canada.

Côté, D., Kuszczak, J., Lam, J.-P., Liu, Y., & St. Amant, P. (2006). A comparison of twelve macroeconomic models of the Canadian economy. *Journal of Policy Modelling, 28*, 523–562.

Conference Board of Canada (1976). *The AERIC short-term quarterly forecasting model of the Canadian economy* (revised ed.). Ottawa: The Conference Board of Canada.

de Bever, L., Foot, D. K., Helliwell, J. F., Jump, G. V., Maxwell, T., Sawyer, J. A., & Waslander, H. E. L. (1979). Dynamic properties of four Canadian macroeconomic models: a collaborative research project. *Canadian Journal of Economics, 12*, 133–139.

de Leeuw, F., & Gramlich, E. (1968). The Federal Reserve-MIT econometric model. *Federal Reserve Bulletin, 54*, 11–40.

Documentation of the DRI model of the U.S. economy (1993). http://tonto.eia.doc.gov/FTPROOT/modeldoc/m061.pdf.

Duessenberry, J. G., Eckstein, O., & Fromm, G. (1960). A simulation of the United States economy in recession. *Econometrica, 28*, 749–809.

Duessenberry, J. G., Fromm, G., Klein, L. R., & Kuh, E. (Eds.) (1965). *The Brookings quarterly econometric model of the United States*. Chicago and Amsterdam: Rand McNally & Co. and North-Holland.

Duessenberry, J. G., Fromm, G., Klein, L. R., & Kuh, E. (Eds.) (1969). *The Brookings model: some further results*. Chicago and Amsterdam: Rand McNally & Co. and North-Holland.

Duggal, V. G., Klein, L. R., & McCarthy, M. D. (1974). The Wharton Model Mark III: a modern IS-LM construct. *International Economic Review, 15*, 572–594.

Duguay, P., & Longworth, D. (1998). Macroeconomic models and policymaking at the Bank of Canada. *Economic Modelling*, *15*, 357–375.

Dungan, P. (1998). *Macroeconometric models and forecasting: focus manual*. Toronto: Institute for Policy Analysis, University of Toronto.

Eckstein, O. (1978). *The Great Recession, with a postscript on stagflation*. Amsterdam: North-Holland.

Eckstein, O. (1983). *The DRI model of the U.S. economy*. New York: McGraw-Hill.

Eckstein, O., Green, E. W., & Sinai, A. (1974). The data resources model: uses, structure and analysis of the U.S. economy. *International Economic Review*, *15*, 595–615.

Elliot, D. A. (1985). *The St. Louis Fed's monetary model: whence it came; how it thrived, 1970–1983*. New York: Garland.

Evans, M. K., & Klein, L. R. (1967). *The Wharton econometric forecasting model*. Philadelphia: Economics Research Unit, University of Pennsylvania.

Fair, R. C. (1971). *A short-run forecasting model of the United States economy*. Lexington: Heath.

Fair, R. C. (1974). *A model of macroeconomic activity, Vol. I: the theoretical model*. Cambridge: Ballinger.

Fair, R. C. (1976). *A model of macroeconomic activity, Vol. II: the theoretical model*. Cambridge: Ballinger.

Fair, R. C. (1984). *Specification, estimation and analysis of macroeconometric models*. Cambridge: Harvard University Press.

Fair, R. C. (1994). *Testing macroeconometric models*. Cambridge: Harvard University Press.

Fair, R. C. (2004). *Estimating how the macroeconomy works*. Cambridge: Harvard University Press.

Fair, R. C., & Alexander, S. L. (1986). *A comparison of the Michigan and Fair models: further results*. http://fairmodel.econ.yale.s.lny.RAYFAIR/pdf/1986B200.PDF.

Fisher, F. M. (1965). Dynamic structure and estimation in economy-wide econometric models. In J. G. Duessenberry et al. (Eds.), *The Brookings quarterly econometric model of the United States* (pp. 589–625). Chicago and Amsterdam: Rand McNally & Co. and North-Holland.

Fromm, G. & Klein, L. R. (Eds.) (1975). *The Brookings econometric model: perspective and recent developments*. Amsterdam: North-Holland.

Global Insight Co. (2002). D.R.I.—WEFA's macroeconometric models. http://www.iccfglobal.org.pdf/DRI-WEFAMacromodel.pdc.

Goldberger, A. S. (1964). *Econometric theory*. New York: Wiley.

Helliwell, J. F., Officer, L. H., Shapiro, H. T., & Stewart, I. A. (1969). *The structure of RDXI* (Staff Research Study No. 3 of the Bank of Canada). Ottawa: Bank of Canada.

Helliwell, J. F., MacGregor, M. E., McRae, R. N., Plourde, A., & Chung, A. (1987). Supply-oriented macroeconomics: the MACE model of Canada. *Economic Modelling*, *4*, 318–340.

Henry, S., Ho, W. M., & Moran, K. (2003). *Simple monetary policy rules in an open economy, limited-participation model* (Bank of Canada Working Paper No. 38).

Hickman, B. G., & Coen, R. M. (1976). *An annual growth model of the U.S. economy*. Amsterdam: North-Holland.

Hirsch, A. A. (1972). Price simulations with the OBE econometric model. In O. Eckstein (Ed.), *The econometrics of price determination conference* (pp. 237–276). Washington: Board of Governors of the US FRS.

Hirsch, A. A., Grimm, B. T., & Narasimham, G. L. V. (1974). Some multiplier and error characteristics of the BEA quarterly model. *International Economic Review*, *15*, 616–631.

Howrey, E. P., & Hymans, S. H. (1995). An open economy analysis of the dynamic properties of the Michigan quarterly model in the U.S. economy. In M. Dutta (Ed.), *Economics, econometrics and the LINK, Essays in honor of Lawrence R. Klein* (pp. 129–158). Amsterdam: North-Holland.

Hymans, S. H., & Shapiro, H. T. (1974). The structure and properties of the Michigan quarterly econometric model of the U.S. economy. *International Economic Review*, *15*, 632–653.

Hyndman, R. M. (1977). *The data resources model of the Canadian economy*. Toronto: Data Resources of Canada.

Institute for Policy Analysis (1977). *FOCUS: quarterly forecasting and user simulation model of the Canadian economy*. Toronto: Institute for Policy Analysis.

Jorgenson, D. W. (1965). Anticipation and investment behaviour. In J. G. Duessenberry et al. (Eds.), *The Brookings quarterly econometric model of the United States* (pp. 35–92). Chicago and Amsterdam: Rand McNally & Co. and North-Holland.

Jump, G. V. (1972). *The University of Toronto quarterly econometric forecasting model: a description of its real sector*. Toronto: Institute for Policy Analysis of the University of Toronto.

Klein, L. R. (1950). *Economic fluctuations in the United States, 1921–1941*. New York: Wiley.

Klein, L. R. (1964). A postwar quarterly model: description and applications. In *Studies in income and wealth: Vol. 28. Models of income determination*. Princeton: Princeton University Press for NBER.

Klein, L. R. (Ed.) (2009). *The making of national economic forecasts*. Cheltenham Glos: E. Elgar.

Klein, L. R., & Goldberger, A. S. (1955). *An econometric model of the United States, 1929–1952*. Amsterdam: North-Holland.

Klein, L. R., Eskin, V., & Mak, W. (2007). *Current quarterly model of the United States economy*. University of Pennsylvania, Forecast Summary, Monday, December 3, 2007.

Kmenta, J. & Ramsey, J. B. (Eds.) (1981). *Large-scale macro-econometric models. Theory and practice*. Amsterdam: North-Holland.

Kuiper, J. (1970). *Model XVI—an econometric model of the Canadian economy* (Mimeo). Dpt of Finance of the Government of Canada.

Liebenberg, M., Hirsch, A. A., & Popkin, J. (1966). A quarterly econometric model of the United States: a progress report. *Survey of Current Business*, *46*, 13–39.

Liebenberg, M., Green, G., & Hirsch, A. (1971). *The Office of Business Economics 70 quarterly econometric model*. Washington: US Department of Commerce, Econometric Branch.

Liu, T. C. (1963). An exploratory quarterly econometric model of effective demand in the post-war U.S. economy. *Econometrica*, *31*, 301–348.

Liu, T. C., & Hwa, E. C. (1974). Structure and applications of a monthly econometric model of the U.S. economy. *International Economic Review*, *15*, 328–365.

London, A., & Stokes, E. (1982). *The medium-term forecasting model (MTFM): an overview*. Ottawa: The Conference Board of Canada.

Lucas, R. E. Jr. (1976). Econometric policy evaluation. A critique. In K. Brunner & A. H. Meltzer (Eds.), *The Phillips curve and labor markets* (pp. 19–46). Amsterdam: North-Holland.

Marwah, K. (1978). Towards money, output and prices: a capacity view of inflation (MOP-CAP). *Empirical Economics*, *3*, 1–29.

May, S. (1966). Dynamic multipliers and their use for fiscal decision making. In *Conference on stabilization policies* (pp. 155–187). Ottawa: Queen's Printer.

Maxwell, T. (1978). A primer on RDX2. *Bank of Canada Review*, 3–10.

McCarthy, M. D. (1972). *The Wharton quarterly econometric forecasting model, Mark III*. Philadelphia: Economics Research Unit, University of Pennsylvania.

Murchison, S. (2001). *NAOMI—a new quarterly forecasting model. Part II. A guide to Canadian NAOMI* (Dpt. of Finance Canada Working Paper No. 25).

Officer, L. H. (1968). *An econometric model of Canada under the fluctuating exchange rate*. Cambridge: Harvard University Press.

Phillips, A. W. (1958). The relation between unemployment and the rate of change of money wage rate in the United Kingdom 1861–1957. *Economica*, *25*, 283–299.

Poloz, S., Rose, D., & Tetlow, R. (1994). The Bank of Canada's new quarterly projection model (QPM): an introduction. *Bank of Canada Review*, 23–28.

Preston, R. S. (1972). *The Wharton annual and industry forecasting model*. Philadelphia: Economics Research Unit, University of Pennsylvania.

Preston, R. S. (1975). The Wharton long-term model: input-output within the context of a macro forecasting model. *International Economic Revue*, *16*, 3–19.

Preston, R. S., Braithwaite, C., Cain, B., Desaulniers, D., Eyford, B. L., Gilby, S., Lodh, B. K., Nevin, P., Ramin, J. E., Rao, P. S., Saiyed, H. M., Jean, C. S., Schweitzer, T. T., Whillans,

J. D., & Willis, M. (1979). *CANDIDE Model 2.0: model description* (Vols. I and II). Ottawa: Economic Council of Canada.

Rasche, R. H., & Shapiro, H. T. (1968). The FRB-MIT econometric model: its special features. *American Economic Review*, *58*, 123–149.

Rhomberg, R. R. (1964). A model of the Canadian economy under fixed and fluctuating exchange rates. *Journal of Political Economy*, *72*, 1–31.

Robidoux, B., & Wong, B.-S. (1998). *CEFM96, part I, Model structure* (Dpt of Finance Canada Working Paper No. 105).

Sawyer, J. A., Carr, J. L., Choudhry, N. K., Jump, G. V., Kotowitz, Y., & Winder, J. W. L. (1976). The TRACE Mark IIIR annual model of Canadian economy. In J. Waelbroeck (Ed.), *The models of Project LINK* (pp. 115–160). Amsterdam: North-Holland.

Schmidt, P., & Waud, R. N. (1973). The Almon lag technique and the monetary versus fiscal policy debate. *Journal of the American Statistical Association*, *68*, 11–19.

Sims, C. A. (1980). Macroeconomics and reality. *Econometrica*, *48*, 1–48.

Suits, D. B. (1962). Forecasting and analysis with an econometric model. *American Economic Revue*, *52*, 104–132.

Taylor, J. B. (1993). Discretion versus policy rules in practice. *Carnegie-Rochester Conference Series on Public Policy*, *39*, 195–214.

Taylor, J. B. (2001). The use of the new macroeconometrics for policy formulation. *American Economic Association*, *83*, 300–305.

Tsurumi, H. (1973). A survey of recent Canadian macro-econometric models. *Canadian Journal of Economics*, *6*, 409–428.

Valavanis-Vail, S. (1955). An econometric model of growth: USA, 1869–1953. *American Economic Review, Papers and Proceedings*, *45*, 208–221.

Waelbroeck, J. (Ed.) (1976). *The models of Project LINK*. Amsterdam: North-Holland.

Waslander, H. E. L. (1979). The dynamic properties of CANDIDE Model 1.2. *Canadian Journal of Economics*, *12*, 139–150.

Wold, H. (1960). A generalization of causal chain models. *Econometrica*, *28*, 443–463.

Yaari, M. E. (1965). Uncertain lifetime, life insurance, and the theory of the consumer. *Review of Economic Studies*, *32*, 137–150.

Chapter 4
Modelling Activities in Europe. Macroeconometric Models of the United Kingdom

4.1 Introduction

Several years from the construction of macroeconometric models in the USA and Canada, other developed market economies followed in their footsteps. Most important developments could be observed in the United Kingdom, France and the Netherlands, the cradle of macroeconometric modelling. From there modelling activities spread over other West-European countries. In the centrally planned economies model building was of special character; these countries started to move towards the market economy structure after 1990, when they launched the transition processes.

Generally, macroeconometric modelling in European countries developed following the US pattern. Modelling activities concentrated first in the academic centres (except for France and the Scandinavian countries) and then within the central administration and central banks.

The mainstream models prevailed. In the 1980s attempts were made, particularly in the UK, to depart partly from these models. Rational expectations were explicitly introduced and neoclassical concepts were applied to model the production sector as well as the household sector. New estimation techniques were used.[1]

In the early 21st century, modelling activities tended to concentrate within research institutions and research centres run by central administration bodies and central banks. The institutions developed multicountry macroeconometric models of the world economy and more recently of the European Union and Eurozone. In many countries, the individual country models became the major components of the multicountry models, such as NIGEM in the UK.

Many models were structured following the neoclassical guidelines. Their blocks of equations did not contain equations explaining the demand and supply sectors, but the activities of households, enterprises, public institutions, etc., as well as markets, for instance, a labour market. The equations were specified according to microeconomic theory.

---

[1] The description of this period in the development of macroeconometric activities rests on the papers published in Bodkin et al. (1991) and Whitney (1994).

W. Welfe, *Macroeconometric Models*,
Advanced Studies in Theoretical and Applied Econometrics 47,
DOI 10.1007/978-3-642-34468-8_4,

The above approach to modelling initiated a tendency promoted by the central banks' research centres towards reducing the differences between the macroeconometric models and the CGE models, especially the DSGE models. As a result, the hybrid models were constructed. The academic centres responded by developing small submodels making explicit use of the cointegrating techniques.

4.2 First Macroeconometric Models of the UK Economy

The first quarterly model of the UK economy was built under the leadership of L.R. Klein and J. Ball at the Oxford University in the years 1958–1959. As the quarterly national accounts were not available then, industrial output, or more precisely its monthly index developed by R. Stone, was chosen to serve as the major endogenous variable. The model had a few equations and was used to produce first forecasts extending to the year 1959 (Klein et al. 1961). The modelling activities were discontinued after Klein departed for the USA in 1959.

4.3 The Annual Multisectoral Model of the UK Economy

In 1960, Richard Stone and his team began to construct a large annual model of the British economy within the Growth Project, initially called "Rocket". The model was founded on the extensive database of national accounts and its special properties followed from R. Stone's earlier contributions. It was highly disaggregated. It made use of the concept of linear expenditure system (LES) developed by R. Stone in the block explaining consumption decomposed into several commodity groups. It also used the I-O relationships. As a result, its static version had several thousands of equations, around 2000 in the 1960s (Cambridge Department of Applied Economics 1962) and ca. 5000 in the 1980s (Whitley 1994). The aggregation process was based on the "bottom up" approach.

The linear expenditure system was the first system of demand functions based on consumer demand theory that was jointly estimated. It distinguished the indispensable expenditures and other expenditures financed from excess income.[2] Consumption was represented by a system of equations meeting particular side conditions, such as the additivity of expenditures to income (including savings). In the model, real expenditures depended on own prices and the prices of substitutes and complementary commodities. This significantly increased the number of the parameters to be estimated; to reduce it, the Slutsky theorem assuming a symmetric matrix of substitution elasticities was applied. As a result, the number of parameter estimates was reduced by half.

In this system of demand functions the number of the jointly estimated equations could not be large (e.g. from 6 to 8). However, in the equation system expenditures

[2]More in Suchecki (2006).

had to be decomposed into 40–60 commodity groups. To achieve this, a multi-step estimation procedure was applied, where the parameters of the major (say 8) groups of commodities were estimated in the first step, and then the parameters of the subgroups formed by decomposing the major groups. The results were controlled by means of parallel investigations using panel data on UK households (Welfe 1964).

The linking of the I-O submodels showing the interindustry relationships between production and prices provided an opportunity for analysing and solving numerous problems. This research was conducted by the team led by J.A.C. Brown.

Solving a system with many thousands of equations was not an easy task. Some of the procedures used to this end had been already applied to the Brookings Model. The system was decomposed into recursively arranged blocks that were weakly dependent on each other.

From the moment it was constructed, the model was used in forecasting and policy simulations. The precision of the medium-term forecasts it generated improved after the equations were dynamized. Its deep disaggregation enabled highly specialist, sectoral simulation analyses (Cambridge Department of Applied Economics 1963, 1964).

The model was initially maintained from public funds. In the mid-1970s public funds ran out, but the newly established commercial institution Cambridge Econometrics led by Terry Barker, a former collaborator of R. Stone, maintained the Cambridge Model all the time (Barker 1976). It became a unique, large, multi-sectoral dynamic model of the UK economy, which was used for more than 50 years in producing numerous sectoral analyses dealing with energy issues, actions to combat air pollution, etc.

In the mid of the 1970s the Cambridge Economic Policy Group (CEPG) led by W. Godley developed an annual model mainly aimed at studies of the impacts of public sector deficit on the fluctuations in current account of the balance of payment (Cripps and Godley 1976). The CEPG systematically published projections for the UK economy on annual basis. Because of funding problems it stopped the activities around 1983.

4.4 The Quarterly Models

In the second half of the 1960s and in the early 1970s many research centres in Britain began to construct macroeconometric models. Six quarterly models were built, 5 of which are still maintained.[3] Their structures were influenced by L.R. Klein. They were mostly demand determined and explained consumer demand, investment demand and employment. (See Table 4.1 for their concise descriptions.)

[3] In the early 1970s, an attempt was made at the Southampton University to construct a multi-sectoral, quarterly model. The project was discontinued in the mid-1970s and was never completed.

4.4.1 The London Business School (LBS) Model

London Business School promoted the praxis-oriented research. It had on its staff outstanding scholars in the field of macro-modelling, who afterwards strengthened other, mainly public research centres. Under the leadership of J. Ball, a quarterly model of the UK was built with the data compiled for the national accounts (Ball et al. 1975). Its original structure was influenced by the Klein-Goldberger and early Wharton models. The model was used in regular forecasting and the first forecast it produced was presented in 1966. Although most of the forecasts were short term, their horizon was sometimes extended to 10 years. The model served the purpose of forecasting for the next 30 years.

Other, mainly public institutions cooperating with the LBS scholars constructed macroeconometric models almost at the same time. The institutions successfully trained their own staff and continue to maintain their models (Ball and Holly 1991; Whitley 1994).

4.4.2 The National Institute of Economic and Social Research (NIESR) Model

The NIESR became involved in forecasting the UK economy early on, as it started to publish forecasts based on individual equations in 1959. Its subsequent forecasts were produced using a quarterly model, which was built in the early 1970s (Byron 1970). The model earned very high reputation and was, and still is, used in many policy simulations and in making regular quarterly forecasts which are published in British journals.

4.4.3 The H.M. Treasury Model

In the 1970s, public institutions built new quarterly models serving practical purposes. The most important of them was the H.M. Treasury model, constructed mainly in order to prepare forecasts. R.J. Ball and J. Eaton from LBS helped to build a large macroeconometric model, which was used in regular forecasting since the mid-70s (Shepherd et al. 1975). Later on, the model was involved in the preparation of fiscal and monetary policy simulations. An iterative procedure was applied, as in the Dutch case. Policy assumptions defined by the top administration were "sifted" through the model and the outcomes were reported to the officers in the major departments. Corrections were invited and following their implementation the final result was obtained, either as updated model-based forecasts or policy simulations subject to further analyses, if necessary. At the end of 1970s, the model had approx. 600 equations and was larger than the quarterly LBS model (280) and the National Institute model (150) (Melliss 1986).

4.4.4 The Bank of England Model

In the middle of 1970s, the Bank of England made a long postponed decision to build its own model in order to analyse the monetary policy effects (Patterson et al. 1987). Because its monetary policy in the earlier years was neutral, it did not have any major impact on economic growth; however, the shocks in the early '70s showed that monetary policy may be of significance. Many studies investigating the role of interest rates indicated that their adjustments were an important tool that could either stimulate or curb consumer and investment expenditures, so ignoring them was not reasonable. In the same period the UK's exchange rate policy radically changed with the replacement of a rigid exchange rate linked to the US dollar for a flexible exchange rate. It was realized then that the exchange rate movements could be explained by the domestic interest rate changes vis-à-vis the changes in the interest rates in other countries. A relative growth of the domestic interest rate attracts foreign speculative capital, which puts the domestic exchange rate under pressure. Because of the findings, the UK Central Bank became interested in having an effective instrument for analysing the likely impacts of monetary policy.

In the 1970s, the "traditional" British economists frequently questioned, like their US colleagues, the accuracy of the model-produced forecasts. The reservations particularly concerned the forecasts' low predictive value in the period of oil shocks. As a result, the institutions using the models ran into financial troubles, which affected their research activity and the capabilities of maintaining (the availability of efficient computers) and using their models. These difficulties were overcome when the Social Science Research Council established the Macroeconomic Modelling Consortium. The Consortium assumed responsibility for financing research dedicated to macromodelling. This solution allowed the existing centres to maintain their models, as well as enabling the establishment of new centres.

4.4.5 The Liverpool University (LPL) Model

In the same years, a new macromodel was built at the Liverpool University. It was a small quarterly neoclassical model accentuating the money sector where rational expectations were used to explain the UK money market. Constructed for the purpose of monetary policy analyses (Minford et al. 1984), the model is used still today.

In the 1980s a small annual model of the UK economy with 60 endogenous variables called CUBS (City University Business School) was built by M. Beenstock. It was oriented to deep specification of the supply sector of the British economy (Beenstock et al. 1986). Its critical analysis can be found in Whitley (1994).

4.5 From Keynesian to Neoclassical Specification

4.5.1 Supply Side and Production Functions

The decade from the mid-1970s to the mid-1980s was characterised by a slow departure from macromodelling based on the Keynesian approach. Indeed, while the majority of macromodels retained blocks of equations generating consumer demand, investment demand and employment, it was emphasised that the supply side was weakly represented. As a result, the model builders decided to extend the blocks describing the supply sector. The neoclassical views were used as the theoretical underpinning, particularly in relation to the concept of production function as the major instrument for generating production supply. However, the direct use of the production function was very rare in the models, because no characteristics were available to generate potential output and the corresponding capacity utilization rates. It took time to overcome this difficulty, so the production functions were explicitly specified in a few models only. The NIESR model was the only one where the production function concept was broadly applied. Its production function was built using a vintage approach. The consecutive vintages of fixed capital were distinguished, with productivity declining with passing time. This allowed the NIESR model to estimate the differences between the total demand for domestic products and their supply, i.e. the occurring market tensions and their impact on price determination, imports and inventory changes (Wren-Lewis 1988).

In the HMT and OEF models, the Cobb-Douglas production function was used indirectly and in the LBS model the capacity utilization rate was calculated using fixed capital-output ratios.

4.5.2 Modelling Labour Markets

In the mid-70s researchers realized that using the production function was not sufficient to characterize the supply side of the economy and that a key role should be given to a system of equations explaining prices, wages and exchange rates. Price equations were initially specified following the American models; prices were determined by unit costs, mainly the costs of labour. At later time, when information on the capacity utilization rates became available, this variable was introduced to serve as a determinant of mark-up changes. The most important contributions of the English econometricians concerned the modelling of subsystems explaining the labour market changes and especially their implications for wages.

A.W. Phillips has shown in his seminal paper that the rate at which wages grow depends on the rate of unemployment. Higher unemployment rate weakens the claims for wage rises. Named the Philips curve after its author, this specification has become quite common. However, subsequent studies on labour markets changed this perspective. In the middle of 1980s, the labour market modelling activities were developing at the Centre for Labour Economics, under the leadership of R. Layard

(Layard and Nickell 1985). After long discussions it was agreed that the changes in inflation rates were fully absorbed by changing nominal wages, meaning that nominal wage elasticities with respect to consumer prices are equal 1. Therefore, models should explain real wages and not nominal wages.

As a result of the above studies, it became necessary to recognize that real wages are formed in the course of negotiations involving employers and trade union representatives. The negotiations are affected by factors determining both labour demand and labour supply, unemployment rate being one of the most important among them. Growing unemployment rate weakens the bargaining position of the trade unions. The demand-side factors are output and labour productivity, and the supply-side factors include immigration. This new understanding resulted in major modifications to the specification of the equations explaining real wages: their long-run level was made dependent on labour productivity and the rate of unemployment. In the 1980s, the changes were being introduced into the British models on a rising scale.

Besides, the above developments made possible the determination—after appropriate theoretical transformations—of the potential unemployment rates, mainly of the natural rate of unemployment related to workers' changing jobs, places of residence, etc. The concept of non-accelerating rate of unemployment (NAIRU) was formulated after the relations determining wages were linked to price changes. With the development of techniques for calculating the NAIRU, the concept spread all over the world.

4.5.3 Rational Expectations

The Lucas critique was followed by a departure from the classical Keynesian approach to model building. Its manifestation was the attempts to utilize the information on economic agents' expectations, particularly in order to explain prices, wages and exchange rates. Most promising was the concept of rational expectations formulated at that time. The concept assumed that economic agents behaved rationally and were forward looking, i.e. that their expectations were model consistent. This assumption seemed realistic as long as it concerned large corporations with research departments that could optimize their behaviour. Its applicability to small firms and households seemed rather doubtful, though. As a result, many model builders remained sceptical. They were of the opinion that adaptive expectations, i.e. expectations formed using past experiences, were more justified on the empirical grounds. The experience gained recently has greater weight than one gained in more distant past. This regularity justified the use of lag distributions with declining weights.

In the 1980s, expectations were widely introduced into the British models, but their character and use differed significantly, mainly reflecting the model builders' preferences. Most public institutions were sceptical about rational expectations. The HMT and Bank of England models incorporated adaptive expectations. The academic institutions' models, such as the Liverpool, LBS and NIESR models, used the hypothesis of rational expectations. The expectations played an important role in modelling the financial markets. They had an immediate impact on the market for bonds and shares, determining also effective changes in the exchange rates. The situation in the other markets is considerably different. Prices in the commodity markets, likewise wages in the labour market, change with long lags when affected by expectations. To account for this, many models reached for the adaptive expectation hypothesis and introduced appropriate lag distributions. In the NIESR model, the expectations codetermined also output. The effective output depended on producers' expectations of the demand for their products, but not on demand itself. Thus generated output determined investments, inventory changes, etc., increasing the accuracy of forecasts (Holden and Peel 1985).

The above changes were implemented in the successive versions of the LBS model (Budd et al. 1984). Much attention was given to the introduction of rational expectations (Hall and Henry 1986). The next respecification of the model took place at the end of the 1980s (Dinesis et al. 1989).

4.6 Developments in the Last Twenty Years

4.6.1 The Oxford Economic Forecasting Model

In the late 1980s, the quarterly Oxford Economic Forecasting (OEF) model maintained by a commercial institution entered the group of models characterised in the section above. It was a medium-sized model comprising 325 equations, with adaptive expectations, whose structure followed one of the versions of the HMT model. The model was used to generate forecasts and to perform policy simulations (Burridge et al. 1991).

4.6.2 Reconstruction of the LBS Model

In the second half of the 1990s, macroeconometric modelling changed even more significantly. The blocks of equations in the supply sector were extended and estimation procedures making use of the cointegration analysis were applied more widely. The most important was the reconstruction of the quarterly LBS model, as a result of which a new version of the model was built (Allen et al. 1994). The equations in the supply sector found a new theoretical underpinning. It was the results of cost minimization. A system of equations with jointly estimated parameters was developed. The system contained producer prices, prices and the demand for production factors. This version of the LBS model included the learning mechanism in addition to rational expectations (see also Garratt and Hall 1997).

It is worth noting that an alternative specification of the consumption function was introduced into the new LBS model. In the traditional specification, household demand depends on real disposable income, personal financial wealth and interest rates. The alternative approach uses a special notion of permanent income that is close to the long-life hypothesis and distinguishes between credit-constrained households and households having free access to credit. For the first group, consumer demand is determined by real incomes. In the second group, consumption depends on the expected, discounted, current and future earned income (expected human wealth), the stock of real financial and physical wealth (asset wealth) and interest rates.

Following the analyses of the alternative concepts of monetary policy rules, it was decided that the nominal interest rates should be treated as functions of the rates of inflation.

Although long-standing and successful, the LBS model was abandoned in the early years of the 21st c., the main reason being financial problems.

4.6.3 The COMPACT Model

In the mid-90s, a small, quarterly model of the UK economy, COMPACT, was constructed at the University of Strathclyde. Its 3.0 version had more than 25 stochastic equations whose parameters were estimated using OLS with ECM specifications, as appropriate. The main goal underlying its construction was to implement macroeconomic theory to facilitate policy analysis (Wren-Lewis et al. 1996). This was a neo-Keynesian model where the nominal rigidities were linked to model consistent wage and price determination. In the short-run, output depended on final demand. The consumption function was derived from intertemporal optimisation to explain the behaviour of the credit-unconstrained households, whereas the other households' behaviour depended on real incomes. In the medium-term, the model referred to the vintage production technology. The foreign trade equations were also extended, using variables showing the relative quality and variety of the traded products. The forward expectations variables treated as rational or model-consistent expectations were widely used. The model was mainly used to produce various policy simulations (Darby et al. 1999).

4.6.4 The Bank of England Models

Towards the end of the 20th c., a new quarterly model MM (Macroeconomic Model) was built at the Bank of England (Harnett and Patterson 1989). It was to serve as an instrument supporting regular forecasting activities and simulation analyses of monetary policy outcomes (Bank of England 1999, 2000). The model distinguished long-term relations from short-term adjustments. Expectations were introduced only occasionally. In the long-run output was determined from the Cobb-Douglas production function and in the short-run it was determined by final demand. Consumer

Table 4.1 Models of the United Kingdom

Model Version Year	Institution Author	Frequency Sample	Equations total stochastic	Decomposition	Estimation method	Properties	Applications
Oxford model 1958–1959	Institute of Statistics, Oxford University L.R. Klein	Quarterly		–	OLS	Explained industrial output, no quarterly NA data	Forecast 1959
Cambridge model Growth Project Rocket 1960	Dpt. of Applied Economics, Cambridge University R. Stone	Annual	2000	Several industries, I-O submodels	OLS	Keynesian, LES consumption functions	Forecasts and policy analyses
Middle of 70s	Cambridge Econometrics T. Barker	Annual	5000	39 industries, I-O submodels	OLS	Neo-Keynesian	Forecasts and policy analysis
LBS model 1965	London Business School	Quarterly	650	Industry, other sectors	ECM	Imperfect competition	Forecasts and simulation analyses
after 1977 after 1980						Rational expectations after 1985; expanded financial sector	Forecasts and simulation analyses
after 1994		Quarterly	ca. 350	Energy, other industries	ECM	Extensions of supply sector; cost function with restrictions; learning process alternative to rational expectations	Forecasts and policy simulations

Table 4.1 (Continued)

Model Version Year	Institution Author	Frequency Sample	Equations total stochastic	Decomposition	Estimation method	Properties	Applications
NIESR 1960	National Institute of Economic and Social Research (NIESR) London		20 11	Industry, other sectors	Cointegration	Financial wealth in consumption function, imperfect competition Layard, Nickell wage negotiations	Forecasts and policy simulations
1970			73 22				
after 1980 after 1985			176			Rational expectations after 1985	
OEF	Oxford Economic Forecasting	Quarterly	325	Industry, other sectors		Imperfect competition, adaptive expectations	Forecasts and simulation analyses
HMT 1965 after 1980	HTM Treasury	Quarterly	ca. 600 1200	Industry, other sectors		Eclectic, adaptive expectations, negotiated real wages	Forecasts and simulation analyses
SLIM 1988		Quarterly	509 332	Industry, other sectors		Extended financial sector	Forecasts and policy simulations
BE	Bank of England	Quarterly	270	Industry, other sectors		Eclectic, adaptive expectations, extended specification of money markets	Short-term forecasts, policy simulations

Table 4.1 (Continued)

Model Version Year	Institution Author	Frequency Sample	Equations total stochastic	Decomposition	Estimation method	Properties	Applications
MTMM (Medium Term Macro Model)	Bank of England	Quarterly					Short-term forecasts, policy simulations
BEQM		Quarterly 1978				Hybrid model of dynamic adjustments	Quarterly forecasts and simulations for Monetary Policy Committee
LPL 1980	Liverpool University	Quarterly	11			Neoclassical, monetarist, rational expectations	Forecasts, policy simulations

demand depended on real disposable income, personal wealth and real interest rate and in the short-run on the rate of unemployment. Investment was dependent on GDP and user costs lagged by 4 quarters. The model had extended blocks of equations that explained the monetary and fiscal sectors, as well as prices and wages. Used as an important monetary policy instrument, the model remained in the mainstream macroeconomic modelling for many years (cf. Taylor 1993).

At the beginning of the 21st c., a new quarterly model was constructed at Bank of England. It substituted the medium-term model MTMM and represented a radical departure from the mainstream models used so far. This new construct was designed for the Monetary Policy Committee, as a refined instrument serving macroeconomic analyses and forecasts (Bank of England 2004).

Because its builders intended to take account of the most recent contributions to macroeconomic theory, the model was split into two submodels. The first submodel represented the steady state constituting a theoretical basis for specifying the long-term relationships, while the second one was a system of dynamic adjustment equations, where public interventions could be accounted for with appropriate parameter changes. The model broadly dealt with expectations, mainly with those related to expected incomes, prices, exchange rates, etc. They were assumed to be rational, whenever appropriate.

The point of departure for the theoretical submodel was the maximisation of economic agents' goals under the relevant dynamic restrictions. The submodel can therefore be viewed as a special form of a dynamic general equilibrium model. It seems that its weak point was the calibration of the parameters of the major equations, likely to move the long-term trajectories away from the trajectories representing observations. It is noteworthy that in all previous models the parameters of the long-term relationships were estimated from relevant samples. The dynamic submodel was to explain why the paths generated by the theoretical submodel and the observed paths diverged from each other, so lagged and/or additional variables can be found in its equations, as well as indicators reflecting the impacts of changing economic policy.

The model consisted of several blocks of equations explaining the behaviour of households, enterprises, public institutions and foreign agents, of the commodity and labour markets, and of broadly treated financial markets (Harrison et al. 2005).

4.6.5 ESRC Macroeconomic Modelling Bureau

The Warwick University's Bureau of Macroeconomic Modelling led by K.F. Wallis has significantly contributed to the development of macromodelling activities in the UK. Between 1983 and 2005, the Bureau successfully carried out many research projects involving the comparative analyses of the forecasts generated by the models run by 6 modelling groups in the UK (Fisher et al. 1989). It also conducted numerous studies into the effects of economic policy. The studies were based on simulations performed with the above models that were deposited at the Bureau.

The major findings from these research projects were presented in a series of monographs that K.F. Wallis edited in the second half of 1980s (Wallis et al. 1987, 4th Report). Computer also the paper by Church et al. (2000).

4.6.6 The Long-Run Structural Macroeconometric Model

At the beginning of the 21st c., an original research project was undertaken at the Department of Applied Economics, University of Cambridge. Its purpose was to establish the rules for constructing structural macroeconomic models that would apply the VAR methodology using the cointegration relationships. This effort brought many methodological achievements (Garratt et al. 2003). The research results were described in an impressive monograph (Garratt et al. 2006).

The above methodology was applied to construct a small structural model of the UK economy—SVPR. The model was based on the theoretical considerations and distinguished five cointegrated (long-term) relations, as well as short-term relations. The long-term relations included equations explaining prices, exchange rates, interest rates and production. The explained variables were defined in terms of the differences between the values of the above variables for the UK and the rest of the world, which implied that the "world" variables were also endogenous. This specification was rather peculiar. The real money stocks and real interest rates were also explained in the model. Contrary to its authors' intentions, the system represented a model explaining the money markets rather than the UK economy.

The way of estimating the model parameters was highly sophisticated and employed the state-of-the art econometric techniques. However, the approach was only applicable to the small systems. Further methodological and numerical contributions are necessary before it can be used in constructing large macroeconometric models.

The properties of the above structural model SVPR were compared with the properties of the structural model COMPACT, a macroeconometric model of simultaneous equations. No evidence was found pointing to one of the models being superior to the other (Jacobs and Wallis 2005).

References

Allen, C., Hall, S. G., & Nixon, J. (1994). The new London Business School model of the UK economy (Center for Economic Forecasting Discussion Paper No. 18-94a).

Ball, R. J., & Holly, S. (1991). Macroeconometric model-building in the United Kingdom. In R. G. Bodkin, L. R. Klein & K. Marwah (Eds.), *A history of macroeconometric model-building* (pp. 195–230). Aldershot: E. Elgar.

Ball, R. J., Boatwright, B. D., Burns, T., Lobban, P. W. M., & Miller, G. W. (1975). The London Business School quarterly econometric model of the United Kingdom. In G. A. Renton (Ed.), *Modelling the economy*, London: Heinemann.

Bank of England (1999). *Economic models at the Bank of England*. London: Bank of England.

Bank of England (2000). *Economic models at the Bank of England, September 2000 update*. London: Bank of England.

Bank of England (2004). The new Bank of England quarterly model. *Bank of England Quarterly Bulletin, 188–193*.

Barker, T. S. (Ed.) (1976). *Cambridge studies in applied econometrics: Vol. 2. Economic structure and policy*. London: Chapman and Hall.

Beenstock, M., Warburton, P., Levington, P., & Dalziel, A. (1986). A macroeconomic model of aggregate supply and demand for the UK. *Economic Modelling*, *3*, 242–268.

Bodkin, R. G., Klein, L. R., & Marwah, K. (1991). *A history of macroeconometric model-building*. Aldershot: E. Elgar.

Budd, A., Dicks, G., Holly, S., Keating, G., & Robinson, B. (1984). The London Business School econometric model of the UK. *Economic Modelling*, *1*, 355–420.

Burridge, M., Dhar, S., Mayes, D., Meen, G., Neal, D., Tyrell, N., & Walker, J. (1991). Oxford economic forecasting's system of models. *Economic Modelling*, *8*, 227–413.

Byron, R. (1970). Initial attempts in model-building at NIESR. In K. Hilton & D. Heathfield (Eds.), *An econometric study of the United Kingdom*. London: Macmillan.

Cambridge Department of Applied Economics (1962). A computable model of economic growth, no. 1. In *A programme for growth*. London: Chapman and Hall.

Cambridge Department of Applied Economics (1963). Input-output relationships 1945–1966, no. 3. In *A programme for growth*. London: Chapman and Hall.

Cambridge Department of Applied Economics (1964). The model in its environment; a progress report, no. 5. In *A Programme for Growth*. London: Chapman and Hall.

Church, K. B., Sault, J. E., Sgherri, S., & Wallis, K. F. (2000). Comparative properties of models of the UK economy. *National Institute Economic Review*, *171*, 106–122.

Cripps, T. F., & Godley, W. A. H. (1976). A formal analysis of the Cambridge economic policy group model. *Economica*, *43*, 335–348.

Darby, J., Ireland, J., Leith, C., & Wren-Lewis, S. (1999). COMPACT: a rational expectations, intertemporal model of the United Kingdom economy. *Economic Modelling*, *16*, 1–52.

Dinenis, E., Holly, S., Levine, P., & Smith, P. (1989). The London Business School econometric model: some recent developments. *Economic Modelling*, *6*, 243–251.

Fisher, P. G., Tanna, S. K., Turner, D. S., Wallis, K. F., & Whitley, J. D. (1989). Comparative properties of models of the UK economy. *National Institute Economic Review*, *129*, 69–87.

Garratt, A., & Hall, S. G. (1997). E-equilibria and adaptive expectations, output and inflation in the LBS model. *Journal of Economic Dynamics and Control*, *21*, 1149–1171.

Garratt, A., Lee, K., Pesaran, M. H., & Shin, Y. (2003). A long run structural macroeconometric model of the UK. *Economic Journal*, *113*, 412–457.

Garratt, A., Lee, K., Pesaran, M. H., & Shin, Y. (2006). *Global and national macroeconometric modelling: a long-run structural approach*. Oxford: Oxford University Press.

Hall, S. G., & Henry, S. G. B. (1986). A dynamic econometric model of the UK with rational expectations. *Journal of Economic Dynamics and Control*, *10*, 219–233.

Harnett, I., & Patterson, K. (1989). An analysis of changes in the structural and simulation properties of the Bank of England quarterly model of the UK economy. *Economic Modelling*, *6*, 20–55.

Harrison, R., Nikolov, K., Quinn, M., Ramsay, G., Scott, A., & Thomas, R. (2005). *The Bank of England quarterly model*. London: Bank of England.

Holden, K., & Peel, D. A. (1985). An evaluation of quarterly National Institute forecasts. *Journal of Forecasting*, *4*, 227–234.

Jacobs, J. P. A., & Wallis, K. F. (2005). Comparing SVARs and SEMs: two models of the UK economy. *Journal of Applied Econometrics*, *20*, 209–228.

Klein, L. R., Ball, R. J., Hazelwood, A., & Vandome, P. (1961). *An econometric model of the United Kingdom*. Oxford: Basil Blackwell.

Layard, P. R. G., & Nickell, S. J. (1985). The causes of British unemployment. *National Institute Economic Review*, *111*, 62–85.

Melliss, C. L. (1986). HM treasury macroeconomic model 1986. *Economic Modelling*, *5*, 237–259.

Minford, A. P. L., Marwaha, S. S., Matthews, K. G. P., & Sprague, A. (1984). The Liverpool macroeconomic model of the United Kingdom. *Economic Modelling*, *1*, 24–62.

Patterson, K., Harnett, I., Robinson, G., & Ryding, J. (1987). The Bank of England quarterly model of the UK economy. *Economic Modelling*, *4*, 348–549.

Shepherd, J. F., Evans, H. P., & Riley, C. J. (1975). The Treasury short-term forecasting model. In G. A. Renton (Ed.), *Modelling the economy* (pp. 38–65). London: Heinemann.

Suchecki, B. (2006). *Kompletne modele popytu (Complete systems of demand)*. Warszawa: PWE.

Taylor, J. B. (1993). Discretion versus policy rules in practice. *Carnegie-Rochester Conference Series on Public Policy*, *39*, 195–214.

Wallis, K. F., Fisher, P. G., Longbottom, J. A., Turner, D. S., & Whitley, J. D. (1987). *Models of the UK economy: a fourth review by the ESRC macromodelling bureau*. Oxford: Oxford University Press.

Welfe, W. (1964). *Changes in consumer behaviour. Observations based on the analysis of family budgets for Great Britain* (Growth Project Papers No. 165). Cambridge: University of Cambridge, Department of Applied Economics.

Whitley, J. (1994). *A course in macroeconomic modelling and forecasting*. New York: Harvester Wheatsheaf.

Wren-Lewis, S. (1988). Supply, liquidity and credit: a new version of the institutes's domestic econometric macromodel. *National Institute Economic Review*, *126*, 32–43.

Wren-Lewis, S., Darby, J., Ireland, J., & Ricchi, O. (1996). The macroeconomic effects of fiscal policy: linking an econometric model with theory. *Economic Journal*, *106*, 543–559.

Chapter 5
Models of the French Economy

5.1 Annual Models and Indicative Planning

5.1.1 FIFI Model

In France, macromodelling activities had different origins than in the USA and the UK. Their development was a response to the clearly articulated needs of indicative planning and forecasting of the major components of national accounts that were to be used to prepare the government budget. The initially constructed five-year plans used the time series of the relevant variables and the input-output tables. Between 1966 and 1968, the INSEE team led by R. Courbis constructed a large, multisectoral model FIFI (Physical-Financial), which was dedicated to the VI and VII Plans (Aglietta and Courbis 1969). The model was subsequently extended to 2000 equations to support the preparation of the VII Plan (Aglietta et al. 1973; Bussery et al. 1975).

The theoretical underpinning of the FIFI model was the theory of "competitioned economies" (économies concurrencées) developed by Courbis (1973). The theory distinguishes between two major sectors—a sheltered sector and a sector exposed to foreign competition. In the sheltered sector of the model, production was determined by demand; it affected the required investments and profits that influenced prices, being otherwise cost determined. This approach revealed the domination of the Keynesian mechanism. Production in the exposed sector depended on production capacity that expanded following new investments. The balance between production and demand was obtained from relevant changes in foreign trade (mainly imports). Prices were determined by the world prices. Rising prices constrained profits and consequently investments activities and production increase. The decisive role in the system was played by the dynamics of the exposed sector.

W. Welfe, *Macroeconometric Models*,
Advanced Studies in Theoretical and Applied Econometrics 47,
DOI 10.1007/978-3-642-34468-8_5, © Springer-Verlag Berlin Heidelberg 2013

5.1.2 The Annual Forecasting INSEE Models

The relevant elements of the national accounts were forecasted using a series of annual models. The first was ZOGOL that P. Herzog and G. Olive built at the INSEE and the Direction de la Provision in the years 1965–1966 (Herzog and Olive 1966). It was a one-sector, medium-size static model displaying Keynesian orientation. It was dynamised in the next annual model called DECA (Demande et Comportment d'Autofinancement), which contained around 200 equations (Billaudot 1971; Malgrange 1972). Between 1969 and 1972, the DECA model was used in forecasting the components of the national accounts. Because the model assumed that firms tended to self-finance their investment outlays, the specification of the investment functions had to be appropriately changed. Wages were endogenized using the Klein-Phillips relationship. The model was mainly used in short-term forecasting stressing the role of final demand. Compare also Nataf and Thonet (1962).

The next annual model STAR (Schema Theorique d'Accumulation et de Repartition) was constructed in order to handle the medium-term problems becoming a dominating element of economic policy. The model was operational in the years 1974–1977. It supported the forecasting of the relevant components of the national accounts (Boulle et al. 1974). Its authors laid emphasis on the specification of financial flows. Firms' profits affected investment activities that depended on output and the initial stock of fixed capital. Respective price adjustments were clearing the commodity markets.

Following the change in the system of national accounts, the STAR model was replaced by the annual model COPAIN (Comportements Patrimoniaux Intégrés), which was constructed in 1981 (Dehove et al. 1981). In response to the need to analyse the capital accumulation problems arising from the 1973 oil shock, the model was enhanced, so that an integrated analysis of financial flows became possible. The specification of the real sector remained Keynesian. However, several modifications were made compared with its predecessor. Personal wealth was introduced to the consumption function, the imports function was modified, interest rates were endogenized and the financial sector was integrated. It was essentially a one sector model, but its equations explaining production and foreign trade distinguished between the manufacturing industry and other industries.

5.2 Annual Models of the Second Generation

In the mid-70s, annual models of the second generation were built. Those were the DMS model constructed at the INSEE in the years 1974–1978, the MOGLI model built between 1974 and 1978 at GAMA, and the OFCE and HERMES-France models developed in the early 1980s (Fonteneau 1983).

5.2.1 The Annual DMS Model

The DMS (Dynamic Multi-Sectoral Model) was designed to replace the FIFI model (Fouquet et al. 1976). Its successive versions were used for economic analyses and contributed to the preparation of the VII, VIII and IX Plans (see Table 5.1). The depth of its disaggregation was comparable with the FIFI model and corresponded to the structure of national accounts. The model stressed the role of profits as a factor in financing investment and price formation. It also accentuated the role of production potential and supply, as the determinants of net exports. An increase in domestic demand resulted—given the domestic output—in the adjustments of exports and imports.

In general, the DMS was a Keynesian-oriented model, in both short and medium term. Output and employment were determined by effective demand. The structure of prices enabled the generation of appropriate profits and investment funds (Charpin and Fouquet 1982).

In the 1980s, the DMS model supported the attempts aimed at optimising the preparation of the VIII and IX Plans.

With a view to facilitating simulation analyses, two simplified versions of the model were built: the MINI DMS with ca. 200 equations and the MICRO DMS containing 30 equations. These abridged models were also used to run long-term simulations (Brillet 1981, 1993). The DMS model served as the major instrument assisting policy simulations and medium-term forecasts for many years, until it was abandoned at the beginning of the 21st c. Some of its functions were taken over by the French economy model being part of the world model built at the INSEE.

5.2.2 The Annual GAMA Models

The MOGLI model (Modèle Glissant) was constructed in the years 1974–1978 at GAMA (Group d'Analyses Macroeconomic), Paris-Nanterre University, under the direction of R. Courbis. It was an annual, multisectoral, dynamic model of the French economy (Courbis 1982; Courbis et al. 1982).

Unlike the DMS model, MOGLI showed the Keynesian orientation only in the short run. In the long run, it followed the concept of économies concurrencées.

Hence, in the long run production was determined by production potential, i.e. by the expected investments. Prices depended on world prices and not on the unit costs. The author of the model made some attempts to endogenize its public and banking sectors. The model supported regular forecasting and simulation analyses preceding the construction of consecutive Plans.

Towards the end of 1970s, the regional REGINA model was built at GAMA (Courbis 1979). It was used in preparation of the VII Plan. Its simplified version, mostly dynamic called REGIS was constructed for 8 regions. In the next years, the multisectoral I-O model ANAIS with 90 branches was added to the group of models characterised above (Courbis and Sok 1983). It was supplemented by a quarterly I-O model TAIS with 36 branches.

Table 5.1 Models of the French economy

Model Version Year	Institution Author	Frequency Sample	Equations total stochastic	Decomposition	Estimation method	Properties	Applications
FIFI 1966–1968 VI Plan VII Plan 1978	INSEE Direction de la Prevision R. Courbis	Annual	1300 2000	29 blocks I-O, 68 industries	OLS	Based on the theory of économies concurrencées	Simulations for the VI and VII plans
ZOGOL 1965–1966	P. Herzog, G. Olive	Annual	107 10	–		Keynesian, static	Short-term forecasts for 1 year; assumptions to the budgets 1966–1968
DECA 1967–1968		Annual	200	–		Keynesian, dynamic	Simulations for the budgets 1969–1973
STAR 1971–1973 1978	Group de Recherche Macroe-conomique (GRM)	Annual		–			Simulations for the budgets 1974–1977
COPAIN 1977–1979	Direction de la Prevision	Annual	350 100	Industry, other branches		Neo-Keynesian	Simulations for the budget
DMS I 1978 II IV 1987 MINI DMS MICRO DMS	INSEE Dynamic Multisectoral Model	Annual	1000 1900 2900	11 industries of the production sector Reduced version		Neo-Keynesian in the short- and in the long-run	Forecasts and simulation analyses

Table 5.1 (Continued)

Model Version Year	Institution Author	Frequency Sample	Equations total stochastic	Decomposition	Estimation method	Properties	Applications
MOGLI 1974–1978	GAMA (Model Glissant) R. Courbis	Annual	1300	Sectoral disaggregation	OLS	Neo-Keynesian in the short-run; dynamic version of économies concurrencées in the long-run	Forecasts and simulation analyses
OFCE 1982–1983	OFCE	Annual				Specification based on MINI DMS, extended financial sector	Simulation analyses
HERMES FRANCE	Ecole Central de Paris	Annual	ca. 1500			Specifications followed HERMES models	Simulation analyses
SIMPLET 1972	Banque de France J.H. David	Quarterly	7	–		Monetarist model	Simulation analyses
METRIC 1977 1981	INSEE	Quarterly	400 900	Sectoral disaggregation		Neo-Keynesian	Forecasts, versions MINI-METRIC, MICRO-METRIC used in simulation analyses
METRIC 1988	INSEE	Quarterly	950			Extended financial sector	Forecasts, policy simulations
ICARE 1982–1983	IPECODE	Quarterly	Large	6 industry groups		Extended supply sector	Forecasts and simulations analyses
PROTEE 1982–1986	GAMA R. Courbis	Quarterly	200			Structure close to MOGLI	Short-term forecasts, monthly updated

Table 5.1 (Continued)

Model Version Year	Institution Author	Frequency Sample	Equations total stochastic	Decomposition	Estimation method	Properties	Applications
OFCE 1983–1984		Quarterly				Model based on MINI METRIC	Forecasts and simulations analyses
Model with Quantity Rationing 1984–1985	P. Artus et al.	Quarterly				Model of disequilibrium	Simulation analyses
AMADEUS 1998	Direction de la Prevision, INSEE	Quarterly					Forecasts
MESANGE 2002	Direction de la Prevision, INSEE	Quarterly 1978–1979	558 40	3 industry groups, 4 economic agents	ECM	Demand neo-Keynesian, neoclassical output, adaptive expectations	Short- and medium-term forecasts; simulation analyses
MANEGE 2001	Direction de la Prevision N. Carnot	Quarterly	36 12		ECM	Keynesian in the short-run, neoclassical supply sector	Small model to be used in policy simulations

5.2.3 Other Annual Models

The annual OFCE model that drew on the MINI DMS model was built in the years 1982–1983. It was extended with large production and financial sectors and served in many simulations of the money market policy (Boutillier and Durand 1986). The annual model HERMES-FRANCE was constructed between 1980 and 1984 at the Ecole Centrale de Paris as a component of a multicountry model being developed for the European Commission. Built after the oil shocks, the model stressed the role of the energy sector and accounted for the impacts of energy prices on economic development. Its general structure resembled the DMS model. It was highly disaggregated and had approximately 1500 equations (Faubry et al. 1984). Compare also the structure of the model AGORA (Peaucelle et al. 1981).

5.3 Quarterly Models of the French Economy

At the beginning of the 70s a small quarterly model SIMPLET was built at the Bank of France. Showing similarities to the American St. Louis model, SIMPLET had a monetarist orientation and only 7 equations (David 1975). The large quarterly models of the French economy were constructed in the mid-70s, nearly 10 years after their annual predecessors.

5.3.1 The METRIC Model

With the quarterly national accounts of the French economy becoming available, a large quarterly macroeconometric model of the French economy METRIC (Modèle Économétrique Trimestriel de la Conjuncture) was constructed at the INSEE (Nasse et al. 1977). Because of sectoral disaggregation, the first version of the model had 400 equations already in 1977. Following the extension of its supply sector in the mid-80s, the number of equations grew to 950 in the model version called METRICX (Bloch et al. 1988).

The model had a neo-Keynesian structure. In the short-run, output and employment were determined by expected demand. An important role in the specification of its equations played consumer and producer expectations that were based on the survey results. The model acknowledged the existence of disequilibria in the commodity, labour and financial markets (Artus and Volle 1982).

To facilitate policy simulations, the reduced versions of the model were constructed, which were called MINI-METRIC and MICRO-METRIC (Deleau et al. 1984).

After the METRIC model was constructed, other large quarterly models of the French economy were built in the 1980s (Artus et al. 1986). The OFCE model following the MINI-METRIC model was added an elaborated block of financial flows

(Sterdyniak et al. 1984). The ICARE (Ipecode—Cadre pour l'Analyse Réflexion Economiques) model was a large quarterly model, which was mainly designed to describe the real sector (6 industries were distinguished). It additionally included an extended block of equations explaining tax incomes in relation to the financial behaviour of enterprises (Dumazet and Khong 1989). Compare also the financial model DEFI (Villa 1982).

5.3.2 *The PROTEE Model*

The PROTEE (Projection Trimestrielle de l'Évolution Économique) model was built at GAMA (Courbis and Salmon 1986). It was used since 1987 mainly to prepare short-term forecasts that were updated on a monthly basis. Its structure resembled the MOGLI model. Demand determined in the short-run, in the long run the model showed supply orientation.

A system of short and hypershort forecasts (monthly and weekly) was developed at GAMA being based on a monthly model EDMEE of GAMA. It uses monthly indicators and also monthly variables in terms of national accounts. This model supplied the necessary information that was next used in quarterly forecasting based on the PROTEE model. A medium-term version of the model was also available (Courbis 1997).

5.4 Disequilibrium Models

In the second half of the 1980s the first attempts were made to construct the disequilibrium models. M.J. Vilares built his annual model of the French economy, assuming constraints in the production sector (Vilares 1986). The quarterly disequilibrium models that were constructed afterwards allowed distinguishing periods with prevailing either classical unemployment or Keynesian unemployment (Artus et al. 1984). The contributions were crucial for the development of macroeconometric disequilibrium models in Europe, which are characterized in the next chapters.

A comparative analysis of the five models of the French economy maintained in the 1980s was presented in a special issue of "Economie et Prevision" (1998).

5.5 Models Built in the Last Twenty Years

In the 1990s, the Ministry of Finance, Division de la Prevision, and the INSEE continued their research on the quarterly models (Courbis 1991). By the end of the decade a new quarterly model AMADEUS drawing on the METRIX model was constructed (Michaudon and Prigent 1998).

5.5.1 The MESANGE Model

The large quarterly MESANGE (Modele Econometrique de Simulation et d'Analyse Genérale de l'Economie) model was built at the threshold of the 21st c. (Allard-Prigent et al. 2002). It was a large model with a small number of stochastic equations, where four groups of economic agents and three groups of industries in the producer sector were distinguished. Its short-run dimension was Keynesian: production and employment were determined by final demand. The model had an extended supply sector with equations based on profit maximization. Prices were determined by labour and capital costs, and wages by retail prices, labour productivity and unemployment rate.

The MESANGE model has been systematically used in forecasting and simulation analyses for government institutions. Because it belongs to this class of the medium-size quarterly models that require highly-specialized staff to be run, it has been only accessible to the research departments of government agencies. It is still in use.

5.5.2 The MANEGE Model

To facilitate the potential business users or users at research institutions to take advantage of the macromodels, at the beginning of the 21st c. N. Carnot built a small quarterly model of the French economy called MANEGE (Modele Agrege National pour l'Etude General de l'Economie) (2002). The model can be viewed as counterpart to the earlier mini-models of the French economy. Its equations explaining consumer demand and investment demand display Keynesian orientation, however those determining the supply side have been derived from the neoclassical concept of profit maximization under monopolistic competition. The model includes equations determining producer prices, wages and demand for production factors, assuming the CES technology and constant returns to scale. The wage equations represent the negotiations between producers and trade unions. The rates of unemployment are determined by both prices and wages. The expectations are adaptive.

References

Aglietta, M., & Courbis, R. (1969). Outil pour le plan: le modèle FIFI. *Economie et Statistique*, *1*, 45–65.

Aglietta, M., Courbis, R., & Seibel, C. (1973). *INSEE collections, series C: Vol. 22. Le modéle FIFI, tome 1, Présentation générale et utilization*. Paris: INSEE.

Allard-Prigent, C., Audenis, C., Berger, K., Carnot, N., Duchene, S., & Pesin, F. (2002). *Presentation du modele MÉSANGE. Modèle économétrique de simulation et d'analyse générale de l'economie*. Paris: Direction de la Prevision, Ministere de l'Economie, des Finances et de l'Industrie.

Artus, P., & Volle, M. (1982). The METRIC model: presentation, simulation and multipliers. In J. Plasmans (Ed.), *Econometric modelling in theory and practice* (pp. 123–158). The Hague: Martinus Nijhoff.

Artus, P., Laroque, G., & Michel, G. (1984). Estimation of a quarterly macroeconomic model with quantity rationing. *Econometrica*, *52*, 1387–1414.

Artus, P., Deleau, M., & Malgrange, P. (1986). *Modélisation macroeconomique*. Paris: Economica.

Billaudot, B. (1971). Le modèle Deca. *Statistique et Etudes Financiers, Series Orange*, *1*, 5–46.

Bloch, L., Kadjar, A., Rabenanarjara, R., Ralle, P., Allard, D., Boeuf, P., Fraichot, J.-P., & Loue, J.-F. (1988). Preséntation du modèle METRICX. *Economie et Prevision*, *85*, 5–26.

Boulle, J., Boyer, R., Mazier, J., & Olive, G. (1974). Le modèle STAR. *Statistique et Etudes Financiers, Series Orange*, *15*, 1–68.

Boutillier, M., & Durand, B. (1986). Investigations in the causal structure of the yearly OFCE model. *Journal of Economic Dynamics and Control*, *10*, 131–137.

Brillet, J. L. (1981). *Collection archives et documents: Vol. 35. MINI-DMS: modèle macroéconomique de simulation*. Paris: INSEE.

Brillet, J. L. (1993). *MICRO-DMS: présentation et proprietes* (Working Paper G 93 06). Paris: INSEE.

Bussery, H., Courbis, R., & Seibel, C. (1975). *INSEE collections, series C: Vols. 37–38. Le modèle FIFI, tome II, Les équations*. Paris: INSEE.

Carnot, N. (2002). MANEGE: a small macro-econometric model of the French economy. *Economic Modelling*, *20*, 69–92.

Charpin, J. M., & Fouquet, D. (1982). The DMS model, version 2. In J. Plasmans (Ed.), *Econometric modelling in theory and practice* (pp. 53–81). The Hague: Martinus Nijhoff.

Courbis, R. (1973). La théorie des "économies concurrencées" fondement du modèle FIFI. *Revue Economique*, *24*, 905–922.

Courbis, R. (1979). The REGINA model, a regional-national model for French planning. *Regional Science and Urban Economics*, *9*, 117–139.

Courbis, R. (1982). La teorie des économies concurrencées et ses applications: de FIFI à MOGLI. In J. L. Reiffers (Ed.), *Economie et finances internationales* (pp. 395–424). Paris: Dunod.

Courbis, R. (1991). Macroeconomic modelling in France. In R. G. Bodkin, L. R. Klein & K. Marwah (Eds.), *A history of macroeconometric model-building* (pp. 231–266). Aldershot: E. Elgar.

Courbis, R. (1997). Quarterly models of the world and French economy. In W. Welfe (Ed.), *Economies in transition and the world economy. Models forecasts and scenarios* (pp. 393–424). Frankfurt/Main: Peter Lang.

Courbis, R., & Salmon, P. (1986). *Le modele PROTEE, un modèle trimestriel de l'economie française. Présentation générale et utilisation* (GAMA Paper 544). In Second French-Polish seminar on modelling, Nanterre, 29–30 May.

Courbis, R., & Sok, H. (1983). Le modèle ANAIS, un modèle inter-sectoriel détailée de l'economie française. *Prevision et Analyse Economique (Cahiers du GAMA)*, *1*, 73–101.

Courbis, R., Fonteneau, A., Le Van, C., & Voisin, P. (1982). The MOGLI model: a pluri-sectoral econometric dynamic model of the French economy. In J. Plasmans (Ed.), *Econometric modelling in theory and practice* (pp. 27–52). The Hague: Martinus Nijhoff.

David, J.-H. (1975). A monetarist model of the French economy. In F. Masera, A. Fazio & T. Padoa-Schioppa (Eds.), *Econometric research in European central banks* (pp. 45–95). Rome: Banca d'Italia.

Dehove, M., Fauqueur, A. Q., Gaudemet, J.-P., Husson, M., Mathis, J., de Monchy, G., & Vallet, D. (1981). Le modèle COPAIN: comportements patrimoniaux et intégration financière. *Economie et Prevision*, *48*, 1–70.

Deleau, M., Malgrange, P., & Muet, P.-A. (1984). A study of short-run and long-run properties of macroeconometric models by means of an aggregative core model. In P. Malgrange & P.-A. Muet (Eds.), *Contemporary macroeconomic modelling* (pp. 215–246). Oxford: Basil Blackwell.

Dumazet, D., & Khong, V. (1989). ICARE 1980. Le modèle ICARE en nouvelle base de comptabilité nationale. *Revue de l'IPECODE*, *22*, 11–43.

Economie et Prevision (1998). Structure et proprietes de cinq modèles macro-economiques française. *Economie et Prevision, 134*, 1–96 (special issue).
Faubry, E., Moncoble, J.-E., Vidal de la Blache, O., & Zagame, P. (1984). Le modèle Hermès-France. *Economie et Prevision*, *66*, 3–29.
Fonteneau, A. (1983). Le modèle OFCE annuel. *Prevision et Analyse Economique (Cahiers du GAMA), 4*.
Fouquet, D. J., Charpin, J. M., Guillaume, H., Muet, P.-A., & Vallet, D. (1976). DMS, modèle de prévision à moyen term. *Economie et Statistique*, *79*, 33–48.
Herzog, P., & Olive, G. (1966). *Le modèle de projection à court terme ZOGOL I* (INSEE-DP Working Paper).
Malgrange, P. (1972). Etude analitique du modèle DECA. *Annales de l'INSEE*, *11*, 85–139.
Michaudon, H., & Prigent, C. (1998). *Présentation du modèle AMADEUS* (Document de travail, G 9801). Paris: INSEE.
Nasse, P., de Menil, G., Artus, P., Bournay, J., Morin, P., Porcher, R., Yohn, F., & Zaidman, C. (1977). METRIC, modèle économétrique trimestriel de la conjuncture. *Annales de l'INSEE*, *26–27*, 1–364.
Nataf, A., & Thonet, P. (1962). Le modèle a moyen terme à prix variables SEEF. *Etudes de Comptabilite National*, *3*, 35–86.
Peaucelle, L., Petit, P., & Saillard, Y. (1981). *Le modèle AGORA. Présentation succincte et analyse de quelques multiplicateurs* (CEPREMAP Working Paper No. 8122).
Sterdyniak, H., Boudier, M.-A., Boutillier, M., Charpin, F., & Durand, B. (1984). Le modèle trimestrielle de l'OFCE. *Observations et Diagnostics Economiques (Revue de l'OFCE)*, *9*, 137–178.
Vilares, M. J. (1986). *Structural change in macroeconometric models. Theory and estimation*. Dordrecht: Martinus Nijhoff.
Villa, P. (1982). Modèlisation macroéconomique des structures financiers. In *Collection archives et documents: Vol. 53. Le modèle DEFI*. Paris: INSEE.

Chapter 6
The Dutch Macroeconometric Models

6.1 Introduction. J. Tinbergen's Contributions

The Dutch macroeconometric modelling started with J. Tinbergen's early contributions. In 1936, Tinbergen constructed the first model of Netherland's economy and described its business cycle (Tinbergen 1937). The model was a precursor of Tinbergen's famous model of the USA that he built for the League of Nations (see Chap. 2).

It was not until 1953 that the Central Planning Bureau headed by J. Tinbergen constructed the first new-generation model of Netherland's economy. The institution continued to build macromodels that were maintained into the mid-70s. These were small, annual models whose structures were influenced by Tinbergen's contributions. From their beginning through the 1970s, they served as major instruments supporting the construction of successive development plans for Dutch economy.

6.2 Annual Models of the Central Planning Bureau (CPB)

6.2.1 The Verdoorn-Koyck Model

The annual CPB models can be identified by the years of their construction, starting with 1953. The annual model of 1955 constructed by the team led by P.J. Verdoorn and L.M. Koyck used a sample containing mainly the pre-war data, likewise the next models of 1961 and 1963D (Verdoorn et al. 1970). All the models initially had similar specifications. Consumer demand was determined by real labour income and other incomes, lagged consumption (Brown's habit-formation approach), prices, as well as by deposits constituting a composite characteristic of households' financial assets. In the early investment function investment activity depended on output (flexible accelerator), but the next model versions used the neoclassical specification where investments depended on financial resources (profits, etc.), prices

W. Welfe, *Macroeconometric Models*,
Advanced Studies in Theoretical and Applied Econometrics 47,
DOI 10.1007/978-3-642-34468-8_6, © Springer-Verlag Berlin Heidelberg 2013

of investment goods and financial assets, and in the long-run on the rate of capacity utilization represented by the rate of unemployment. Prices were determined by the major components of costs. It must be emphasised that all relationships in the model were expressed through rates of growth, as opposed to the initial versions, where relationships were defined in terms of first differences.

6.2.2 *The CS, VINTAF and FREIA Models*

In the second half of the 1960s C.A. van den Beld constructed a medium-term CS model with explicit production functions that were used to generate production capacity utilization rates (Van den Beld 1968). The model represented an intermediate stage towards the building of the second-generation models, whose major representative was the annual VINTAF model that H. Den Hartog and H.S Tjan constructed in the mid-1970s, which was afterwards updated to VINTAF II. The characteristic property of the VINTAF models was that the vintage approach was introduced into the production functions. Fixed capital was obtained as a weighted sum of machinery and equipment, the weights representing declining rates of growth. The demand for labour was determined by rising fixed capital-labour ratios (Den Hartog and Tjan 1976).

In early 1980s the VINTAF models were substituted by an annual model called FREIA, where the VINTAF's real sector was combined with a new submodel of financial flows (Hasselman et al. 1983). The data used in the FREIA model covered a period going back to 1954. The new thing about this medium-size model (exceeding 300 equations) was that employment in the real sector was determined as a weighted sum of labour demand and supply. The disequilibria in the financial sector were also allowed for (interest rates did not bring the money markets into balance).

6.3 The Quarterly CPB Models

6.3.1 *The Driehuis Model*

The necessity to construct quarterly models was recognized as early as the beginning of the 1970s. The first small quarterly model built by W. Driehuis used the 1951–1964 data (Driehuis 1972). Being generally a demand-oriented model, it used many neoclassical elements to specify the investment and employment equations. For details cf. Table 6.1.

6.3.2 *The KOMPAS, FREIA-KOMPAS and FKSEC Models*

In the early 1980s, the Driehuis model was substituted by a large, quarterly KOMPAS model built under the direction of P.J.C.M. Van den Berg. The model had a vintage production function and a large block of equations to explain financial flows, including the securities and insurance markets. The model was very dynamic and used many lag distributions, also the Koyck transformation.

The KOMPAS model was merged with the FREIA model around the mid-1980s, as a result of which the large quarterly model FREIA-KOMPAS having more than 500 equations came into being. The specification of its particular equations was generally improved in some cases. For many years, the model served as a forecasting tool and an efficient policy simulation instrument (Van den Berg et al. 1987).

The quarterly FKSEC model constructed in the early 1990s was a continuation of the FREIA-KOMPAS model.

6.4 The CPB Models in the 1990s and Later

In the 1990s new versions of the CPB models were prepared (Broer et al. 1998). It is worth noting that CPB's macromodelling activity was characterised by evolutionary improvement of model specifications that responded to new needs, and by limited absorption of innovations introduced by foreign research centres (Barten 1991). In the second half of the 1990s several new models of various orientations were built at the CPB (Broer et al. 1998).

6.4.1 The JADE Model

In the last years of the 1990s the new, large annual model JADE (Joint Analysis of Dynamics and Equilibrium) was introduced, which was primarily treated as an instrument serving policy simulations. The model distinguished the exposed and sheltered industries, and the building industry. Employees were divided into two groups by their educational attainment. Its short-run dimension was demand determined, while in the long-run the specification was based on the neoclassical theory of production. In the estimation process, the long-term relations were treated separately from the short-term adjustments; the ECM techniques were employed (Huizinga 1998).

6.4.2 The ATHENA and MIMIC Models

The multisectoral model ATHENA constructed in the mid-1980s (1900) expanded over the next 15 years to ca. 7500 equations (CPB 1990). It had an extended I-O block and a block for the financial flows of the institutional sectors. The model included sectoral vintage production functions. Used mainly for sectoral analyses, it also helped prepare the medium- and long-term scenarios of economic development (Vromans 1998).

Table 6.1 Models of the Dutch economy

Model Version Year	Institution Author	Frequency Sample	Equations total stochastic	Decomposition	Estimation method	Properties	Applications
Model 1955 1953–	Central Planning Bureau P.J. Verdoorn	Annual 1949–1966	27	–	First difference	Extended specification of Tinbergen model	Simulations for the Central Plans, short-term forecasts
Model 1961 1958–	P.J. Verdoorn, E.M. Koyck	Annual 1923–1938 1948–	36	–			
63D 1964–1969	C.A. Van den Beld	Annual 1949–1966		–	OLS TSLS	Medium-term model	Simulation for the Central Plan, short-term forecasts
CS model 1967							
Driehuis model 1970	W. Driehuis	Quarterly 1951–1964	68 71				Short-term forecasts
VINTAF 1974–1975	H. Den Hartog, H.S. Tjan	Annual	112		TSLS	Vintage production function	Forecasts and policy simulations
FREIA 1983	Central Planning Bureau P.J.C.M. van den Berg	Annual 1954–1975	332		Rates of growth		Forecasts and simulation analyses

Table 6.1 (Continued)

Model Version Year	Institution Author	Frequency Sample	Equations total stochastic	Decomposition	Estimation method	Properties	Applications
KOMPAS 1983		Quarterly	851			Integration of real and financial sectors	Forecasts and simulation
FREIA-KOMPAS	Central Planning Bureau	Quarterly	500			Combining models FREIA-KOMPAS	Forecasts and simulations
FKSEC	Central Planning Bureau	Quarterly		–		FREIA-KOMPAS models continued	Forecasts and simulations
JADE (Joint Analysis of Dynamics and Equilibrium) 1997	Central Planning Bureau	Annual	2200	6 groups of industries, 2 groups of employees by education levels	ECM	In the short-run demand determined, in the long-run neoclassical	Mainly scenario analyses, medium-term forecasts
ATHENA (Multi-sector Model)	Central Planning Bureau						
1988		Annual			Calibration, OLS	Sectoral analyses	Scenarios, forecasts

Table 6.1 (Continued)

Model Version Year	Institution Author	Frequency Sample	Equations total stochastic	Decomposition	Estimation method	Properties	Applications
1996		Annual	7500 2500	15 industry groups, I-O matrix, 6 institutional sectors	Calibration, OLS	Vintage production function	Scenarios medium and long-term forecasts
MIMIC 1997	Central Planning Bureau	Annual	15000	6 industry groups, 5 factors of production, 3 categories of employees	Calibration	Computational general equilibrium model	Simulation analyses, mainly regarding tax and insurance systems
SAFE 1998	Central Planning Bureau	Quarterly				Substituted FKSEC, specification similar to JADE	Short-terms forecasts simulation analyses
GRECON	Econometric Institute, University of Groningen	Annual			Rates of growth, TSLS		Forecasts
1977			18				
1986			9	–			
SECMON 1978–1980 1981 1986	University of Amsterdam W. Driehuis	Annual	700	18 groups		Multisectoral, including I-O submodel	Simulation analyses

Table 6.1 (Continued)

Model Version Year	Institution Author	Frequency Sample	Equations total stochastic	Decomposition	Estimation method	Properties	Applications
MORKMON	Central Bank M.M.G. Fase					Extended monetary sector	Forecasts and simulation analyses
I 1981		Quarterly 1970–1979	360 60				
II 1992		Quarterly			ECM		
CESAM 1990	CSSO, University Groningen S.H. Kuipers et al.	Annual				Mixed model: Keynesian demand, production based on profit maximization	Forecasts, simulations analyses
IBS-CCSO 1995	CCSO University of Groningen and Twente, IBS Amsterdam, Brokerage firm	Quarterly	172		ECM	Distinguished financial and real blocks, adaptive expectations.	Forecasts and simulations analyses by the brokerage firm

Another noteworthy model is the long-term general equilibrium model MIMIC. It contained extended submodels explaining the labour market, financial flows and particularly the social insurance system. The model was used during studies in preparation of the fiscal system reform and within the unemployment studies (Graafland and de Mooij 1998).

6.4.3 The Model SAFE

In the late 1990s the FK SEC model was replaced by the quarterly model SAFE designed for the purpose of short-term forecasting (up to 3 years) and economic analyses (Donders and Lunsing 1999). The structure of the SAFE model resembled that of JADE's, but sectoral decomposition and the estimation of the production function were abandoned. However, the model generated the capacity utilization rate affecting prices and exports. The demand for the production factors was obtained by solving the enterprises optimization problem. The investment function was specified taking account of real user costs and profits, in addition to output. The demand for labour was determined by output (weighted by labour-output ratios) and the relative labour costs.

6.5 The MORKMON Model of the Central Bank

At the beginning of the 1980s the quarterly model MORKMON of the Dutch economy was built at the Central Bank of Netherlands under the direction of M.M.G. Fase. In the first version of the model the data sample covered the years 1970–1979. The model integrated the real and financial sectors and contained submodels for the monetary and real sectors (Fase 1985). In the first sector, the equations explaining interest rates and exchange rates were specified. In the real sector, the equations were specified like in the CPB models. Consumer demand was determined by real labour incomes and other incomes, and investment by available incomes (mainly profits) and long-term interest rates. Prices depended on unit costs and a mark-up, and production was dependent on expected sales and inventory changes. Employment was determined by output and fixed capital.

Although the MORKMON model's structure resembled that of the CPB models in many respects, it also contained many original solutions, particularly regarding the modelling of the financial sector.

6.6 The Models of the Academic Institutions

At the end of the 1970s, preconditions appeared for the Universities of Amsterdam and Groningen to build the macroeconometric models.

6.6.1 The SECMON Model

In the early 1980s, the large, annual model SECMON was constructed at the University of Amsterdam under the direction of W. Driehuis (Driehuis et al. 1983). It was a multisectoral structure with around 700 equations, which distinguished 18 industry groups and an I-O submodel. It was mainly used to run medium-term policy simulations on behalf of large corporations. At that University the research on Hermes model for the Netherlands was developed (Mot et al. 1989).

6.6.2 The Groningen University Models

In the second half of the 1970s a small, annual model called GRECON was constructed at the Econometric Institute of the Groningen University. The parameters of its equations were estimated with data in the form of the rates of growth, using the TSLS procedure (Voorhoeve 1986).

Toward the end of the 1980s, macromodelling activities developed also at the Groningen University's Centre for Cyclical and Structural Research (CCSO). The first to be built was the CCSO annual model of the Dutch economy, which had an extended production sector (Kuipers et al. 1987, 1990). In the next years the annual model CESAM and the quarterly model IBS-CCSO having similar structures were constructed in collaboration with the Twente University. The quarterly model was used commercially by the brokerage firm IBS that made forecasts and simulation analyses for its clients with it (Jacobs and Sterken 1995).

The models were Keynesian regarding the generation of the short-term demand, but neoclassical with respect to the generation of production. Potential output and potential employment were determined using a vintage CES production function. The quarterly model had an extended sector of financial flows based on the post-Keynesian portfolio theory. The equation parameters were estimated with the ECM models.

The Groningen University models supported regular forecasting and numerous policy simulations.

References

Barten, A. P. (1991). The history of Dutch macroeconometric modelling, 1936–1986. In R. G. Bodkin, L. R. Klein & K. Marwah (Eds.), *A history of macroeconometric model-building* (pp. 153–194). Aldershot: E. Elgar.

Broer, P., de Mooij, R., & Okker, V. R. (1998). CPB models and their uses. *GBP Report*, *3*, 17–20.

CPB (1990). *CPB monograph: Vol.* 30. *ATHENA, een bedrijfstakken model voor de Nederlandse economie (ATHENA, a sectoral model of the Dutch economy)*. The Hague: CPB.

Den Hartog, H., & Tjan, H. S. (1976). Investment, wages, prices and demand for labour (A Clay-Clay vintage model for the Netherlands). *De Economist*, *124*, 32–82.

Donders, J., & Lunsing, H. (1999). SAFE: a model for short-term analyses and forecast for the Dutch economy. *CPB Report, Quarterly Review of CPB, 3*, 42–44.

Driehuis, W. (1972). *Fluctuations and growth in a near full employment economy: a quarterly econometric analysis of the Netherlands*. Rotterdam: Universitaire Pres.

Driehuis, W., van Ierland, E. C., & van den Noord, P. J. (1983). *A sectoral model for the Netherlands economy, SECMON-C* (Mimeo). Foundation for Economic Research of the University of Amsterdam.

Fase, M. M. G. (1985). *Monetary monographs: Vol. 2. MORKMON, a quarterly model of the Netherlands economy for macroeconomic policy analysis*. Dordrecht: Nijhoff.

Graafland, J., & de Mooij, R. (1998). An applied general equilibrium model for the Netherlands. *GPB Report, 3*, 27–31.

Hasselman, B. H., Okker, V. R., & den Haan, R. J. A. (1983). *FREIA, een macroeconomisch model voor middellange termijn (FREIA, a macroeconomic model for the medium-term)*. ‘s-Gravenhage: Centraal Planbureau.

Huizinga, F. (1998). JADE a model for the joint analysis of dynamics and equilibrium. *GPB Report, 3*, 21–26.

Jacobs, J., & Sterken, E. (1995). The IBS-CCSO quarterly model of the Netherlands. Specification, simulation and analysis. *Economic Modelling, 12*, 111–163.

Kuipers, S. H., Kuper, G. H., & Sterken, E. (1987). *The CCSO annual model of the Dutch economy, specification, estimation and forecasting* (Mimeo). Faculty of Economics, University of Groningen.

Kuipers, S. H., Jongbloed, B. W. A., Kuper, G. H., & Sterken, E. (1990). The CCSO annual model of the Dutch economy. *Economic Modelling, 7*, 202–250.

Mot, E. S., van den Noord, P. J., van der Stelt-Scheele, D. D., & Koning, M. A. (1989). *HERMES, the Netherlands, SEO*. Amsterdam: Foundation for Economic Research, University of Amsterdam.

Tinbergen, J. (1937). *An econometric approach to business cycle problems*. Paris: Hermann et Compagnie.

Van den Beld, C. A. (1968). An experimental medium-term macro model for the Dutch economy. In M. G. Kendall (Ed.), *Mathematical model building in economics and industry* (pp. 31–48). London: Griffin.

Van den Berg, P. J. C. M., Gelauff, G. M. M., & Okker, V. R. (1987). *The FREIA-KOMPAS model for the Netherlands: a quarterly macro-economic model for the short and medium term* (Occasional Paper 39). ‘s-Gravenhage: Centraal Planbureau.

Verdoorn, P. J., Post, J. J., & Goslinga, S. S. (1970). *The 1969 re-estimation of the annual model: model 69-C* (Mimeo). Central Planning Bureau.

Voorhoeve, W. (1986). *Short-term economic forecasting in practice: the GRECON case. Econometric fundamentals and empirical results using data of the Dutch economy* (Mimeo). Groningen: University of Groningen.

Vromans, M. (1998). ATHENA. The multi-sector model. *CPB Report, 3*, 32–36.

Chapter 7
The Models of the Nordic and Baltic Countries

7.1 Introduction

In Scandinavian countries, modelling started to develop in 1970s, mainly at public institutions such as the Danish and Norwegian central statistical offices and the Danish and Finnish Ministries of Finance. The models were annual and supported economic planning. Later on, quarterly models were constructed at the central banks of Denmark, Finland, Norway and Sweden to aid regular forecasting and policy simulations.

The above models were intended to characterise the functioning and growth of small open economies. They mainly had the Keynesian orientation. In the short-run, production and employment were determined by demand. The large models included I-O submodels and distinguished industries exposed to foreign competition and the sheltered ones. The price and foreign trade equations were differently specified for these groups of industries (Bergman and Olsen 1992; Whitley 1992). In the Baltic countries (Estonia, Latvia, Lithuania), modelling activities intensified with the countries' becoming independent, supported by the Danish and Finnish econometricians.

The models' properties are discussed in the sections below and are summarised in Table 7.1.[1]

7.2 The Models of the Danish Economy

7.2.1 The ADAM Model

The model ADAM (Aggregated Danish Annual Model) was constructed at the Danish Central Statistical Office, the country's economic modelling centre since the

[1] In writing the text of this chapter the following monographs were consulted: Lybeck et al. (1984), Bjerkholt and Rosted (1987), and Bergman and Olsen (1992).

Advanced Studies in Theoretical and Applied Econometrics 47,
DOI 10.1007/978-3-642-34468-8_7,

1970s. The model was chiefly used by the Ministry of Finance as an analytical tool supporting the development of financial plans and of the government budget. In 1981, it became a member of Project LINK (Dam 1986).

The ADAM model was started as a large annual model with an I-O submodel that distinguished 20 industries and ca. 72 commodity groups representing final demand.

The model had a Keynesian orientation. Consumer demand was determined by real income and real personal financial and physical wealth. Investment demand depended on an increase in production, on the relative prices of production factors, including interest rates, and on the Tobin "Q" coefficients. Final demand determined output and employment in most of the industries. The production functions were introduced to generate potential output and the rates of capacity utilization used to determine exports.

The supply side was represented by the wage and price equations. It was assumed that wage changes resulted from bargaining, so wages depended on prices, shares in profits and the rate of unemployment. Prices were dependent on the domestic unit costs and import costs plus a mark-up.

The model had an extended financial sector with equations generating budget revenues and expenditures and financial flows. Interest rates being the major channels of transmission from the financial sector to the real sector were assumed to clear the financial markets (Hansen and Smidt 1992).

The model was revised on a regular basis. The data were updated 3 times a year and the equations were respecified every second year. The most recent samples cover the years 1966–2009 for most series. The model had 2038 equations (90 were stochastic) and a large number of exogenous variables. Its parameters were mostly estimated with OLS.

The model ADAM was, and still is, frequently used by the Ministry of Finance and other non-public economic organisations for forecasting and policy simulation purposes.

7.2.2 The SMEC and CLEO Models

At the end of the 1970s, the model SMEC (Simulation Model of the Economic Council) was built to serve the needs of the Economic Council of Denmark (Hansen and Paldam 1973). It was an annual, medium-size model with a Keynesian structure. The supply-side components such as prices, wages, and labour productivity were exogenous. The financial sector was omitted. In the 1990s, the computational general equilibrium model GESMEC was built for the same institution (Frandsen et al. 1994). In the same decade, a special, annual model CLEO was constructed at the Copenhagen University for the long-run analyses (Kaergärd 1991).

7.2.3 The MONA Model

The quarterly model MONA built at the Central Bank of Denmark at the end of 1980s was to support short-term forecasting and economic analyses. It was a medium-size model with an extended financial sector. It had a Keynesian orientation. Its equations explaining financial markets used model consistent expectations (Christensen and Knudsen 1992).

7.3 The Models of the Estonian Economy

7.3.1 The ESTMOD Model

In Estonia, macromodelling activities started in the transition period based on an agreement that the Estonian Ministry of Finance signed with its Finnish counterpart. A team directed by A. Leppä built the quarterly model MODEST (Leppä 1995), afterwards renamed ESTMOD. The original model used the 1994–1995 quarterly data and had three versions, each one using more data, which ended in 1998. Many parameters of the model's 16 stochastic equations were calibrated.

The model had a mixed structure. Final demand determined production and employment, while consumer demand depended on real disposable income and nominal, short-term interest rates. Investment demand was dependent on output, capital costs and the real interest rate. The model included equations that explained producer and retail prices, as well as wages determined by retail prices and GDP. The equations accounting for foreign trade transactions and prices were extended; an additional equation was specified to explain the inflow of foreign capital, an important factor stimulating the growth of the Estonian economy.

Most equation parameters were estimated using the ECM procedure. Several policy simulations were performed to test the model for accuracy (Leppä et al. 2000).

At around the same time, a model of the Estonian economy started to be constructed at the Central Bank of Estonia.

7.4 The Models of the Finnish Economy

7.4.1 The KESSU Model

The annual model KESSU was built in the late 1970s at the Finnish Ministry of Finance. It was intended to be used in planning processes and medium-term simulation analyses supporting the construction and execution of government budgets (Mannermaa and Kaski 1980). Because of its intended use, it was a large macromodel (see Table 7.1) decomposed into 25 industries, with an I-O submodel. It had several versions, but the first version defined its basic structure that was maintained in the next versions. In the KESSU III model several changes were introduced into the specifications of equations explaining the behaviour of households and enterprises (Leppä 1987). The changes were retained in the KESSU IV version that was provided with an extended submodel of financial flows; the interest rates were endogenized (Hetemäki 1992).

Consumer demand was decomposed and made dependent on real disposable income, financial wealth and real interest rates. Investment demand was generated according to profit maximization and was dependent on the user costs of capital and output. The equations explaining production and employment were separately specified for the exposed industries (mainly the manufacturing industry) where profit maximization and relative prices prevailed and for the sheltered industries with employment determined by output and relative prices.

Prices in the exposed industries depended on the world prices, whereas in the sheltered industries they were determined by unit costs. Wage rates were shaped by exogenous, negotiated rates and by factors generating deviations, such as changes in labour productivity.

7.4.2 The BOF and QMED Models

In the 1970s, the Bank of Finland's research teams started to construct a quarterly model of the country's economy, whose successive versions were subsequently elaborated in the 1980s and 1990s (Tarkka and Willman 1985). The most significant change in the builders' approach took place in the mid-80s, when they abandoned the Keynesian orientation (models BOF1 to BOF3) to replace it with the neoclassical one. For the BOF4 and BOF5 models it was assumed that in the short-run final demand would be determining output and employment, while in the long-run the decisive factor would be production formation. Besides, rational expectations were introduced into both versions. The modifications were associated with the change in the estimation procedure, involving mainly the replacement of OLS-based estimation with ECM.

As its predecessors, the model BOF 4 was a large quarterly system. It distinguished 5 industries and used I-O relationships (Männistö et al. 1992). The model BOF 5 constructed in the mid-90s was even larger (see Table 7.1) due to an extended block of equations explaining the financial sector (Willman et al. 2000).

The consumer demand function in the above models initially drew on Friedman's permanent income hypothesis. In BOF 5, the point of departure was intertemporal utility maximization under a given budget constraint. In both cases, consumer demand depended on real current and expected incomes and personal wealth of households that initially covered financial wealth only, but in BOF 5 also physical wealth. In the latter model, the approach justified the attempt to specify the equations explaining changes in the real estate markets. The other consumer demand determinants in these models were real interest rates and a rate of inflation.

Table 7.1 Models of Nordic and Baltic countries

Country Version Year	Institution Author	Frequency Sample	Equations total stochastic	Decomposition	Estimation method	Properties	Applications
DENMARK							
ADAM 1990	Central Statistical Office, Ministry of Finance	Annual	927 92	20 industries, I-O submodel		Keynesian in the short-run adaptive expectations	Forecasts and policy simulations by Ministry of Finance
1992			139 19				
2010			2308 90				
SMECIII 1979	Economic Council of Denmark	Annual 1952–1979	323 39		OLS	Keynesian, prices, wages exogenous	Forecasts and simulation analyses
MONA	National Bank of Denmark	Quarterly	16			Keynesian in the short-run	Forecasts and simulation analyses
ESTONIA							
ESTMOD 1996	Ministry of Finance A. Leppä	Quarterly		–	OLS ECM	Experimental, mixed structure	Simulations
Model BANK of ESTONIA 2001	Central Bank of Estonia O. Basdevant	Quarterly 1993–2001			TSLS ECM	Mixed structure	Forecasts

Table 7.1 (Continued)

Country Version Year	Institution Author	Frequency Sample	Equations total stochastic	Decomposition	Estimation method	Properties	Applications
FINLAND							
KESSU	Ministry of Finance	Annual		20 industries, I-O submodel		Neo-Keynesian adaptive expectations	Forecasts and simulation analyses
I 1980							
IV 1990			569 240				
BQF	Central Bank A. Willman et al.	Quarterly					
BOF 1 1970					OLS		
BOF 4 1990			272 70	5 industries, I-O submodel	OLS	Ne-Keynesian	Forecasts and simulation analyses
BOF 5 1994–1995			360 60			Neo-Keynesian	
QMED	Central Bank Uman	Quarterly	70 20			Neoclassical rational expectations	Forecasts, policy simulations
LITHUANIA							
LITMOD 2000	Ministry of Economy A. Kazlauskas, A. Leppä	Quarterly				Rational expectations	

Table 7.1 (Continued)

Country Version Year	Institution Author	Frequency Sample	Equations total stochastic	Decomposition	Estimation method	Properties	Applications
2004	Institute of Economics, Academy of Science F.M. Andersen et al.	Quarterly 1995–2002	205 180	12 industries, I-O submodel	OLS ECM	Demand oriented	Annual forecasts and simulation analyses
NORWAY							
MODIS 1960	Central Statistical Office Statistic NORWAY	Annual					
MODAG 1980		Annual	1325 183	31 industries, I-O submodel	OLS FIML		Forecasts and simulation analyses
KVARTS		Quarterly	300 90	12 industries, I-O submodel		Adaptive expectations	Forecasts and simulation analyses
RIMINI	Central Bank of Norway	Quarterly	30				Short-term forecasts and simulation analyses mainly of inflation processes

Table 7.1 (Continued)

Country Version Year	Institution Author	Frequency Sample	Equations total stochastic	Decomposition	Estimation method	Properties	Applications
SWEDEN							
KOSMOS	National Institute of Economic Research	Midyear	467 21	8 industries	OLS	Description of transmission and inflationary processes	Forecasts and simulation analyses
STEP 1979	Stockholm School of Economics	Quarterly 1963–1974	362 125		OLS TSLS	Adaptive expectations	Simulation analyses
SNEP 1980	University of Uppsala	Annual 1955–1978	32 14		OLS TSLS	Structure similar to early BOF models Keynesian, wages depend on labour productivity and unemployment rate	Simulation analyses
GUESS 1981	University of Gothenburg	Annual 1954–1977	97 36		OLS		Policy simulations
RIKSMOP	Riksbank	Quarterly				Demand determined, potential output	Simulation analyses, forecasts
BOE (Small Open Economy) 2005		Quarterly			Bayesian estimation	Modelled after Canadian QMP	Policy simulations, forecasts

Enterprise investment demand was derived from intertemporal profit maximization. As a result, its specification assumed that it depended on the demand for fixed capital. The latter was determined by the marginal productivity of fixed capital that was related to the current and expected user costs, mainly determined by interest rates. In industry, value added remained the major explanatory variable.

The demand for domestic production was obtained from the difference between final demand and imports. It determined employment (working time) using an inverted CES production function with considerable lags.

Prices were determined taking into account marginal production costs that included labour costs and imports. The links among final demand, gross output deflators and valued added deflators were found using the I-O price equations. The wage equations were built in line with the principles of the extended Phillips curve.

The BOF models contained extended blocks of equations that explained financial flows in both the money market and the financial sector. They were regularly used in forecasting and in many policy simulations.

In the early 1990s, a small quarterly model QMED was constructed at the Central Bank of Finland. Rational expectations of prices, interest rates and incomes were consistently introduced into the respective equations. The model was used for forecasting and policy simulation purposes.

7.5 The Models of the Lithuanian Economy

7.5.1 The LITMOD Model

The first version of the model LITMOD was created at the Lithuanian Ministry of Economy by the team directed by A. Kazlauskas and A. Leppä in the year 2000. Its next version was built four years later by a Danish-Lithuanian team whose members represented the Rise National Laboratory in Roskilde and the Institute of Economics, the Lithuanian Academy of Science (Andersen et al. 2005). The second version of LITMOD was a large quarterly system disaggregated into 12 industries interconnected by a system of I-O equations.

The model was demand determined and addressed the specific properties of an economy in transition towards a market system (the sample covered the years 1995–2002). It contained equations explaining final demand, i.e. consumer and investment demand, public institutions' demand and exports. Final demand was assumed to determine production and employment. The foreign trade equations were specified under the Armington concept. The demand for domestic output included intermediate demand, making use of the I-O relationships. Domestic production determined the demand for production factors and generated the incomes of the production sector. Prices were formed by unit costs, while wages by labour productivity and the rate of unemployment. The fiscal and monetary parameters (interest rates) were exogenous.

The equation parameters were estimated with the OLS and ECM methods. The model was used for medium-term forecasts and policy simulations.

7.6 The Models of the Norwegian Economy

7.6.1 The MODIS and MODAG Models

The Norwegian Central Bureau of Statistics (Statistics Norway) became involved in macromodelling activities as early as the 1960s. As a result, the large annual model MODIS containing an extensive block of I-O equations was constructed. It was used for more than 20 years in forecasting and policy simulations. The description of its last version MODIS IV can be found in a publication by Statistic Norway (Bjerkholt and Longva 1980).

MODIS IV provided the basis for constructing a large annual system called MODAG, whose operational version was ready in 1983.

The MODAG model was highly disaggregated into more than 30 industries and 40 consumer commodity groups. It had an extended I-O submodel. As a result, the number of its equations was ca. 1300. Their parameters were initially estimated with the OLS method and then with IIV and FIML. The data sample started in 1960 (in special cases in 1970). The model was re-estimated every year. The Ministry of Finance used it for short- and medium-term forecasting and policy simulations (Cappelen and Longva 1987).

A special feature of the model was that it distinguished between the exposed and sheltered industries. Consumer demand was analysed separately for durables and non-durables using a LES submodel. In the production sector, the raw-materials industry and other industries were distinguished. For each industry production, imports and demand for labour were generated (by inverting the production functions). Gross investments depended on gross output and profitability (including interest rates) and on the prices of production factors.

Potential output was generated from the production functions built for particular industries. Price equations were formulated assuming monopolist competition. Prices were determined by unit costs, world prices and the rate of capacity utilization. Industrial wages were dependent on prices, labour productivity, tax rates and unemployment rate, following the Phillips curve concept. The financial flows sector in the model was not extensive (Cappelen 1992).

The most recent versions of the MODAG model retained their major properties. They were highly disaggregated like their predecessors and distinguished 21 different sectors and 45 different products. However, the economic structure was analysed within a different framework, distinguishing three markets: products, labour and finances. In the product market the I-O submodel played a significant role. To determine the demand for input factors in 11 industries a Cobb-Douglas production function was used for materials, a CES-aggregate for energy and a CES-aggregate for high- and low-educated labour. Statistics Norway used a special version of the model to analyse the labour market and presented predictions extending to the year 2030 (Bjornstad et al. 2010).

7.6.2 The Quarterly KVARTS Model

In the 1980s a quarterly model of the Norwegian economy was built at Statistics Norway. It was a large model containing approximately 800 equations structured following the annual model. It was used for forecasting and short-term analyses (Biorn et al. 1987). Compare also Bjerkholt (1998).

7.6.3 The Quarterly RIMINI Model

In the middle of the 1990s a quarterly model of the Norwegian economy was constructed at the Central Bank of Norway to support short-term forecasting. It was mainly applied to analyse the monetary policy effects. The model turned out to be instrumental in describing the transmission processes in the context of inflation analyses (Olsen and Wulfsberg 2001).

7.6.4 Modelling Inflation in an Open Economy

In the first years of the 21st c. a group of Norwegian scholars set out to improve the procedures used for constructing macroeconometric models employing the multivariate cointegration methods. Given the then stage of knowledge, the methods could only be used to estimate the parameters of a limited number of equations. The scholars proposed a piecewise procedure implying the construction of small, weakly-dependent segments of a macromodel to be linked into a large system afterwards.

An application of the model of inflation in Norway intended for the analyses of monetary policy effects was demonstrated. Its core was a two-equation model explaining average wages and prices in the classical manner. Its long- and short-run parameters were estimated. The explanatory variables that included import prices (more precisely, the exchange rates), GDP changes, the rate of unemployment and interest rates were assumed weakly exogenous. The respective equations' parameters were estimated using the cointegrating techniques. This system showing the dynamics of inflationary processes was used in monetary policy simulations (Bårdsen et al. 2005). However, the road to including it into a complete model of the Norwegian economy, such as RIMINI, seemed very distant.

7.7 Models of the Swedish Economy

7.7.1 The KOSMOS Model

In the early 1980s, the large model KOSMOS was constructed at the National Institute of Economic Research (NIER) in Stockholm. Its builders aimed to improve the quality of the Institute's forecasts, including the extension of the forecasting horizon from 1 to 3 years. The Swedish Ministry of Finance also used the model. For the limited reliability of the Swedish national accounts, KOSMOS was a semi-annual model. It was decomposed into 8 industries and had over 460 equations. In the estimation process the ECM techniques were used (Ernsäter and Nordström 1992).

The model had a Keynesian orientation and applied adaptive expectations. In its first version only prices and wages were endogenized, while interest rates and exchange rates remained exogenous.

Consumer demand was linearly dependent on household real incomes, nominal interest rate and the rate of inflation. Investments in the production sector were separately specified for buildings and machinery and equipment. The growth rate of fixed capital was determined by the growth rate of value added and a profitability indicator representing the difference between the profit-fixed capital ratio and the real interest rate on bonds. Gross investment adjusted for changes in the rate of capacity utilization included also depreciation.

The model accounted for changes in the share of the Swedish exports in total OECD imports, the major explanatory variables being the relative prices. An increase in imports was explained in terms of rising domestic demand and relative prices.

In the model, production was derived from the difference between final demand and imports. It determined employment in manufacturing industries. The demand for working hours was influenced by value added and real labour costs. In the other industries employment was arrived at by dividing output by exogenous labour productivity.

Prices were dependent on unit costs, among which only labour costs were endogenous. In the manufacturing industries prices depended also on a mark-up and/or world prices.

Wage dynamics in the manufacturing industry was assumed to determine wage dynamics in other industries. In the manufacturing industries, wage changes resulted from a bargaining process involving the entrepreneurs and the trade unions. As a result, the wage equations depended in the long-run on the ratio between the wage fund and value added and on the rate of unemployment. In the short-run, they were also determined by labour productivity, retail prices and/or world prices. This specification was rather far from the standard specification, where labour productivity and price impacts enter the long-run relationship.

A block of equations explaining financial flows was added in the later versions of the model.

In the late 1980s and in the early 1990s three macroeconometric, demand-oriented models were constructed in the academic institutions. These were the quarterly forecasting model STEP built at the School of Economics and annual models designed rather for policy simulations—the smaller model SNEP at the Uppsala University and a larger model at the Gothenburg University (Lybeck et al. 1984).

7.7.2 The STEP Model

The quarterly, medium-size model STEP constructed at the Stockholm School of Economics was structured following the early version of the Finnish model BOF 3. It was demand oriented, but restrictions were introduced into the consumption and investment functions to account for credit constraints. The model also assumed foreign trade disequilibria. The rate of capacity utilisation affected exports and imports, as well as inventory changes. The model supported regular forecasting and numerous simulation analyses and participated in the Project LINK (Ettlin et al. 1979).

7.7.3 The SNEP Model

The annual SNEP model built at the Uppsala University had a Keynesian structure, with final demand determining production and employment. Prices depended on unit costs, while wages on labour productivity and unemployment rate. Interest rates were exogenous and served as monetary policy instruments (Berg et al. 1981).

7.7.4 The GUESS Model

The annual GUESS model built at the Gothenburg University was a demand-determined construct with supply restrictions entered into the equations explaining imports and inventory changes. The model generated potential output. Prices depended on unit costs. The impacts of the world prices were transmitted through the costs of both imports and exports (Lybeck and Carlsson 1981).

7.7.5 The Riks Bank RIXMOD and SOE Models

The quarterly model RIXMOD of the Swedish economy was built at the Riks Bank mainly for monitoring the monetary policy effects. It drew on the models constructed by the central banks of Canada, USA and the United Kingdom (Nilsson 2002). In the next years, the dynamic stochastic general equilibrium (DSGE) model was constructed at the same institution. The model was called the Small Open Economy (SOE) model. It consisted of two segments, i.e. long-run relationships with calibrated parameters and short-run adjustments with parameters estimated with the Bayesian techniques (Adolfson et al. 2007).

A similar DSGE model was constructed at the Institute of International Studies at the Stockholm University (Curdia and Finocchiaro 2005).

References

Adolfson, M., Laséen, S., Linde, J., & Villani, M. (2007). Evaluating an estimated new Keynesian small open economy model. *Sveriges Riksbank Working Paper Series, 203*.

Andersen, F. M., Celov, D., Grinderslev, D., & Kazlauskas, A. (2005). A macro-econometric model of Lithuania LITMOD. *Economic Modelling, 22*, 707–719.

Bårdsen, G., Eitrheim, Ø., Jansen, E. S., & Nymoen, R. (2005). *The econometrics of macroeconometric modelling*. Oxford: Oxford University Press.

Berg, L., Bergström, R., Bergström, V., & Nilsson, C. (1981). *Svenska modellen (SNEP)*. Lund: LIBER.

Bergman, L., & Olsen, Ø. (1992). *Economic modeling in the Nordic countries*. Amsterdam: North-Holland.

Biorn, E., Jensen, M., & Reyment, M. (1987). KVARTS: a quarterly model of the Norwegian economy. *Economic Modelling, 4*, 77–109.

Bjerkholt, O. (1998). Interaction between model builders and policy makers in the Norwegian tradition. *Economic Modelling, 15*, 317–339.

Bjerkholt, O., & Longva, S. (1980). MODIS IV. A model for economic analysis and national planning. *Social Economic Studies from the Central Bureau of Statistics, 43*. Oslo.

Bjerkholt, O., & Rosted, J. (1987). *Contributions to economic analysis: Vol. 164. Macroeconomic medium-term models in the Nordic countries*. Amsterdam: North-Holland.

Bjornstad, R., Gjelsvik, M. L., Goday, A., Holm, I., & Stolen, N. M. (2010). Demand and supply of labor by education towards 2030. Linking demographic and macroeconomic models for Norway (Reports 39/2010). Oslo-Kongsvinger: Statistics Norway.

Cappelen, Å. (1992). MODAG a macroeconometric model of the Norwegian economy. In L. Bergman & Ø. Olsen (Eds.), *Economic modeling in Nordic countries* (pp. 55–93). Amsterdam: North-Holland.

Cappelen, Å., & Longva, S. (1987). MODAG A: a medium-term macroeconomic model of the Norwegian economy. In O. Bjerkholt & J. Rosted (Eds.), *Macroeconomic medium-term models in the Nordic countries*. Amsterdam: North-Holland.

Christensen, A. M., & Knudsen, D. (1992). MONA. *Economic Modelling, 9*, 10–74.

Curdia, V., & Finocchiaro, D. (2005). An estimated DSGE model for Sweden with a monetary regime change (Seminar Paper No. 740). Institute for International Economic Studies, Stockholm University.

Dam, P. U. (1986). The Danish macroeconomic model ADAM. *Economic Modelling, 3*, 31–44.

Ernsäter, L., & Nordström, T. (1992). KOSMOS—a short term model for Sweden. In L. Bergman & Ø. Olsen (Eds.), *Economic modeling in Nordic countries* (pp. 185–196). Amsterdam: North-Holland.

Ettlin, F. A., Lybeck, J. A., Erikson, I., Johansson, S., & Järnhäll, B. (1979). *The STEP I quarterly economic model of Sweden: the equation system*. Stockholm: EFI.

Frandsen, S. E., Hansen, J. V., & Trier, P. (1994). A general equilibrium model for Denmark with two applications in economic and financial modelling (Mimeo). Copenhagen: Danish Economic Council.

Hansen, J., & Paldam, M. (1973). *SMEC*. Copenhagen: Economic Council.

Hansen, C. B., & Smidt, J. (1992). Integration of real and monetary sectors in ADAM. In L. Bergman & Ø. Olsen (Eds.), *Economic modeling in the Nordic countries* (pp. 95–113). Amsterdam: North-Holland.

Hetemäki, M. (1992). KESSU IV: an econometric model of the Finnish economy. In L. Bergman & Ø. Olsen (Eds.), *Economic modeling in the Nordic countries* (pp. 115–137). Amsterdam: North-Holland.

Kaergärd, N. (1991). The Danish growth model CLEO and the stability of the economy. *Economic Modelling, 8*, 175–186.

Leppä, A. (1987). KESSU III: an econometric planning model for the Finnish economy. In O. Bjerkholt & J. Rosted (Eds.), *Macroeconomic medium-term models in the Nordic countries*. Amsterdam: North-Holland.

Leppä, A. (1995). MODEST—an econometric macro model for the Estonian economy. In W. Welfe & M. Majsterek (Eds.), *Macromodels and forecasts* (pp. 95–96). Łódź: Absolwent. Abstract. An update: Ministry of Finance Discussion Paper 52. Tallinn, 1996.

Leppä, A., Martihhina, A., & Merikull, J. (2000). The effects of the Russian crisis 1998 on the Estonian economy. Simulations with ESTMOD. In W. Welfe & P. Wdowiński (Eds.), *Modelling economies in transition* (pp. 34–69). Łódź: Absolwent.

Lybeck, J. A., & Carlsson, E. (1981). The use of macroeconometric models, as illustrated by the GUESS model of Sweden. *Statistisk Tidskrift, 5*.

Lybeck, J. A., Bergström, R., Bergström, V., Carlsson, S., Johansson, S., Järnhäll, B., Kukkonen, P., Tarkka, J., Visholm, T., & Willman, A. (1984). A comparison of the dynamic properties of five Nordic macroeconometric models. *Scandinavian Journal of Economics*, *86*, 35–51.

Mannermaa, K., & Kaski, E.-L. (1980). *KESSU: the forecasting and planning system for the Finnish economy*. Helsinki: Economics Dpt., Ministry of Finance.

Männistö, H.-L., Tarkka, J., & Willman, A. (1992). The structure and foundations of the BOF4 model. In L. Bergman & Ø. Olsen (Eds.), *Economic modeling in the Nordic countries* (pp. 139–183). Amsterdam: North-Holland.

Nilsson, C. (2002). RIXMOD—the Riksbank's macroeconomic model for monetary policy analysis. *Sveriges Riksbank Economic Review*, 46–71.

Olsen, K., & Wulfsberg, F. (2001). The role of assessments and judgment in the macroeconomic model RIMINI. *Economic Bulletin Norges Bank*, *72*, 55–64.

Tarkka, J., Willman, A. (1985). *The BOF3 quarterly model of the Finnish economy*. Helsinki: Bank of Finland. D:59.

Whitley, J. D. (1992). Comparative properties of the Nordic models. In L. Bergman & Ø. Olsen (Eds.), *Economic modeling in the Nordic countries* (pp. 3–53). Amsterdam: North-Holland.

Willman, A., Kortelainen, M., Männistö, H.-L., & Tujula, M. (2000). The BOF5 macroeconometric model of Finland, structure and dynamic microfoundations. *Economic Modelling*, *17*, 275–303.

Chapter 8
Models of the Central and South European Countries

8.1 Introduction

In the Central and South European countries macroeconometric modelling developed later than in the United Kingdom, France and Netherlands. Nevertheless, the Austrian, Belgian, German, Greek, Irish, Italian, Spanish and Swiss models were successful enough to join in Project LINK in the late 1980s (Klein et al. 1999). The models will be briefly characterised below, with stress being laid on the developments in macromodelling in Germany and Italy (see Table 8.1).

8.2 The Models of the Austrian National Economy

At the end of the 1980s, an annual model of the Austrian economy was constructed for the purpose of policy simulations, especially in the context of linking the shadow economy and the national economy. The model was demand oriented and generated potential output.

8.2.1 The LIMA Model

The quarterly model LIMA constructed at the Institute of Advanced Studies in Vienna was a major Austrian model. Developed as a medium-size model intended for forecasting and policy analyses, it was systematically improved for many years. At the end of the 1990s its LIMA 97 version had over 90 equations, 23 of which were stochastic. The model distinguished 6 major blocks of equations explaining final demand, foreign trade, labour market, prices and budget revenues. It supported regular forecasting activities and numerous policy simulations investigating the impacts of the country's entering into the European Union. It became a participant of Project LINK as after it was built (Fidrmuc and Pichelmann 1999).

W. Welfe, *Macroeconometric Models*,
Advanced Studies in Theoretical and Applied Econometrics 47,
DOI 10.1007/978-3-642-34468-8_8,

In the 1990s, an annual, multisectoral model of the Austrian economy was constructed at the WIFO Institute under the MULTIMACI project. Its special feature was that it used a top-down approach: variables' values were obtained for industries through disaggregation (Kratena and Wuger 1995).

8.3 The Models of the Belgian Economy

In the 1980s, an annual model of the Belgian economy was built at the Free University of Brussels. It was a medium-size model used in forecasting and policy simulations (D'Alcantara 1983).

In the later years, use was made of a large model of Belgian economy being part of the HERMES model system. This model had a block of equations for the production sector based on the I-O relationships (Bossier et al. 1989). It was used for a time by the Belgian Planning Bureau.

8.3.1 The MIRABEL Models

In the mid-80s, the Belgian Planning Bureau constructed the annual model MIRABEL I, which was replaced after several years by the large disequilibrium model MIRABEL II (Bogaert et al. 1990). The second model was designed to run medium-term analyses of growing unemployment rates. With its 227 equations it was the first large disequilibrium model. It was able to generate potential (unobservable) GDP and employment taking into account not only the restrictions constraining final demand, but also those affecting labour force and fixed capital.

At the end of the 1980s, a small model of the Belgian economy was built at the De Boeck University, which was mainly used for policy simulations (Barten and Dhaene 1990).

8.3.2 The Belgian National Bank Models

In the middle of the 1990s, the National Bank of Belgium built a quarterly model of the country's economy with a view to making it a component of the European System of the Central Banks' models. Its main purpose was to facilitate the analyses of the monetary policy transmission mechanisms in the EURO area and short-term forecasts of the Belgian economy in this environment (Jeanfils 2000).

The model's structure was influenced by the developments concerning the use of economic theory and econometric methodology that took place in the central banks in Canada and the USA. Two segments were distinguished in the model: the long-term (steady state) and dynamic adjustments, where rational expectations played an important role; this approach entailed a broad use of leads and lags. The parameters of the short-run equations were estimated with the Polynomial Adjustment Costs approach, leading to the creation of a special form of ECM. Forecasts were prepared using separate VAR models.

The model was medium sized. Its disaggregated version was built already in 1995. It distinguished a household sector with a special consumption function. Households maximizing expected labour incomes played a key role, while the income-constrained households represented ca. 22 % of the population. The model included also the enterprise sector. The inverted Cobb-Douglas production function determined employment (i.e. time worked), the demand for fixed capital, and thus investment demand. The model had equations explaining the price system, wages and broadly understood financial flows.

The model supported short-term forecasting and many simulation analyses of the impacts generated by monetary and fiscal policy.

8.3.3 The NONAME Model

After several years, the above model was followed by the quarterly NONAME model (Jeanfils and Burgraeve 2005), which was structured like its predecessor. Its long-term (theoretical) segment assumed that the point of departure would be the intertemporal optimization of households and enterprises. Dynamic adjustments allowed for multinomial costs under the assumption of rational expectations. The model utilized a CES production function. The price equation included a mark-up. The model was used as a forecasting and policy simulation tool.

8.4 The Models of the German National Economy

This section first presents the modelling activities in West Germany and then in the whole country, after the unification of two German states. The macroeconometric models constructed in the former German Democratic Republic are discussed in the next chapter that deals with macroeconomic modelling in the former centrally planned economies.

In the Federal Republic of Germany macroeconomic modelling started to develop in academic centres in the 1960s and 1970s. By 1989, more than 200 of mostly ephemeral macromodels were built. Heilemann and Wolters (1997) compiled a list of 100 macromodels that had been constructed between 1973 and 1995. Only a few of the listed models that have significantly contributed to the development of macromodelling in the FRG and have been used for many years as forecasting and policy simulation tools will be described below (see also Gahlen and Sailer 1985).

8.4.1 The BONN Models

A major macroeconometric modelling centre in West Germany was the Bonn University with its modelling team led by W. Krelle. In the 1960s, the team constructed a medium-sized annual model (the Bonn Model) (Krelle et al. 1969). The model was systematically improved and extended, i.a. by including a large financial sector (Krelle 1986). Its 11th version had 470 equations, 157 of which were stochastic. The model had a large number of exogenous variables. The data it used spanned the period 1964–1982. The equation parameters were estimated with the OLS method (Sarrazin 1986).

The model construction process started with the equations explaining final demand. The consumption function was built using T. Brown habit persistent hypothesis (Brown 1952), but the list of the explanatory variables was extended to include a variable approximating household personal wealth. Consumer demand was decomposed into major commodity groups (by applying LES) and by source (domestic, foreign). Investment demand was decomposed into 4 groups. The key role was played by the demand function for machinery and equipment that was obtained using the neoclassical assumptions. The demand for fixed capital was determined first and then the investment demand was derived, after the replacement demand has been allowed for. The model generated potential output from multiplying fixed capital by its exogenous productivity.

The model contained a system of equations that explained the labour market. Unemployment was generated as a difference between labour supply and labour demand. The equations explaining foreign trade had a standard form, but they included variables standing for pressures occurring in the domestic markets.

The price equations constructed for 16 groups of commodities were determined by the major components of unit costs (labour costs and imports), as well as by the capacity utilization rates representing domestic market pressures.

The model had a large public sector (191 equations) whose major component was the social security submodel. Particularly in the 1980s, the pertinent federal institutions used the submodel to run numerous simulation analyses. The model also had a large monetary sector that explained financial flows, including the system of interest rates.

The Bonn model was widely used for policy simulations. In the 1970s it participated in the Project LINK system.

In the 1980s, two I-O submodels were separated from the Bonn model, one covering real flows and the other representing the price system (Nakamura 1986). The technical coefficients in the submodels were endogenized by relating them to relative domestic and foreign prices, fixed capital-labour ratios and rates of capacity utilization.

In the mid-1980s, a quarterly model was constructed. The Bonn Quarterly Model was built to meet the requests reported by both public agencies and business representatives (Müller and Nakamura 1986). This medium sized model was used for business cycle monitoring, short-term forecasting and for the simulation of the impacts of fiscal and monetary policies.

Its structure resembled that of the annual model. Its distinctive feature was the use of the translogarithmic production function. It utilized information on the rate of production capacity utilization collected from IFO surveys, as well as the DIW (German Institute of Economic Research) indicator. As well as being helpful in computing the potential output, this characteristic was also utilised in the equations explaining prices, employment and imports.

The quarterly model had a large financial sector that was linked with the real sector mainly through interest rates, but also via inflation rates.

8.4.2 The SYSIFO Model

The quarterly model SYSIFO of the FRG economy was constructed by G. Hansen and U. Westphal at the Hamburg University in the mid-70s (de Menil and Westphal 1985). In 1982, it replaced the Bonn model in the world economy model maintained by Project LINK.

SYSIFO was a large model of the West German economy. In the 1990s it was enlarged to 1400 equations. It retained separate submodels for East and West German states, even after they unified. The submodel for West Germany was significantly extended and included an I-O model of 13 sectors. It was used in short-term forecasting and in monitoring the developments in the FRG economy, also serving educational purposes.

After U. Westphal's untimely death in 1996, further work on the model extension and use was abandoned.

8.4.3 The RWI Models of Business Cycles

In the mid-70s, U. Heilemann built a small annual model of business cycle at the Rheinisch Westphaelisches Institut fuer Wirtschaftsforschung (RWI) in Essen. After a few years, the model was replaced by a medium-sized quarterly model of Western Germany, which was intended for forecasting and short-term analysis purposes. Consumer and producer expectations derived from surveys (including IFO) were included in this model, however without a major success (RWI 1985).

This medium-sized model became the main model of the German economy. Following the unification, it covered the whole German territory. It substituted the SYSIFO model in the Project LINK' system of the world models (Heilemann and Wolters 1997).

In 1978, the federal authorities implemented a System of Structural Reports that required 5 research institutes, i.e. DIW in Berlin, HWWA in Hamburg, IFO in Munich, IFW in Kiel and RWI in Essen, to produce common annual reports. Some of the institutions prepared their reports using the I-O models (Stäglin 1995). It the next years, RWI in Essen, IWH in Halle, IFW in Kiel, and IFO in Munich were requested to submit half-annual reports prepared in cooperation with other institutes. The recipient of the common reports containing forecasts and opinions on the current economic situation was the Federal Government. The forecasts were partly based on the quarterly macroeconometric models operated by the RWI and IWH and on the annual IFW model.

8.4.4 The Freiburger Model

In 1969, the first version of a quarterly model was constructed under the direction of D. Lüdecke at the University of Freiburg. The model was systematically developed. In the 1980s, it was operated jointly with the financial sector model built at the University of Tübingen. It was extended since 1989, receiving its own financial sector. After the German states unified its equations were respecified—the changes in the regime were mainly forced through the introduction of dummies. The model was used for forecasting and policy simulations.

8.4.5 The Bundes Bank Quarterly Model

The Central Federal Bank of the FRG engaged in modelling activities relatively early. A quarterly model was constructed, which initially supported short-term forecasting and then many simulation analyses of the monetary policy impacts (Deutsche Bundesbank 1994). In the early 1990s, the model was entered into MEMMOD, a system of the world models constructed at the Bundes Bank (Deutsche Bundesbank 2000). At that time, rational expectations were explicitly introduced to the model. The model had over 120 equations whose parameters were estimated with OLS. It had a mixed orientation. Final demand for consumer goods and investments determined imports and indirectly employment. The production function was used to determine potential output and the rate of its utilisation that affected prices. The model had an extended sector explaining financial flows.

Quite recently, the experimental DSGE model was constructed for the Bundes Bank, intended for simulations analysing the monetary policy impacts (Pytlarczyk 2005).

8.4.6 Disequilibrium Models

Towards the end of the 1980s H.-J. Hansen of the J.W. Goethe University constructed a two-market disequilibrium model for the FRG. The model assumed that three types of constraints existed: demand for goods, production capacity and labour supply. It applied the min-condition at the aggregate level. This rationing model was estimated with the 1962–1986 sample (Hansen 1995).

At the end of the 1980s, a quarterly disequilibrium model (MDM) for West Germany was constructed at the University of Konstanz (Franz et al. 1993). The Konstanz disequilibrium model belonged to the third generation of models where the min-condition was assumed to apply to particular markets. The demand for and supply of products and labour were aggregated using a CES-type function. Prices were assumed to adjust only sluggishly in the short-run. In the late 1990s, the quarterly Konstanz model had 68 equations (36 stochastic) and its data sample covered the years 1960–1994 (Franz et al. 1997). The forecasting and simulation analyses run with the model showed that the German economy was predominantly demand constrained (Schellhorn and Winker 1994). The model was also used in other interesting simulation analyses like the studies on the impacts of migration from Eastern Europe into West Germany (Franz et al. 1994), the research on spillovers and feedbacks of international trade (Beck and Winker 2004) and first of all the studies of employment and unemployment.

The 1960–1988 version of the disequilibrium model for Germany was used as a component of an international project that was designed to study unemployment in Western Europe (Dreze et al. 1990). It was chiefly used for simulation analyses focusing on the unemployment problems.

8.4.7 New Quarterly Models

In early the 21st c. two German institutes operated new quarterly models. In the late '90s the Halle Institute for Economic Research (IWH) started to build a macroeconometric model of the German economy distinguishing the East and West German states (Brautzsch and Dreger 1997). Later on, with the data becoming available only for the whole country, the model equations were integrated. In 2008, the model had approximately 150 equations, around 50 of which were stochastic. The estimation procedure was based on the ECM approach. The parameters of the long-run equations were estimated with TSLS and OLS.

The model was structured following the theory of monopolistic competition in the commodity and labour markets. Four sectors were distinguished in the model: the commodity market with foreign trade, the labour market, public institutions and the financial market. A key role in explaining the long-run supply was given to a CES production function with constant returns to scale. The factor demand functions were obtained following profit maximisation. The unemployment rate, both effective and NAIRU, were calculated. It was assumed that wages were determined through bargaining. The model had equations explaining the particular components of final demand in a more or less standard manner. In the long-run, demand and supply achieved balance through price and wage adjustments; in the short run, foreign trade adjustments and policy interventions were used to take account of temporary disequilibria. The model is used today for monitoring the economic situation, forecasting and simulations (Scheufele 2008).

At the Macroeconomic Policy Institute (IMK) in Duesseldorf, a quarterly model of the German economy has been constructed very recently. It is based on the 1980–2006 quarterly data. The parameters of its equations have been estimated with ECM. The model lays stress on the differences between the long- and short-term impacts.

The model has a Keynesian orientation and distinguishes a final demand sector, an income sector and an employment sector, a public sector and a sector for prices, exchange rates and interest rates. Output and employment are determined by final demand components. On the supply side, prices and wages have been specified allowing for the existence of nominal rigidities. The model has been systematically used for forecasting and policy simulations (Duong et al. 2008).

8.5 The Models of the Greek Economy

The macroeconometric models of the Greek economy were built at the National Bank of Greece under the direction of N.C. Garganas. The first, small model was developed in 1975. Its versions were substantially extended and elaborated over the next years. The model was demand determined and had an extensive sector for public institutions. It found application in forecasting and monetary policy simulations (Garganas 1991). At the end of the 1990s it was re-estimated with data extending to the year 1997 and its monetary sector was completely revised. At the same time, a new, compact model of the Greek economy addressing the long-run supply side of the economy was developed (Zonzilos 2000).

In the mid-80s the Centre of Planning and Economic Research in Athens began to construct annual models of the Greek economy. A medium-size model of Keynesian orientation was built first, which supported forecasting activities and numerous simulations. The model participated in the system of world models maintained by Project LINK.

In the mid-90s an annual macroeconometric model of the Greek economy was constructed within the HERMIN system of models.

8.6 The Models of the Irish National Economy

In the 1960s and 1970s the macroeconometric modelling activities in Ireland developed on a large scale. The models that were built then had a Keynesian orientation (Bradley and Fanning 1991).

8.6.1 The Bank of Ireland Model

At the end of the 1970s a medium-sized annual model of the Irish economy was constructed at the Bank of Ireland. The model was used to run simulations of the monetary policy impacts and to forecast the development trends in the Irish economy. It participated in Project LINK.

8.6.2 The HERMES and HERMIN Models

The annual HERMES model constructed in the late 1990s followed the general tendency to extended modelling of the supply sector. It had a neo-Keynesian orientation and was treated as a component of the system of models built in the EU. It was a large model, with more than 700 equations and 10 sectors (Bradley and Fitz Gerald 1971). After it was found to be complicated in use, the decision was made to construct a smaller, 4-sector annual model HERMIN (Bradley et al. 1995b).

The HERMIN model that was characterised by the thorough specification of the supply sector gave birth to a whole system of macroeconometric models for the EU's peripheries (Bradley et al. 1995a, 1995b).

8.7 The Models of the Italian National Economy

8.7.1 The PROMETEIA Model

In the mid-70s a quarterly model of the Italian economy was constructed at the Institute of Economics PROMETEIA in Bologna (D'Adda et al. 1976). Initially called the Bologna Model, the model was progressively developed and became the Modello Trimestrale di PROMETEIA (MTP). This medium-sized model was used for forecasting and policy simulations. In the next decade a flow-of-funds sector was added to it, which was to facilitate the analyses of fiscal and monetary policy impacts (D'Adda and Fomasari 1980; Vincenzi 1989). Systematically extended, in 1991 the model reached around 900 equations (Ferrari et al. 1991).

The model was demand determined. In generating final demand the main roles were given to the consumer demand functions and the enterprise investment functions. Household consumer demand was determined by real personal incomes and financial wealth, as well as capital gains and consumer credits. Investment demand depended on the level of output, the interest rate and the ratio between investment goods prices and production prices.

Final demand determined output and indirectly employment. Producer prices were mainly determined by wage costs and import prices. The wage equation followed the extended Phillips curve concept, including the impact of wage indexation. Import prices depended on the world prices and the exchange rate.

The PROMETEIA model was regularly used to prepare short-term forecasts and manifold policy simulations for many years (Belfiori et al. 1997).

8.7.2 The Bank of Italy Model

In the mid-80s, D. Terlizzese directed the construction of a quarterly model of the Italian economy at the Bank of Italy (Visco et al. 1989). The model was intended

Table 8.1 Models of the Central and South Europe

Country Version Year	Institution Author	Frequency Sample	Equations total stochastic	Decomposition	Estimation method	Properties	Applications
AUSTRIA							
1981	Institute for Advance Studies Vienna	Quarterly	167 37		OLS		Forecasts and simulation analyses
BELGIUM							
	Free University of Brussels	Annual	139 69				Forecasts and simulation analyses
MARIBEL II 1989	Planning Bureau H. Bogaert et al.	Annual	227		OLS	Demand determined, disequilibrium model	Policy simulations of unemployment
Model with explicit expectations 2000	National Bank P. Jeanfils	Quarterly 1981	95 52		Steady state-calibration, OLS; dynamics-polynomial adjustment costs—ECM; expectations, VAR	Rational expectations, extended financial sector	Simulation analyses, forecasts
NONAME 2005	National Bank P. Jeanfils, K. Burggraeve	Quarterly	120 50			CES production function, rational expectations	Forecasts and simulation analyses

Table 8.1 (Continued)

Country Version Year	Institution Author	Frequency Sample	Equations total stochastic	Decomposition	Estimation method	Properties	Applications
GERMANY							
Bonn Model							
1962	Bonn University W. Krelle	Annual	63 28	Model I-O 18 sectors	OLS		Forecasts and policy simulations
1986		Annual 1964	472 157	Model I-O for prices	OLS	Extended financial sector	Medium-term forecasts, simulation analyses
BONN models disaggregated 1986	Bonn University M. Kyi, S. Nakamura	Annual			OLS	Inter-industry analyses, I-O relationships	Medium-term forecasts, simulations analyses, incl. social insurance system
Bonn Quarterly Model 1986	Bonn University C. Müller, S. Nakamura	Quarterly	80 48		OLS	Mixed structure	Short-term forecasts, simulation analyses
Freiburger Model 1995	Freiburg University K. Lüdeke	Quarterly	165 65			Mixed structure, after 1989 extended financial sector	Forecasts, policy simulations
Konstanzer Modell 1995	W. Franz, P. Winker, K. Goggelmann	Quarterly 1960–1994	68 36			Disequilibrium model, min-condition	Simulation analyses

Table 8.1 (Continued)

Country Version Year	Institution Author	Frequency Sample	Equations total stochastic	Decomposition	Estimation method	Properties	Applications
SYSIFO							
1980	Hamburg University	Quarterly	600 220	7		Keynesian	Forecasts and policy simulations
1996	G. Hansen, U. Westphal		1400 280	13 industries, I-O, West Germany		Separated East and West Germany	Forecasts, simulations analyses
1994	Deutsche Bundesbank Jahnke	Quarterly	331 120			Extended financial sector	Forecasts, policy simulation
RWI Business Cycle Model	Rheinisch Westphälisches						
1978	Institut für Wirtschafts-	Annual	45 33			Mixed structure	Simulation analyses, forecasts
1995	forschung Essen U. Heilemann	Quarterly	120 41			Demand determined	Short-term forecasts and policy simulations
IFM	Institut fuer Weltwirtschaft, Kiel	Annual				Demand determined	Forecasts, simulation analyses
IWH 2007	Halle, Institute for Economic Research	Quarterly				Demand determined	Forecasts, simulation analyses

Table 8.1 (Continued)

Country Version Year	Institution Author	Frequency Sample	Equations total stochastic	Decomposition	Estimation method	Properties	Applications
IMK 2008	Macroeconomic Policy Institute, Düsseldorf G.A. Horn	Quarterly				Demand determined	Forecasts, simulation analyses
GREECE							
Model of the Bank of Greece	Bank of Greece N.C. Garganas						
1975		Annual			OLS	Keynesian	Forecasts and simulation analyses
1991		Annual 1960–1988	120		OLS	Keynesian	Forecasts and simulation analyses
1997		Annual 1960–1997	450 50		TSLS ECM	Extended monetary sector	Forecasts and simulation analyses
Simulation 1997		Annual 1960–1997	80 20		ECM	Keynesian in the short-run, neoclassical in the long-run	Forecasts and simulation analyses
MYKL	Center of Planning and Economic Research, Athens Papadopoulos						
1986		Annual	101 44				Forecasts and simulation analyses
1991		Annual 1954–1981	Small model			Extended financial sector	Simulation analyses

Table 8.1 (Continued)

Country Version Year	Institution Author	Frequency Sample	Equations total stochastic	Decomposition	Estimation method	Properties	Applications
IRELAND							
1979	Bank of Ireland	Annual	ca. 100 29				Forecasts and simulation analyses
HERMIN 1993	J. Bradley et al.	Annual		Several sections and I-O submodel			Forecasts and policy simulations
ITALY							
MTP Modello Trimestrale di PROMETEIA	PROMETEIA Bologna C. D'Adda et al.	Quarterly 1980					
1982			229 71			Demand determined	Forecasts, policy simulations
1991			ca. 800 150			Extended financial sector	Forecasts, policy simulations
1986	Banca d'Italia C. Galli et al.	Quarterly 1970–1984	Above 700 ca. 120			Keynesian in the short-run, neoclassical in the long-run, extended sector of financial flows	Forecasts and simulation analyses of impacts of monetary policy
SPAIN							
CEPREDE	Autonomous University of Madrid	Annual	122 50			Demand determined	Simulation analyses, forecasts

Table 8.1 (Continued)

Country Version Year	Institution Author	Frequency Sample	Equations total stochastic	Decomposition	Estimation method	Properties	Applications
Quarterly Model 2002 2004	Banco de Espagña A. Estrada	Quarterly			Two-step ECM	Block of the multicountry ESCB model	Forecasts and simulation analyses
SWITZERLAND							
CREA 1982	University of Lausanne J.C. Lambelet et al.	Annual	ca. 100 50		FIML	Extended production sector	Forecasts, simulation analyses
KOF 1995	Technische Hochschule, Zurich P. Stalder	Quarterly	52 19				Short-term forecasts and policy simulations

for the Bank to make short-term forecasts and simulation analyses of monetary policy effects. Developed over the next years, in the mid-90s it reached the status of an exemplary solution in macroeconometric modelling for other central banks in Europe to follow. It must be stressed that the simulations focused on the major issues in monetary policy. The assumptions underlying the analyses and the analyses' outcomes were discussed at the Research Department and reported to the bank management on a regular basis.

The model had a mixed structure. In the short-run, economic growth was determined by final demand affecting the volumes of output and employment. In the medium- and long-term, the model followed the neoclassical viewpoint, with entrepreneurs' decisions being the major determinants of potential output and the demand for production factors.

Prices depended on unit costs plus a mark-up. The wage equations referred to the extended Phillips curve, allowing for expected inflation and the NAIRU unemployment rate.

The model had extended blocks of equations explaining financial flows related to fiscal activities, but mainly to the money markets.

At the end of the 1980s, a reduced version of the model was built to study the short- and long-term properties of the full model (Galli et al. 1989).

8.8 The Models of the Portuguese National Economy

8.8.1 The HERMIN Model

The annual model of the Portuguese national economy was constructed within the HERMIN system of the macroeconometric models of European peripheral countries (Modesto and Neves 1994). Its structure conformed to the standard typifying this class of models. The equations explaining final demand and output were specified. The supply sector was extended and the domestic and foreign commodity flows were distinguished. The model was mainly used to run simulation analyses preceding the country's entry to the European Union.

8.9 The Models of the Spanish National Economy

8.9.1 The Academic Centres' Models

In Spain, macroeconometric modelling was initiated by academic centres in the 1980s. Worth noting is the annual model constructed at the University Autonoma in Madrid. This medium-sized model had a Keynesian orientation. It was regularly used in forecasting and policy simulations. It entered the Project LINK world model. In the mid-1990s, another annual model was built within the HERMIN system of models (Bradley et al. 1995a).

8.9.2 The Bank of Spain Models

In the first years of the 21st c., the Central Bank of Spain decided to expand its macromodelling activities (Gali and Lopez-Salido 2001). The first model was built in 2002 to be a component of the CB system (Willman and Estrada 2002). A new, improved version of the quarterly model followed soon, replacing its predecessor (Estrada et al. 2004).

The model distinguished two levels. The first level represented the long-term relationships arising from the theoretical assumptions. The second level described the dynamic adjustments with the ECM method. In the short-run, output and employment were determined by final demand; in the long-run, production depended on the supply factors. The supply block was broadly specified. The model's system of equations explaining financial flows was specified to meet the needs of the Central Bank. The model was used in regular forecasting, helping also to run numerous simulation analyses exploring the monetary policy impacts.

8.10 The Models of the Swiss National Economy

8.10.1 The CREA Model

In Switzerland, the mainspring of the development of macroeconometric activities was the team led by Ch. Lambelet at the Lausanne University (Lambelet et al. 1982). The team constructed a medium-size annual model that supported forecasting activities and numerous policy simulations. It also participated in the international Project LINK.

In the short run, the model laid stress on the disequilibria arising in the commodity markets and in the labour market. An important role was given to the price and wage equations. They had a standard structure that allowed for the impacts of market pressures. The equations explaining labour demand and investment demand were derived from cost minimization. The model used an inverted Cobb-Douglas production function. In the consumer demand functions (that distinguished demand for durables), permanent real income and real personal wealth were used as the explanatory variables. The model included an extended foreign trade sector and a financial sector (Bodkin 1988).

8.10.2 The Quarterly KOF Model

In the middle of the 1990s, a small quarterly model was constructed at the Konjunkturforschungsstelle KOF/ETH, the Zurich Polytechnic University (Stalder 1995). The model had a conventional sector for final demand and an extended sector explaining supply. The equations in the latter sector were derived from the neoclassical assumptions about monopolistic competition among firms. The demand for production factors was determined with the vintage production functions, which resulted in a pertinent segmentation of investment demand.

The model generated both effective and potential outputs. They adjusted with considerable lags because of price and wage rigidity. Hence, the model generated excess demand, on one hand, and underutilized productive potential and unemployment, on the other. Financial flows were not covered in the model.

The KOF used its model regularly for forecasting purposes, also in cooperation with the IFO Institute in Munich, and for policy simulations.

8.10.3 The Quarterly Model of the University of St. Gall

The quarterly model of the Swiss economy that was constructed at the University of St. Gall towards the end of the 1980s had 208 equations (38 were stochastic) and its data sample covered the years 1974–1987 (Abrahamsen et al. 1995). The model's demand sector was developed fairly well, including the foreign trade. It had the income distribution, as well as the government sector. The labour market and consequently unemployment were not modelled, because labour supply was exogenous (dependent on the regulated immigration). The supply side was represented by the price and wage equations. Despite its relatively small size, the monetary sector contained all elements reflecting the impacts of the transmission mechanisms. Interest rates were the major channel of transmission from the monetary sector to the real sector.

The distinctive feature of the model was that it was specifically designed for simulation purposes. Its ex-post forecasts covering the period of 1974–1987 were found particularly important. Another important field of the model applications were the simulations of the fiscal and monetary policy impacts.

8.11 The Models of the Turkish National Economy

In the middle of the 1980s, a small model of the Turkish economy containing 28 equations was constructed. It mainly supported the analyses of inflationary pressures (Yagci 1987).

The special, quarterly model of the country's national economy built in the first years of the 21st c. was geared to analyse inflationary processes and the debt level of the Turkish economy (Özatay 2000).

References

Abrahamsen, Y., Huth, C., & Chips, B. (1995). A macroeconomic model for Switzerland used as a training instrument in management courses. In W. Welfe & W. Florczak (Eds.), *MACROMODELS'91* (pp. 41–75). Łódź: Absolwent.

Barten, A., & Dhaene, G. (1990). *Ouvertures economiques, series balises. Pratique des modeles macroeconometrique. Gestion de l'economie et de l'enterprise. L'approche quantitative*. De Boeck Universite.

Beck, M., & Winker, P. (2004). Modeling spillovers and feedback of international trade in a disequilibrium framework. *Economic Modelling*, *21*, 445–470.

Belfiori, L., Ferrari, M., & Tomasini, S. (1997). *Ricerche applicate e modelli per la politica economica. Errori di previsione e impatto strutturale della crisi 1992–1993: una analisi con il modello econometrico di Prometeia*. Banka d'Italia.

Bodkin, R. G. (1988). A survey of non-Dutch European macroeconometric models. Some international perspective. In W. Drehuis, M. M. G. Fase & H. Den Hartog (Eds.), *Challenges for macromodelling* (pp. 221–257). Amsterdam: North-Holland.

Bogaert, H., de Biolley, T., & Verlinden, J. (1990). A disequilibrium macroeconomic model of the Belgian economy. The MIRABEL II model of the Planning Bureau. *Economic Modelling*, *7*, 310–375.

Bossier, F., Martens, B., Strumelle, P., & van der Baken, W. (1989). HERMES: a macrosectoral model for the Belgian economy. *Economic Modelling*, *6*, 374–398.

Bradley, J., & Fanning, C. (1991). Twenty five years of macromodeling the Irish economy. Retrospect and prospect. *Journal of the Statistical and Social Inquiry Society of Ireland*, *24*, 107–131.

Bradley, J., & Fitz Gerald, J. (1971). The ESRI medium-term model. In J. Bradley, J. Fitz Gerald & D. McCoy (Eds.), *Medium-term review 1991–1996*. Dublin: The Economic and Social Research Institute.

Bradley, J., Modesto, L., & Sosvilla-Ribero, S. (1995a). Similarity and diversity in the EU periphery. A HERMIN based investigation. *Economic Modelling*, *12*, 313–322.

Bradley, J., Whelan, K., & Wright, J. (1995b). HERMIN Ireland. *Economic Modelling*, *12*, 249–274.

Brautzsch, H.-U., & Dreger, C. (1997). *Dokumentation: Das makrooekonometrische Modell des IWH—Version 1.0* (IWH-Diskussionpapiere 055).

Brown, T. M. (1952). Habit persistence and lags in consumer behaviour. *Econometrica*, *20*, 355–371.

D'Adda, C., & Fomasari, C. (1980). The flow of funds in the Bologna model. In *Review of economic conditions in Italy*. Roma: Banco di Roma.

D'Adda, C., et al. (1976). *Il Modello Econometrica dell'Universita di Bologna: struttura e simulazioni*. Bologna: Il Mulino.

D'Alcantara, G. (1983). *SERENA: a macroeconomic sectoral regional and national accounting econometric model for the Belgian economy*. Leuven: Vitgeverij ACCO.

de Menil, G., & Westphal, U. (Eds.) (1985). *Stabilization policy in France and Germany*. Amsterdam: North-Holland.

Deutsche Bundesbank (1994). *Macroeconometric model of the German economy*. Frankfurt/Main: Deutsche Bundesbank.

Deutsche Bundesbank (2000). *Macroeconometric multi-country model, MEMMOD*. Frankfurt/Main: Deutsche Bundesbank.

Dreze, J. H., Bean, C., Lambert, J. P., Metha, F., & Snessens, H. R. (1990). *Europe's unemployment problem*. Cambridge: MIT Press.

Duong, M. H., Logeay, C., Stephan, S., & Zwiener, R. (2008). *IMK German macro model. Development of a Keynesian macro-econometric model of Germany* (Mimeo). Dusseldorf: Hans Bockler Stiftung, IMK.

Estrada, A., Fernandez, J. L., Moral, E., & Regal, A. (2004). *A quarterly macroeconometric model of the Spanish economy* (Documents de Trabajo No. 0413). Madrid: Banco de Espana.

Ferrari, M., Onofri, P., & Tomasini, S. (1991). *Modelli econometrici e fluttuazioni cicliche: una analisi col modello trimestriale di Prometeia ed il suo core. Ricerche applicate e modelli per la politica economica* (Banka d'Italia, Nota di Lavoro No. 9102). Bologna: PROMETEIA.

Fidrmuc, J., & Pichelmann, K. (1999). Austrian experience of the accession to the European Union. In R. Courbis & W. Welfe (Eds.), *Central and Eastern Europe on its way to European union. Simulation studies based on macromodels* (pp. 271–293). Frankfurt/Main: P. Lang.

Franz, W., Heidbrink, W., & Smolny, W. (1993). *The impact of German unification on West Germany's goods and labor market: a macroeconometric disequillibrium model in action* (Discussion Paper No. 7). Konstanz: Center for International Labor Economics, University of Konstanz.

Franz, W., Oser, U., & Winker, P. (1994). A macroeconometric disequilibrium analysis of current and future migration from Eastern Europe into West Germany. *Population Economics 7*, 217–234.

Franz, W., Goeggelmann, K., & Winker, P. (1997). Ein makrooekonometrisches Ungleichgewichtsmodell fuer die westdeutsche Volkswirtschaft 1960 bis 1994: Konzeption, Ergebnisse und Erfahrungen. In U. Heilemann & J. Wolters (Eds.), *RWI Mitteilungen: Vol. 47. Gesamtwirtschaftliche Modelle in der Bundesrepublik Deutschland. Erfahrungen und Perspectiven* (pp. 115–165). See also Duncker and Humbolt, Berlin, 1998 (pp. 115–165).

Gahlen, B. & Sailer, M. (Eds.) (1985). *Macroeconometric modelling of the West Germany* (Mimeo). Berlin.

Gali, J., & Lopez-Salido, A. (2001). *A new Phillips curve for Spain* (Documento de Trabajo No. 0109). Madrid: Banco de Espana.

Galli, G., Terlizzese, D., & Visco, I. (1989). Short and long run properties of the Bank of Italy quarterly econometric model. In *IFAC dynamic modelling and control of national economies* (pp. 379–382).

Garganas, N. C. (1991). *The Bank of Greece econometric model of the Greek economy*. Athens: The Bank of Greece.

Hansen, H.-J. (1995). A two-market disequilibrium model for the FRG. In W. Welfe & C. Lipinski (Eds.), *MACROMODELS'88* (pp. 25–33). Łódź: Wydawnictwo UŁ.

Heilemann, U., & Wolters, J. (1997). Gesamtwirtschaftliche oekonometrische Modelle in der Bundesrepublik Deutschland: Erfahrungen und Perpektiven. *RWI-Mitteilungen*, *47*, 31–46.

Jeanfils, P. (2000). *A model with explicit expectations for Belgium* (Working Paper, Research Series). National Bank of Belgium.

Jeanfils, P., & Burgraeve, K. (2005). *NONAME—a new quarterly model for Belgium* (Working Paper Research No. 68). National Bank of Belgium.

Klein, L. R., Welfe, A., & Welfe, W. (1999). *Principles of macroeconometric modeling*. Amsterdam: North-Holland.

Kratena, K., & Wuger, M. (1995). *Das WIFO-modell fuer die oesterreichischen Wirtschaftssektoren* (WIFO Monatsberichte 9).

Krelle, W. (1986). Oekonometrische Pognosemodelle: wissenschaftliche und wirtschaftspolitische Bedeutung. Der Beitrag des Sonderforschungsbereichs 21. In W. Krelle (Ed.), *Oekonomische Prognose-, Entscheidungs- und Gleichgewichtsmodelle* (pp. 59–75). Weinheim: DFG, VCH Verlagsgesellschaft.

Krelle, W., Beckerhoff, D., Langer, H. G., & Fuss, H. (1969). *Ein Prognosesystem fuer die wirtschaftliche Entwicklung der Bundesrepublik Deutschland*. Meisenheim.

Lambelet, J.-C., Bachetta, P., & Campero, L.-F. (1982). *The 'Crea' macroeconometric model of Switzerland's economy* (Cahier de recherche economiqué No. 8210). Université de Lausanne.

Modesto, L., & Neves, P. (1994). HERMIN Portugal. *Economic Modelling*, *12*, 275–294.

Müller, S., & Nakamura, S. (1986). Das Bonnier Vierteljahresmodell fuer die Bundesrepublik Deutschland. In W. Krelle (Ed.), *Oekonomische Prognose-, Entscheidungs- und Gleichgewichtsmodelle* (pp. 107–122). Weinheim: DFG, VCH Verlagsgesellschaft.

Nakamura, S. (1986). Ein Multisektorales Modell zur Erklaerung der Preise und der Verflechtungsstruktur fuer die Bundesrepublik Deutschland. In W. Krelle (Ed.), *Oekonomische Prognose-, Entscheidungs- und Gleichgewichtsmodelle* (pp. 99–106). Weinheim: DFG, Verlagsgesellschaft.

Özatay, F. (2000). A quarterly macroeconometric model for a highly inflationary and indebted economy: Turkey. *Economic Modelling*, *17*, 1–11.

Pytlarczyk, E. (2005). *An estimated DSGE model for the German economy within the Euro area* (Discussion Paper, Series 1: Economic Studies Nr. 33/2005). Frankfurt/Main: Deutsche Bundesbank.

RWI (1985). *RWI-Konjukturmodel. Dokumentation* (Mimeo). Essen: Rheinisch-Westfälisches Institut für Wirtschaftsforschung.

Sarrazin, H. T. (1986). Das BONNER MODELL 11. In W. Krelle (Ed.), *Oekomische Prognose-, Entscheidungs- und Gleichgewichtsmodelle* (pp. 76–88). Weinheim: DFG, VCH Verlagsgesellschaft.

Schellhorn, M., & Winker, P. (1994). Stochastic simulations of a macroeconometric disequilibrium model for West Germany. In W. Welfe & R. Kelm (Eds.), *Macromodels'94, Proceedings* (pp. 31–47). Łódź: Absolwent.

Scheufele, R. (2008). *Das Makrooekonometrische Modell des IWH: Eine angebotsorientierte Betrachtung* (Mimeo). Halle Institute for Economic Research (WWH).

Stalder, P. (1995). *The KOF-macromodel for Switzerland. Theoretical set-up, estimation and simulation*. Zurich: Konjunkturforschungsstelle KOF/ETH Eidgenossische Technische Hochschule.

Stäglin, R. (1995). Input-output models for policy analysis: the structural reporting system in the Federal Republic of Germany. In W. Welfe & M. Plich (Eds.), *Joint conferences on MACROMODELS'87 and MODELS and FORECASTS'87*, Porąbka-Kozubnik, 8–11.12.1987 (pp. 187–194). Łódź: Wydawnictwo UŁ.

Vincenzi, L. (1989). A note on the monetary sector of PROMETEA's model: the flow of funds and the reaction function. In *Dynamic Modelling and Control of National Economies* (pp. 481–487).

Visco, I., Terlizzese, D., & Galli, G. (1989). Un modelle trimestriale per la previsione e la politica economica: le proprieta di breve e di lungo periode del modello della Banco d'Italia. *Politica economica, 1*.

Willman, A., & Estrada, A. (2002). *The Spanish block of the ESCB—multi-country model* (Documents de Trabajo No. 0212). Madrid: Banco de Espana.

Yagci, F. (1987). An econometric model of the Turkish economy: the pivatal role of inflation. In W. Welfe & K. Markowski (Eds.), *Models and Forecasts'84, Macromodels'84* (pp. 259–279). Wrocław: Ossolineum.

Zonzilos, N. G. (2000). Structural econometric modelling at the Bank of Greece. In W. Welfe & P. Wdowiński (Eds.), *Macromodels'99* (pp. 81–94). Łódź: Absolwent.

Chapter 9
The Models of the Middle and East European Countries

9.1 The Models of the Centrally Planned Economies

9.1.1 Introduction

The national economies of the Middle and East European countries were centrally planned since the 1950s. The dominant role in the system was played by the socialized, mainly state-owned sector. In the early years, the command-distributive approach to management prevailed. Enterprises' economic activities were determined by plans that were either agreed with or approved by the central authorities, so the activities were regarded as deterministic. The requirements for commodities and materials presented in the plans were balanced at the central level using the material balances approach that was treated as the major instrument of coordination, while the role of price adjustments was ignored. At the macro scale, this made it possible to use the input-output models. The theoretical basis for applying the models as well as the optimization models was formulated in the numerous O. Lange's and W.S. Nemchinov's contributions on the functioning of socialist economies.

Production plans were drawn up in an iterative process. Enterprises formulated their production plans that were adjusted and approved by the central authorities, allowing for the centrally assigned tasks and taking account of constrained material supplies. The built-in motivation systems were to stimulate the fulfilment of the production plans by their maximisation (initially gross-output and then net output or value added).

Researchers realised after a time that the mezzo- and macro levels of this process could be approximated using the systems of production functions. Noteworthy are the Polish attempts to use the production functions to analyse the national economy (Pajestka 1961) and the attempts made afterwards in the former Soviet Union (Michalevskij and Soloviev 1966). The systems provided the backbone of the macroeconometric models that were constructed for the centrally planned economies in the later periods (Costa and Mienszikow 1979).

The theoretical foundations underlying the use of the production functions at the macro scale were formulated by M. Kalecki in his theory of growth of socialist

W. Welfe, *Macroeconometric Models*,
Advanced Studies in Theoretical and Applied Econometrics 47,
DOI 10.1007/978-3-642-34468-8_9, © Springer-Verlag Berlin Heidelberg 2013

economies (Kalecki 1963). This monograph identified major barriers to growth that became a basis for the concept of a chronic disequilibrium economy that J. Kornai developed several years later (1980). See also Bobińska (1982).

In the 1960s, the continuous development of official statistics enabled the construction of a special system of balances at the macro-scale that included the annual data, which was called a Material Product System (MPS). At that time only desk calculators were available; it was not until the 1970s that computers were introduced.

The above circumstances explain why the first, experimental macromodels were built as late as the mid-60s (Fomin and Tomaszewski 1971). They were small, linear models describing Hungary (Halabuk et al. 1966) and Poland; the latter was constructed by Z. Pawłowski (Barczak et al. 1968). These macromodelling activities were influenced by L. Halabuk's cooperation with H.T. Shapiro (Michigan University) and Z. Pawłowski's contacts with R. Stone (University of Cambridge).

In the 1970s, with the command-distributive management systems being slowly withdrawn to be replaced by market-oriented economies, macroeconometric, annual models were built in many of the socialist countries. This applies to former Czechoslovakia (VVS models), GDR (KP models), Hungary (CSO and INFELOR models), Poland (Planning Office's KP models and IEiS UŁ's W-models) and to the former USSR (Novosibirsk AS). The person that played a crucial role in priming the processes was L.R. Klein, who hosted at the University of Pennsylvania I. Sujan from the VVS in Bratislava and W. Welfe from the IEiS UŁ in the early 1970s.[1] The structure of these models was presented in the relevant chapters of Welfe (1982) and its English version in Welfe (1983). Compare also Welfe (2004b). Concise characteristic is presented in Table 9.1. More can be found in Mienszikow and Costa (1979).

In those years international conferences on macromodelling were organised on a regular basis: "Models and Forecasts of Socialist Countries" was held every two or three years and "MACROMODELS" was an annual event organised by the Institute of Econometrics and Statistics (IEiS) UŁ, Poland.

In the 1980s, a new generation of models was built. Their characteristic feature was that they more and more often attempted to take account of the elements of market economies. This was manifested through the introduction of the demand-determined model versions in addition to the supply-oriented versions, and through the construction of the disequilibrium models. The new models were provided with systems of price equations and of financial flows. To account for disequilibria, attempts were made to extend the production functions by adding the scarcity in the supply of raw materials, especially the imported ones, and to respecify the consumer demand functions in order to reflect the chronic deficits in the supply of consumer goods. Most of these contributions originated from the series of the W models for Poland.[2]

[1] The development of macromodelling activities to the mid-70s can be found in the papers by Shapiro and Halabuk (1976), Shapiro (1977) and in the monograph by Kolek and Šujan (1978).

[2] The detailed characteristics of the structures of the models, especially those for Poland, can be found in our monograph dealing with the modelling of the Polish economy (Welfe 1992).

In the mid-80s, special, small-sized disequilibrium models were constructed, accentuating excess demand in the markets for consumer goods and for investment goods.

At the turn of the century, most Middle and East European countries started their transition towards a market system.

The changes in the economic system had a serious effect on the programmes of the macromodelling centres. Several of them, for instance in Hungary and East Germany, suspended their activities, others carried on, e.g. IEiS UŁ in Poland or INFOSTAT in CSRS. After enough data on the transition period had been accumulated, new modelling centres in Bulgaria, Romania, the Ukraine, the Russian Federation, but also in Poland (the Central Bank, the Ministry of Finance), the Czech Republic (the Central Bank) and Slovakia (SAV) started to operate, mainly constructing the quarterly models.

9.1.2 The Models of the Czechoslovak National Economy

9.1.2.1 The VVS models

The main macroeconometric modelling centre in the Czechoslovak Republic was the Computer Research Centre (VVS) in Bratislava, affiliated with the United Nations. I. Šujan built the first model in 1970. The research team headed by I. Šujan constructed the VVS system of macromodels containing annual models for the CRSR and Slovakia and a quarterly model for the CSRS (Šujan et al. 1974).

The models were supply oriented. Exogenous investments determined fixed capital and exogenous labour supply affected employment (allowing for fixed capital). All these production factors determined gross output by industry that for the most part influenced exports and imports. Given the exogenous average wages, employment had an effect on real incomes that in turn shaped the volume of retail trade. The macroeconomic identities were not present. Because the models were mostly forecasting tools, dynamic, recursive relationships played a dominant role in their structures. The VVS developed a range of computer programmes to support its modelling and forecasting activities. See also Kyn et al. (1967).

9.1.2.2 The VUSEiAR Models

In the late 1970s and in the early 1980 the Bratislava centre that changed its name to the Research Institute of Socio-Economic Information and Management Automation (VUSEiAR) intensively developed its modelling activities. A new system of the second-generation macromodels was constructed. It included the annual medium-sized Central Model (CEM), which was mainly dedicated to policy simulations, the Input-Output model, and the sectoral models of commodity supply and consumption. The system was linked with the model of the CMEA countries (Šujan 1978a,

1978b). The Central Model was built in many variants (Šujan 1986). At the same time, a quarterly model of the CSRS economy was constructed. The model, which also had more than one version, found application in short-term forecasting and policy simulations (Šujan and Kolek 1974; Klas et al. 1979).

The Central Model's structure was characteristic of the supply-oriented models. It started with a two-factor production function (that allowed for technical progress) that generated the potential gross output. The function could be modified by taking account of the likely supply restrictions. Employment was exogenous. Fixed capital increase depended on investment (with lags) and the rate of liquidation. Investments were determined at the central planning level where the supplies of domestic and imported investment goods were taken into account. The demand-determined consumption depended on exogenous wages and prices, being also adjusted for restrictions in the supply of domestic and imported commodities (Štrauch 1990). The Central Model was the main component of the system of models that included also the MSA model of the I-O type and the satellite models of consumption and financial flows.

Independently Brada and King constructed a structural model of Czechoslovakia (1980).

9.1.2.3 The Disequilibrium Models

In the mid-80s, the disequilibrium models were constructed for the CSRS economy, likewise for other centrally-planned countries. The models attempted to estimate the excess consumer demand. They also showed the mechanisms transmitting disequilibria into foreign trade, production and the labour market (Dlouhy 1984; Šujan and Štrauch 1989).

9.1.3 The Models of the German Democratic Republic

9.1.3.1 The DEM Model

In the East German economy, planning processes were mostly based on the balance approach, including the Input-Output models. The activities aimed at building macroeconometric systems started later than in the other CMEA countries. In the early 1970s, a small annual model DEM 1 was built at the Planning Committee, structured following the model for Ukraine (Anders et al. 1971). This fully supply-oriented model was used for preparing forecasts supporting the planning process. Gross output was generated from a production function. It was allocated among investment, material supplies and exports. Personal consumption was determined by real incomes and social consumption was residual.

9.1.3.2 The ISI Model

At the end of the 70s, the team led by M. Wölfling constructed at the Economic Institute of the Academy of Science an annual, macroeconometric model of the GDR economy called ISI 1 (Woelfling 1977). The system was composed of 5 interrelated blocks of equations explaining employment and wages, production, foreign trade, the distribution of national income and fixed capital. It also contained 10 submodels. Its structure was peculiar and the economic mechanisms used were rather strange. The two principal exogenous variables were independently treated: national income and investment. The second variable, i.e. investment, was regarded as a policy (central planner's) instrument. National income determined gross output and indirectly employment in the material product sector (allowing for labour supply). It also made dependent real personal income being a function of labour productivity, taking account of average wages. Real incomes determined retail sales. Investments decided upon the increase in fixed capital. Gross output was arrived at by multiplying employment by labour productivity depending on the ratio between fixed capital and employment.

The ISI 1 model was a fully supply-oriented system designed as the planners' instrument for macroeconomic forecasting and simulation analyses.

9.1.4 The Models of the Hungarian National Economy

9.1.4.1 The CSO's Models

The first macroeconometric M-1 model of the Hungarian economy was built in 1966 as a result of the research conducted by L. Halabuk's team at the Econometric Laboratory of the Central Statistical Office (CSO) in Budapest (Halabuk et al. 1966). This small, annual, linear model was designed for experimental purposes. Its construction was a learning ground that provided the necessary experience to construct an operational version of the multisectoral M-2 model four years later. M-2 was medium-size model with 26 equations; it was linear but dynamic (Halabuk et al. 1973). The model was, in principle, supply determined, with 8 industries and 5 commodity groups within retail sales. Most important were the linear two-factor production functions generating gross output that determined global consumption. However, retail sales were dependent on real personal incomes per capita being a function of average wages and employment. Employment depended on output and exogenous labour force at the same time. Fixed capital was assumed to be exogenous. The model included also the export and import functions of mixed specification (i.e. both demand and supply were allowed for) (Kolek and Šujan 1978).

The equation parameters were estimated with OLS and TSLS, allowing for collinearity. The model was carefully validated. It was used in preparing forecasts for the years 1970–1975.

At the turn of 1969, the CSO entered into cooperation with the VVS Centre in Bratislava that resulted in the construction of a small M-3 model. A parallel model was built for Czechoslovakia (Kolek and Šujan 1978). In the mid-70s, the large annual M-4 model was constructed, which was a macroeconometric model integrated with the Input-Output model (Hulyak 1973).

9.1.4.2 The SZAMKI Models

In the second half of the 1970s, macroeconometric modelling activities were developing in the Budapest research institute INFELOR directed by G. Szakolczai. The SZAMKI models constructed at the Institute included an annual model and the first quarterly model of the Hungarian economy. The models were expanded and used in forecasting and policy simulations until the end of the 1980s (Szakolczai et al. 1985).

9.1.4.3 Models Developed at the Institute of Economics and Market Research

In the late 1970s, an annual model of the Hungarian economy was constructed at the Institute of Economic and Market Research in Budapest by the team directed by A. Simon. It was a demand oriented model (Simon 1978). It participated in the system of world economy models Project LINK. It served for many years as forecasting and policy simulation tool.

9.1.4.4 The Disequilibrium Models

The early 1980s witnessed the development of theoretical contributions to the disequilibrium models in Hungary, among which J. Kornai's contributions are considered the most prominent (Kornai 1980). Empirical research based on the contributions developed in the mid-1980s. It was summarised in the paper by M. Lacko (Davis and Charemza 1989). K. Hulyak built a macroeconometric, multisectoral disequilibrium model of Hungarian economy, presented in the above monograph.

The Hulyak's model distinguished consumer goods, investment goods markets, labour market, as well as exports and imports, specifying the demand and supply functions for each of the markets. Excess demand in the markets was explained assuming quantitative adjustments and the central planner's interventions affecting investments. Consequently, it was assumed that prices did not clear the markets. The disequilibrium elements affected also exports and imports.

The model was small and substantially aggregated. It was mainly used for running ex-post simulation analyses of experimental character (Hulyak 1986).

9.1.5 The Models of the Polish Economy

9.1.5.1 The Model by Z. Pawłowski

The first macroeconometric model of the Polish economy was developed in the mid-60s on the inspiration from Z. Pawłowski (Barczak et al. 1968). It was a small, annual, experimental model, which was primarily intended to pave the way for the construction of macromodels that the centrally planned economies could use as auxiliary tools in preparing central economic plans.

The model used a sample spanning the years 1950–1964 and was very dynamic. This supply-determined system was decomposed into three sections: agriculture, other industries in the material sphere and non-material services. A key role was given to the linear, two-factor production functions that generated net output and national income. National income determined the investments in the material production sector that, allowing for lags, determined an increase in fixed capital. An important feedback was thus established in the model, which we later used to call a supply accelerator (Welfe 1983). It significantly distinguished the above model from those constructed in other countries in later periods, where investment was assumed to be an important policy variable.

Another distinctive feature of the model was the specification of the equation explaining average real wages in industry. They were dependent on labour productivity and followed the rules that the authorities assumed to make decisions about wage rises. The average wage dynamics in industry influenced wage growth in the non-material services. The authors built the model for experimental purposes and did not develop it any further.

New macroeconometric models were built only in the 1970s (cf. Maciejewski 1980). Two of them—one at the Planning Committee (KP) and another at the Institute of Econometrics and Statistics UŁ (W-1)—had operational meaning. The other models were experimental and were not used on a regular basis. This concerns the models constructed at the Institute for Economic Cycles and Prices (Kalisiak and Piaszczyński 1972; Bożyk et al. 1973), at the Department for Statistical and Economic Research at the GUS (CSO) and at PAN (Kudrycka 1974).

9.1.5.2 The Models of the Planning Committee (KP)

The Computational Centre of the Planning Committee started its macroeconometric modelling activities in the years 1972–1973. Its research team led by W. Maciejewski entered into close cooperation with the VVS Centre in Bratislava.

The first to be constructed was the annual model KP-1. This forecasting-simulation model was intended to enhance the analyses preceding the implementation of the 5-year plan 1971–1975 (Maciejewski et al. 1973). The model was medium sized, with 41 equations. It was linear and dynamic, in principle recursive, based on the 1955–1970 data. It was supply determined, with the linear production functions used to generate net output. Lagged fixed capital depending on lagged investment was a key production factor. Investment was determined by the volume of industrial output. These relationships defined a feedback that we called the supply accelerator. At the macro level, the model generated both supply and demand, thus enabling analyses of the potential disequilibria.

The experience gained from running the model led to the construction of its new, consistent version, which was called KP-2 (Maciejewski and Zajchowski 1974). The new, medium-sized model having 188 equations was also linear and dynamic, but jointly dependent. The sample covered the years 1960–1972. The model was fairly disaggregated, distinguishing 15 industries and 7 commodity groups in consumption.

Its characteristic feature was that it defined the strategic industries controlled by the central authorities that determined their investments and wages. Regarding the non-strategic industries, the feedbacks ensured autonomous increase in their output. The net output (net value added) was generated from the linear, two-factor production functions. The model assumed full employment, with employment in the non-material services sector being treated as residual, given the supply of labour force.

Consumer demand depended on real income and relative prices that were treated as exogenous. The model generated also commodity supply, which depended on domestic output and imports. This approach made feasible the analyses of local disequilibria in the commodity markets.

The model was intensively used in forecasting and during numerous policy simulations. Organizational changes caused, though, that this modelling activity was given up at the end of the 1970s.

The team lead by W. Maciejewski and J. Zajchowski constructed also the first, operational, quarterly model of the Polish economy KP-3K (Kalinin et al. 1977). It was used for evaluating the future implementation of the annual plans and as a tool supporting their preparation.

KP-3K was a small, linear, dynamic model based on the 1967–1975 sample. Because the quarterly national accounts did not exist then, the number of its variables was limited to those on which the quarterly data were available, namely industrial output, foreign trade, wages and incomes. The model's production functions explained gross-output in manufacturing and building industries as being dependent on exogenous employment and imports. Retail sales were generated from both the supply and demand side. Exports equations explained the supply of exported commodities. The model was used in short-term forecasting until the economic crisis at the turn of the 1970s.

9.1.5.3 The Macroeconometric, Annual W Models. The W-1 Model

Between 1971 and 1972, macroeconometric models of the Polish economy were also constructed by the research team led by W. Welfe at the Institute of Econometrics and Statistics UŁ (IEiS UŁ) in Łódź, Poland. The models were initially built for the learning purposes—they were intended to facilitate the analysis of the properties of the real functioning of a centrally-planned economy. Unlike the planning models, the macroeconometric models were to replicate the reduced forms of relationships explaining both plan preparation and implementation processes.[3]

W. Welfe constructed his W-1 model in Philadelphia in 1971, where he had the opportunity to consult with L.R. Klein (Welfe 1973). W-1 was a medium-sized,

[3]More detailed characteristics of these models were provided in our monograph (Welfe 1992).

complete, non-linear, and dynamic model with jointly interdependent equations. It was fairly disaggregated, distinguishing 4–5 industries and 6 commodity groups. The model was annual and most data it used came from the years 1955–1969.

The model had an extended production sector. A key role was played by the linear and non-linear production functions that generated net output that determined investments and fixed capital increase. These relationships defined a feedback known as the supply accelerator. For each industry (and market) both supply and demand functions were determined that enabled the estimation of excess demand. The last variable did not have an effect on any other endogenous variables, though. Its changes could be modified with the quantitative adjustments, including the planners' interventions, without using the exogenous price adjustments.

The equation parameters of the W-1 model were estimated with the OLS method, after the equations were linearized. The model was carefully validated and then structural analysis laying emphasis on the stability of the parameters, especially on the relevant elasticities, was performed.

During the next years the W-1 model was frequently modified with updated samples. The 1976 version was based on a sample extending to the year 1974 (Welfe 1975; Czyżewski et al. 1976). This version distinguished three variants: (a) short term, supply determined, where production allocation was assumed to be regulated by inventory changes, (b) medium term, supply determined, where the market was regulated by foreign trade adjustments (these two variants reflected the economic situation in the 1960s fairly well), (c) medium term, demand determined, where final demand determined the demand for production and imports; the demand for production was confronted with potential output generated from the production functions—the resulting discrepancies affected the foreign trade adjustments. This version seemed to provide the most adequate description of the economic situation in the early 1970s.

In the supply-oriented versions of the model, production was allocated among particular groups of users based on the commodity supply equations. In the demand-oriented version, personal consumption, investments, and exports equalled the demand realized by households and other pertinent economic agents (Welfe 1979).

The parameters of the models' equations were estimated with OLS and TSLS and with the iterative techniques (Bodin 1986). The validation procedures included multiplier analyses and stability testing.

9.1.5.4 The Annual W-3 Model

The rapid economic expansion in the early 1970s called for a new instrument of macroeconomic analyses. Accordingly, a new, operational W-3 model drawing on the accumulated experience was constructed at the IEiS UŁ between 1975 and 1976. This model had ca. 300 equations, thus being larger than its predecessors (Czyżewski et al. 1976). The manufacturing industry was decomposed into 6 groups and the foreign trade classification was made more specific. The equation parameters were estimated with OLS and TSLS, using a sample with annual data ranging from 1969 to 1974.

The model's structure was not significantly different from the last version of the W-1 model (of 1976), but many important modifications were made to the specifications of particular equations. The W-3 model was respecified each year using updated samples. Two versions of the model were prepared. The demand-determined version (W3-D) was used until the economic crisis at the turn of the 1970s. The supply determined version (W3-S) was used even after that period. The most important innovation in the model was that the lag distribution parameters in the equations linking fixed capital and investment outlays were estimated (Welfe 1980).

The supply-determined version was rebuilt at the beginning of 80s, mainly by imposing restrictions on economic growth. Updated every six months, the fully supply-determined version was operated until the year 1984, serving both macroeconomic analyses and regular forecasting of economic development (Juszczak and Welfe 1983). It was non-linear and dynamic, and represented a jointly interdependent system of equations. The data sample covered the years 1961–1979. This model used the production functions to generate gross and net output. It was allocated among the relevant intermediate and final users assuming constant allocation coefficients. The feedbacks in the model were the supply accelerator and the foreign trade loop: output determined exports and indirectly imports of machinery and equipment, thus affecting fixed capital increase and consequently output. An important novelty was a system of price equations representing the cost-push approach that was added to the model (Welfe 1982). For the time being, wages remained exogenous.

The W3-S model was replaced by a new version in 1983 (Juszczak 1987). The new version had over 500 equations and was estimated using an updated sample spanning the years 1961–1982. It was used between 1986 and 1987 to produce regular forecasts and policy simulations (Juszczak et al. 1987).

It is worth adding that a minimodel W4 of the Polish economy replicating the properties of the W-3 model was additionally constructed. It was mainly used for testing alternative estimation techniques (Romański and Welfe 1981). Another minimodel of the Polish economy was built by J. Gajda, as a component of the system of models of the CMEA countries (Gajda 1986).

9.1.5.5 The Input-Output and Sectoral Models

In the late 1970s, new research projects were initiated at the IEiS UŁ. An attempt was made to construct a large W-2 model with a view to integrating an econometric model with an I-O model (Tomaszewicz and Welfe 1979). The construction of the W-2 model was never completed, but the experience gained then helped build the I-O and sectoral models of the Polish economy. The input-output models were developed on a large scale in the 1980s by Ł. Tomaszewicz and her research team (Tomaszewicz 1983). It is worth noting that Ł. Tomaszewicz has prepared an original—supply determined version of the I-O model (Tomaszewicz 1983, 1985). Later on the structure of these models returned to the classic, demand oriented models. The above models were incorporated into the system of the world INFORUM models. The I-O models were integrated with macroeconometric submodels explaining the major components of final demand, as well as the demand for the production factors. They were considerably extended afterwards. Their construction was greatly ahead of what a country in transition to a market economy may have needed.

Several sectoral models were developed at the IEiS UŁ. Worth mentioning are the models of the financial sector (Łapińska-Sobczak et al. 1979), of income distribution (Dębski 1987) and of employment, inventories and commodity markets.

9.1.5.6 The W-5 Model

In the early 1980s, the IEiS UŁ team led by W.Welfe constructed the W-5 annual model of new generation, being an attempt at implementing W. Welfe's concept of a complete, integrated model (see Czerwiński and Welfe 1982 and Welfe in Klein and Welfe 1982). The model contained blocks of equations generating both demand for and supply of production and production factors, financial flows, as well as wages and prices. The number of the equations grew to 800 and their parameters were estimated using OLS with a data sample covering the years 1964–1979. In the next years, the model was extended to around 1100 equations, as well as being systematically updated. It was a non-linear, dynamic and highly simultaneous model. After access was gained to efficient simulation packages developed under Project LINK, in 1987 the model started to be used for regular medium-term forecasting and numerous policy simulations (Welfe 1985).

The structure of the W-5 model based on the solutions developed for the previous versions of the W models, particularly regarding its supply sector. The Cobb-Douglas production functions generating net output remained the most important. The functions contained the primary production factors—fixed capital and employment—as well as the indicators of their utilization, such as the numbers of shifts and average time worked by employees, the impact of technological progress and finally the restrictions in supply of raw materials, mainly from import.

Fixed capital increase was dependent on investment and the rates of scrappings. Investment demand (by industry) could be modified to allow for the central planners' preferences. Employment was determined by demand and depended on net output and the impact of technological progress. It was not assumed, however, that the demand would be balanced with labour force supply.

Severe restrictions in the supply of raw materials were common in the 1980s. To account for them, the excess demand indicators representing the ratios between the potential demand for imported raw materials and their supply had to be constructed in the production functions. The restrictions in the supply of the domestic raw materials could be confined to those affecting fixed capital or employment in the industries that produced the materials. This specification was an important novelty in the construction of the production function.

The structure of the supply sector was concluded by specifying the equations explaining commodity supply to the particular groups of users; attempts were made to present central planners' varying preferences, with either investments or consumption being assumed as residual.

The W-5 model generated the final demand of the major final users, including foreign agents, as well as the industries' demand for domestic output. The new problems that had to be dealt with resulted from the frequently appearing deficits, i.e. excess demand in particular markets. They were solved in the following manner. The initial (unobservable) demand was specified first. Then the effective demand equal to the realized (observable) supply was obtained by subtracting the excess demand from the initial demand.

The initial consumer and investment demand functions were specified following the standard rules applying to market economies. To arrive at the effective demand functions several modifications were necessary, regarding the designation of demand and incomes. For instance, the modifications involved the introduction of the notion of postponed demand, household incomes enlarged for forced savings, etc. Excess demand was generated from the functions of special disequilibrium indicators constructed using the observed data (Welfe 1992). These elements were new to the consumer demand functions and the reason for adding them were the chronic disequilibria in the 1980s.

The W-5 model included extended blocks of equations generating prices, wages, and financial flows (Juszczak et al. 1997a).

Beginning from 1988, the W-5 model was used for many years to prepare regular forecasts and numerous policy simulations jointly with the Planning Committee. The commencement of the transition period to a market economy caused that the model's demand-determined version assuming an equilibrated market economy with prices clearing the markets was found particularly valuable. The W-5 model was entered into the Project LINK world economy model (Welfe 1987).

In the mid-80s, the W-6 minimodel replicating the W-5 model was constructed. This model stressed the disequilibria in the Polish economy (Romański and Welfe in Kotyński 1990). See also Romański and Welfe (1991).

9.1.5.7 The Disequilibrium Models

The IEiS UŁ research activities aimed to measure and analyse the impacts of the disequilibria resulted in contributions involving the specification of production functions, as well as of the consumer demand functions. They contributed to the emergence of the W-A market models for the Polish economy constructed by Welfe (1984, 1986). A generalized notion of consumer demand, including postponed demand, was introduced into the models, and real incomes were extended to allow for forced savings. Special indicators of excess demand were constructed and widely used.

In the mid-80s, several centrally-planned countries developed special macroeconometric disequilibrium models (Charemza and Quandt 1982; Welfe 1991). They mostly stressed the role of the consumer goods markets and imperfect price adjustments (Portes et al. 1987). This class of models included a small disequilibrium model of experimental character, which was developed by Charemza and Gronicki (1985). The model reproduced relationships prevailing in the period of chronic disequilibria affecting the commodity markets, production and foreign trade. Its properties were described in full vis-à-vis the properties of the W-5 model (Welfe 1989).

The paper provided a starting point for the discussion about the rules to be applied to construct large macroeconometric models for the East European countries, including the role of economic theory, the equation specification rules, data mining, etc. The discussion was initiated by Charemza (1989, 1991) who was answered by Welfe, Gajda and Żółtowska (1991, 1992). The last word belonged to Klein (1991).

9.1.5.8 Other Models of the Polish National Economy

Between 1981 and 1982, J. Pawilno-Pacewicz constructed the medium-term model SAPO. It had a supply orientation linked to the planning process (Garbicz et al. 1985, 1986). The Polish Planning Committee used it as an auxiliary instrument enhancing the planning processes and the monitoring of their implementation.

The models built at the Institute for Economic Cycles and Prices drew on the input-output models, but they showed a supply orientation.

The Poznań research centre constructed multisectoral dynamic planning models. The models were demand determined and supported policy simulations (Czerwiński et al. 1981, 1998).

In the mid-70s, the Planning Institute formed the concept of constructing a comprehensive system of macroeconometric models based on the input-output models with a view to making the planning process more efficient. Its realization was the macroeconometric model MEDIG, which was built at the Institute (Bocian and Burża 1978; Bocian 1984). The Institute of National Economy being the successor of the Planning Institute continued this research, as a result of which a new system of macromodels was produced (Barteczko et al. 1985). Attempts to ease system analysis were undertaken (Kulikowski et al. 1979).

The last supply-determined model to be mentioned here was constructed by B. Kłos from the National Bank of Poland. The model was intended to reproduce the development of the Polish economy in the years 1971–1985 (Kłos 2002).

Towards the end of the 1980s, the approaching transformation attracted researchers' attention to the short-term analyses. Unfortunately, I. Kudrycka from the ZBSE GUS & PAN was the only one to address issues in the short-term forecasting and analyses. However, the attempt to construct a system of quarterly models turned out to be unsuccessful. The major quarterly model she constructed was highly disaggregated and distinguished 26 industries. It revealed its supply orientation by stressing the description of the production process. It was planned that the quarterly model would be accompanied by a submodel of the monthly production of selected commodities. After preliminary work, the research was discontinued in year 1985 (Kudrycka 1990).

9.1.6 The Models of the United Soviet Socialist Republics (USSR)

9.1.6.1 The Early Models

The first macroeconometric models were constructed for particular republics—Ukraine (1974) and the Baltic countries (Bielkin et al. 1978). Best known were the UKR1 and UKR2 models of the Ukrainian economy built by A.S. Jemielianov and F.I. Kushnirskij (Jemielianov and Kushnirskij 1974; Maciejewski 1976; Kolek and Šujan 1978). The models were clearly supply determined. In the larger model UKR2 that was constructed for particular industries the linear, two-factor production functions played a key role. A feedback mechanism resembling the supply accelerator was built into the models—gross output was assumed to determine accumulation that affected fixed capital increase, allowing for up to 2-year lags, and then fixed capital determined output again. Consumption was residual.

For the USSR as a whole only small macroeconometric models were constructed at the Institute of Industry Economics and Organization, the Novosibirsk Branch of the Academy of Sciences. They were used in the comparative studies on the Soviet, American and Japanese economies.

9.1.6.2 The Input-Output Systems

At the end of the 1970s researchers concentrated their efforts on attempts involving the application of the input-output systems. The leading institution in this field was the Central Economic-Mathematical Institute of the Academy of Science (CEMI) in Moscow. The point of departure assumed was the planning system (Michalevskij 1967, 1972). Simulations making use of extended I-O systems were to support the preparation of economic plans as well as their implementation (Jaremienko 1984).

In the next years, the research efforts aimed to transform the balance models into macroeconometric models allowing for disequilibria took place. In particular, the assumption about the technical coefficients being constant was given up. Many attempts were made to account for technological changes by means of the regression equations. Further, equations explaining the use of raw materials in particular industries were constructed. They were specified to allow for factors determining both demand and supply (cf. Jaremienko 1982).

9.1.6.3 The Macroeconometric Models

Interestingly, the large macroeconometric models of the USSR were constructed outside the Soviet Union. The most renowned model was the annual multisectoral SOVMOD model, which was built at the Wharton Econometric Forecasting Association (WEFA) in Philadelphia (Green and Higgins 1977).

Towards the end of the 1980s, two macroeconometric models were constructed at the CEMI in Moscow. The first model, SUN, built by the team led by E.E. Gavrilenkov (1990) had several versions. It was an annual, medium-sized model with a supply orientation—the major role was given to production functions generating gross output. It was used for medium-term forecasting and simulation analyses that preceded the preparation of economic plans. The second model was intended to support structural analyses, especially those concerning the disarmament policies (Bessarab and Dadajan 1991). It was an annual medium-sized model that practically only had the production sector.

9.1.7 The Models of the Yugoslav National Economy

In the middle of the 1970s, macromodelling activities in former Yugoslavia were started by the team led by J. Mencinger. The quarterly EIPF model they initially constructed (in 1975) consisted of 51 equations, growing to 115 by 1981. At that time, an input output submodel was added. This demand-determined model was mainly used in short-term forecasting.

In the first half of the 1980s, an annual model of the Yugoslav economy constructed by Pfajfar was operated. Initially consisting of 34 equations (1980), within the next four years it was enlarged to 174 equations. The model was used as a tool supporting medium-term analyses and forecasts.

9.1.8 The Models of the Council of Mutual Economic Assistance (CMEA) Countries

The CMEA was an association of the centrally-planned countries in Europe, which coordinated their economic activities. The mutual economic relationships between the countries were analysed with the support of the system of the annual models of the CMEA members, which was constructed by J. Gajda's team at the IEiS UŁ in 1979. Particular country models had similar structures and were linked together via the foreign trade transactions. The models served the preparation of medium-term forecasts and policy simulations. This system was in use until the end of the 1980s (Gajda 1986).

9.1.9 The Models of the Centrally Planned Countries. Recapitulation

Macroeconometric modelling developed in the centrally planned countries in a somewhat peculiar way. Initially, when the centrally planned command-and-distribution system prevailed, the decisive role was played by the planning models,

Table 9.1 Models of Central and Eastern Europe in the period of centrally planned economies

Country Model Version Year	Institution Author	Frequency Sample	Equations total stochastic	Decomposition	Estimation method	Properties	Applications
CZECHOSLOVAKIA							
VVS 1 1971	VVS Bratislava I. Šujan et al.	Annual 1956–1968	12 8		OLS, TSLS	Linear model, supply oriented	Experimental, forecasts
VVS 2 1972		Annual 1955–1970	27 17	4 sectors	TS IIV, principal components	Extended for a block of fixed capital, investment, and household incomes	Experimental, forecasts
VVS 3 1973		Quarterly 1961–1971	52 37	8 industries, 6 groups of commodities in retail sales	OLS recursive	Supply determined, disaggregated	Short-term forecasts; large number of predetermined variables
VVS 4 Model Slovakia 1974		Annual 1963–1969	12 8		OLS	Supply oriented	Medium-term forecasts
VVS 5 1977		Annual	135		OLS	Supply oriented	Simulation analyses, forecasts
CEM Models	VUSEiAR						
CEM1 1979	I. Šujan et al.	Annual	111	5 industries, 8 groups in foreign trade, 3 in retail sales	OLS	Supply oriented	Forecasts and simulation analyses

Table 9.1 (Continued)

Country Model Version Year	Institution Author	Frequency Sample	Equations total stochastic	Decomposition	Estimation method	Properties	Applications
CEM2 1981		Annual	150		OLS	Supply oriented	Simulation analyses
CEM3 1983		Annual	187		OLS	Extended foreign trade	Policy simulations
CEM 3.2		Annual 1960–1983	192 26	4 industries	OLS recursive	Prices and financial flows absent	Simulation analyses, medium-term forecasts
CEM4 1986		Annual	264	5 industries	OLS		
MSA 1983		Annual		I-O, 14 industries	OLS	Component of the models system CEM	Simulation analyses
Quarterly 1982		Quarterly	130			Supply determined	Short-term forecasts
Model CSRS 1978–1980	A.S. King, J.C. Brada	Annual	From 42 to 75			Supply determined	Simulation analyses
Disequilibrium model 1983	V. Dlouhy, K. Dyba	Annual				Small disequilibrium model	Policy simulations
Disequilibrium model 1989	VUSEiAR I. Šujan, D. Štranch	Annual		–		Disequilibrium model	Forecasts and policy simulations

Table 9.1 (Continued)

Country Model Version Year	Institution Author	Frequency Sample	Equations total stochastic	Decomposition	Estimation method	Properties	Applications
GERMAN DEMOCRATIC REPUBLIC							
DEM 1 1971	Planning Committee P. Franken et al.	Annual 1960–1970	18 9	–	OLS	Supply determined	Forecasts supporting central planning
ISI 1 1977	Institute of Economics A.W. DDR M. Woelfling	Annual	207 146	13 industries	OLS	Supply determined, hierarchical structure (10 submodels)	Forecasts, simulation analyses
HUNGARY							
M-1 1965	Central Statistical Office, Econometric Laboratory L. Halabuk et al.	Annual	9 5		OLS	Supply determined	Experimental
M-2 1970		Annual 1950–1964	26 23	8 industries, 5 groups of commodities	OLS	Supply determined, consumer demand function	Forecasts and simulation analyses
M4 1974	K. Hulyak	Annual		I-O	OLS	Integrated model (I-O including)	Forecasts and simulation analyses

Table 9.1 (Continued)

Country Model Version Year	Institution Author	Frequency Sample	Equations total stochastic	Decomposition	Estimation method	Properties	Applications
SZAMKI 1985	INFELOR L. Hunyadi, G. Szakolczai	Annual 1960–1985	Above 600 ca. 150	19 industries, 5 sectors	OLS	Mixed structure	Forecasts and policy simulations
1985.	G. Szakolczai et al.	Quarterly			OLS	Financial flows	Short-term forecasts and policy simulations
Models of Hungary	Institute of Economics and Market						
1978	Research A. Simon	Annual 1960–1972	171	15 industries I-O	OLS	Mixed, mostly demand oriented	Forecasts and simulation analyses
1985		Annual	81 35	4 sectors I-O	OLS	Proexports oriented	Forecasts and simulation analyses
Disequilibrium model 1985–1986	Institute of Economic Planning, Budapest K. Hulyak	Annual 1965–1984	22 12		OLS TSLS	Disequilibrium model, demand and supply functions	Simulation analyses
POLAND							
1968	Model Z. Pawłowski	Annual 1950–1964	17 12	3 sectors	OLS	Supply oriented Supply accelerator	Experimental model

Table 9.1 (Continued)

Country Model Version Year	Institution Author	Frequency Sample	Equations total stochastic	Decomposition	Estimation method	Properties	Applications
KP-1 1972	Computational Center, Planning Committee W. Maciejewski, J. Zajchowski	Annual 1955–1970	41 32	4 industries	OLS	Supply determined, linear, recursive, disequilibria in consumer goods markets	Forecasts and simulation analyses for elaborating the 5-year plan 1971–1975
KP-2 1974		Annual 1960–1972	188 127	15 industries, 7 groups of commodities	OLS	Supply determined, supply of consumer goods confronted with demand	Forecasts and policy simulations
KP3K 1977	D. Kalinin, W. Maciejewski, J. Zajchowski	Quarterly 1967–1972 Quarterly 1967–1975	17 13	3 sectors	OLS	Supply determined, explained activities of manufacturing, building industries and agriculture	Short-term forecasts
Multisectoral Model 1974	Dept. of Statistical and Economic Research GUS PAN I. Kudrycka	Annual	60 51		OLS	Supply determined	Experimental
W models[a]	Institute of Econometrics and Statistics UŁ W. Welfe et al.						
W-1 1971		Annual 1960–1969	243 158	5 industries, 6 groups of commodities	OLS TSLS	Disequilibrium model	Experimental Structural analyses

Table 9.1 (Continued)

Country Model Version Year	Institution Author	Frequency Sample	Equations total stochastic	Decomposition	Estimation method	Properties	Applications
W3 1975–76		Annual 1961–1976	960 268	11 industries, 7 groups of commodities	OLS TSLS	Mixed orientation	Structural analyses
W3S 1980–1983		Annual 1961–1976	512 195	11 industries, 7 groups of commodities	OLS	Supply determined	Medium-term forecasts up to 1990, simulation analyses
W5 1980–1983		Annual 1963–1982	1055 436	11 industries, 7 groups of commodities	OLS	Submodels of supply and demand orientation, financial sector, participated in Project LINK	Forecasts up to 1995, simulation analyses
W6 1986–1991	J. Romański, W. Welfe	Annual 1961–1985	111 46	2 sectors	OLS TSLS	Minimodel with disequilibrium components	Simulation analyses
MEDIG 1978	Planning Institute A. Bocian et al.	Annual 1965–1978	ca. 100	I-O	OLS	Supply determined	Component of a system of planning models
SAPO 1981–1983	SGPiS J. Pawilno-Pacewicz	Annual		I-O	OLS	Supply determined	Forecasts and simulations for planning purposes
Model I-O 1982–1985	IEiS UŁ Ł. Tomaszewicz	Annual		I-O	OLS	Supply oriented, returned to demand determined in the 90s	Structural and simulation analyses

Table 9.1 (Continued)

Country Model Version Year	Institution Author	Frequency Sample	Equations total stochastic	Decomposition	Estimation method	Properties	Applications
Quarterly model	Dpt. for Statistical and Economic Research GUS PAN I. Kudrycka	Quarterly		26 sectors	OLS	Supply determined	Experimental
Disequilibrium model 1985	U. Gdański W. Charemza, M. Gronicki	Annual 1960–1980	35 13		OLS TSLS	Disequilibrium model	Simulation analyses
USSR							
UKR 1 1970	Ukraine Branch of the Institute of the Planning Committee of USSR A.S. Jemielianov, H.F. Kushnirskij	Annual 1965–1984	22 12	I-O 12 industries, 5 sections	OLS TSLS	Supply determined	Experimental
UKR 2 1971		Annual	89		OLS	Supply determined, decomposition by industries	Structural analyses used in 5-year planning

Table 9.1 (Continued)

Country Model Version Year	Institution Author	Frequency Sample	Equations total stochastic	Decomposition	Estimation method	Properties	Applications
SOVMOD	Wharton Economic Forecasting Association (WEFA), Philadelphia D. Green, C. Higgins						
1 1977		Annual	113 81	5 sections	OLS	Multisectoral model I-O added	Simulation analyses
3			295 189	6 sections	OLS	Submodel I-O added	Simulation analyses, forecasts
Model 2000 1998–1999	CEMI, Moskva C. Bessarab, V. Dadayan	Annual 1965–1988	68 46	5 sections	OLS	Supply determined, demand absent	Simulation analyses, including impacts of disarmament
SUN 88 1989	CEMI, Moskva E. Gavrilenkov	Annual 1970–1988	112	5 sections	OLS	Supply determined, production functions generating gross-output	Medium term forecasts, simulation analyses
SUN 89, 90 1990–1991		Annual 1970–1989	189	6 sections	OLS		Medium term forecasts, simulation analyses
YUGOSLAVIA							
EIPF							
1975	J. Mencinger	Quarterly	51 40	–	OLS	Demand oriented	Forecasts
1980	Bole, J. Mencinger	Quarterly	115 91	–	OLS		Forecasts and simulations

Table 9.1 (Continued)

Country Model Version Year	Institution Author	Frequency Sample	Equations total stochastic	Decomposition	Estimation method	Properties	Applications
CMEA	IEiS UŁ Łódź						
EIS1[b] 1979	J. Gajda et al.	Annual	80 42		OLS	Supply determined linked with foreign trade equations	Simulation analyses, forecasts
EIS2[b] 1983		Annual			OLS	Foreign trade adjustments assumed balanced growth	Forecasts and policy simulations

[a]Full characteristic of the consecutive versions of the W model in Welfe (1992)

[b]Similar structure for the following countries: Bulgaria, ČSRS, GDR, Hungary, Poland, Romania and USSR

which were supported at the macro-scale by the input-output models. The production functions were only applied to analyse how the plans had been implemented.

The macroeconometric models of the 1960s and 1970s that were constructed on L. Halabuk's initiative were supply determined. The only models to include the demand functions that explained consumer demand in the first place were built for the CSRS (I. Šujan) and Poland (W. Maciejewski and W. Welfe). The W-1 model built at the IEiS UŁ paved the way for the disequilibrium models.

Commodity deficits that affected all centrally planned countries in the 1980s induced the tendency towards building disequilibria into the large macromodels and constructing special, small disequilibrium models. R. Portes, W. Charemza and M. Gronicki proposed new notions of consumer demand for the disequilibrium models, and A. Welfe for the W-A models for Poland. The notions of the production functions in the large multisectoral macroeconometric models, e.g. W-3 and W-5 for Poland, were enlarged by imposing restrictions on the supply of raw materials, especially of the imported ones, as suggested by W. Welfe.

In the next years, macroeconometric models grew even larger. In the models of the CSRS and Polish economies the demand-determined and supply-determined submodels were distinguished, which established a foothold necessary to construct macroeconometric models of economies in transition to a market economy.

The models were systematically used for preparing regularly updated short- and medium-term forecasts and policy simulations that supported the making of the short-term policy decisions and the formulation and implementation of central authorities' five-year plans.

9.2 The Models of the Transition Economies

9.2.1 Introduction

Regarding the activities of the macroeconometric centres, the early 1990s, when the central-command countries started their transition to market economies, were a highly diversified period. Czechoslovakia (VSEiAR) and Poland (IEiS UŁ) were the only countries to carry on with constructing and maintaining macromodels. In the beginning of the 1990s operational rules were developed for dealing with essential changes in the economic regimes, which amounted to full or partial parameter changes (Hall 1994; Welfe 1993, 1995). Despite these theoretical contributions, macroeconomic activities in other countries began several years later.

The changes in the economic system and the ensuing reforms built a bridge for the market mechanisms. After the transformation recessions that occurred at the beginning of the '90s, the economic situation in the Central and East European countries stabilized only in the mid-90s. Hence many research centres decided to undertake macromodelling activities in the second half of the 1990s. They focused on the short-term models based on the quarterly data covering the transition period, i.e. starting from 1992 or 1993. Only a few modelling centres, such as VSEiAR in Slovakia and IEiS UŁ in Poland, decided to respecify their annual models, shifting the equation parameters appropriately to account for the economic regime changes.

In all these countries the statistical information was deeply revised. They accepted the SNA framework and in most cases the macrodata was recalculated from the MPS to SNA system beginning with the year 1981. This also applies to quarterly time series, being the base of modelling the transition period. Concise characteristic of this development is provided in Table 9.2.

The above models were built using modern econometric techniques. The long- and short-term relationships were distinguished and the ECM procedure was applied. The general rules guiding the construction of the macroeconometric models for economies in transition were laid down in the monograph by Klein et al. (1999).

9.2.2 The Models of the Bulgarian National Economy

In the mid-90s, an annual model of Bulgarian economy was constructed at the Economic Institute, Bulgarian Academy of Sciences, under the leadership of G. Minassian (1996). The model was systematically extended and updated over the next years.

Its first version that was used to analyse and forecast the major GDP components and its sectoral structure distinguished industry, agriculture and other sectors. Because Bulgarian statistics took a fairly long time to adjust to the European standards, many variables, for instance deflators, including CPI, exchange rate and the rate unemployment rate, were initially treated as exogenous. On the other hand, the blocks of equations explaining financial flows were expanded.

A small macroeconometric model has been put into use in Bulgaria very recently. It has 15 equations, 7 of which are stochastic. It reveals a supply orientation. GDP is generated from the production function that takes account of TFP impacts. Consumption is obtained as a residual. Investment depends on domestic credit, exports are supply determined. Much emphasis has been given in the model to the monetary market variables (Minassian 2012).

The above models have been systematically used to make forecasts of the Bulgarian economy and in numerous simulation analyses investigating monetary and fiscal policy issues.

9.2.3 The Models of the Czechoslovak National Economy

In the initial period of transition the Research Institute INFOSTAT in Bratislava (M. Olexa and his team) played a crucial role in adjusting the Czechoslovak economy models to the transition economy conditions. Additional support was given by I. Šujan, the then President of the Federal Statistical Office in Prague.

The demand-oriented models developed in the early 1990s anticipated the formation of market mechanisms, particularly increased price sensitivity among economic agents. The models had expanded blocks of equations explaining financial flows that were intended for analysing the fiscal and monetary policy impacts.

An important role was played by a quarterly medium-sized model that was constructed for preparing short-term forecasts (Šujan and Olexa 1984; Šujan et al. 1995). Even larger significance was attributed to the annual models. Constructing its first version in 1991, the researchers carefully modified the specification of the equations to ensure a "smooth" passage from the pre-transition period (1980s) to the transition period. The model was extended after a year to include a detailed specification of the foreign trade equations (Šujan et al. 1993). The models were used to produce medium-term forecasts and simulation analyses, i.a. in the last year of the Federal Czech and Slovak Republic to identify the consequences of the divorce between the two countries (Šujan 1997).

9.2.4 The Models of the National Economy of the Czech Republic

The above activities were continued despite the split and a separate model for the Czech Republic was constructed. It was a demand-determined model with an expanded foreign trade sector (Šujanova and Šujan 1994).

In the late 1990s, the Czech National Bank (CNB) extended its macroeconometric activities. The first model was built by a CNB team with D. Vàvra and L. Beneš (Vàvra et al. 1998) in cooperation with the French experts (representing INSEE: J.L. Brillet et al.). It was a quarterly model using a quarterly data sample starting in 1993. The sample used in this experimental version ended in the year 1998 and was extended at a later time.

The model was demand oriented, with GDP determined by domestic final demand adjusted for net exports. Its special feature was that it attempted to estimate potential industrial output and the rate of its utilization assuming that fixed capital was the only limiting factor. An inverted production function was used to generate the demand for investments and employment. The model had equations explaining average wages and prices. Prices depended on labour costs, prices of imports and a mark-up. Its first version only analysed household personal incomes that depended on labour incomes. The model was subsequently extended by adding the equations explaining financial flows.

Many stochastic equations in the model were estimated with ECM. The model was employed to prepare short-term forecasts and medium-term policy simulations facilitating the study of the economic policy impacts, such as growing budget expenditures, and the impacts of the shocks generated by wage or exchange rate increases.

A new initiative was started recently to build a dynamic stochastic general equilibrium (DSGE) model as a component of a system for forecasting and economic analyses (Beneš et al. 2005). The model distinguished a theoretical plane assuming balanced growth and a short-term adjustments plane, allowing for Keynesian rigidities and lags. All equation parameters were calibrated, which is to the model's disadvantage compared with other models in the same class. The model has been designed for simulation analyses of the monetary policy impacts.

9.2.5 Models of the Hungarian National Economy

In the first transition years macroeconometric activities Hungary came to a freeze. Research was continued at a small scale only at the National Bank of Hungary. Towards the end of the 1990s, the Bank entered into cooperation with NIESR in London, which resulted in the construction of a macroeconometric quarterly model of the Hungarian economy. It became a component of the world model NIGEM that included also the models for Estonia, Poland and Slovakia (Barrel et al. 2002).

The model's structure was typical of the country models incorporated into the NIGEM system, as shown by the specification of the long-run equations whose parameters were either calibrated or estimated jointly for the four countries. The parameters of the short-run equations were estimated with ECM separately for particular countries, allowing i.a. for the specific properties of the Hungarian economy.

The model was demand determined; consumer demand depended on real disposable income and household financial wealth. Investment demand was assumed to equal the expected increase in fixed capital generated from the production function. The model had an extended supply sector. Potential GDP was generated using a CES production function with constant returns to scale and technical progress embodied in FDI. The parameters of the equation were estimated with the functions of labour productivity and of fixed capital productivity. The functions were used to determine the demand for production factors.

CPI dependent on labour costs, import prices and the utilization rate of potential GDP was taken as a point of departure for the price equations. The equations explaining real wages took account of the impacts of labour productivity and of the rate of unemployment. Many equations were specified in order to explain financial flows.

The model was mainly used for running numerous simulations of the impacts of the National Bank's monetary policy and for regular forecasting.

9.2.6 The Model of the Montenegrin National Economy

The Slovenian econometricians led by F. Štiblar have built a quarterly model of the national economy of Montenegro in the last years. The model is demand oriented, with final demand determining output and employment. It distinguishes exposed and sheltered industries and includes a block of equations that explain financial flows. The model has been created to run simulations of the economic policy impacts (Štiblar et al. 2006).

9.2.7 The Macroeconometric Models of the Polish National Economy

In the first years of the transition period the only research centre in Poland that was engaged in macroeconometric modelling was the IEiS UŁ in Łódź, where W. Welfe led his team. The demand version of the large, annual W-5 model was maintained there. The attempt in the mid-90s to replace it with the large, multisectoral W-10 model was not successful (Welfe 1997; Juszczak et al. 1997b). Its functions were taken over by the integrated model IMPEC, which was based on an I-O model linked to the SAM. This unsuccessful attempt caused that a new, one-sectoral, annual medium-term model W-8 was constructed instead (Courbis and Welfe 1999). It had many versions. In the first years of the new century its long-term version W-8 D was built, paving the way for the construction of a model of knowledge-based economy (Welfe 2007, 2009a, 2009b).[4]

With the economies in transition realising that the instruments enabling short-term analyses and forecasting were the most useful for them, the tendency towards building the quarterly models became widespread. In the mid-90s, a series of the WK quarterly models of the Polish economy was built at the IEiS UŁ by the team headed by W. Welfe. At the end of the same decade and in the first years of the new century the models were substantially developed and analysed with modern cointegrating techniques by the teams led by A. Welfe. They were used for preparing regular short-term forecasts and policy simulations.[5]

In the middle of the 1990s, the Faculty of Economics, University of Warsaw, embarked on the initiative to build a Macroeconomic Data Base containing mainly the quarterly forecasts of the Polish economy. W. Maciejewski' research team regularly organised meetings of the forecasting institutions, which were supported under the Phare Programme. The team successfully integrated the projections and then presented the findings to the government (Grajek 2000; Greszta 2002).

Following the commencement of the transition period, attempts were made at the Central Planning Office (CUP) to build a system of macromodels, able to support the preparation of the medium-term plans. The short-term SAS model was constructed and put into operation in the years 1992–1993. The liquidation of the CUP put an end to these ambitious endeavours. They were continued to some extent by K. Barteczko and A. Bocian from the Institute of National Economy (IGN) in Warsaw and then at the University of Białystok. The authors proposed a system of annual simulation macromodels for running medium-term policy analyses and for constructing long-term scenarios of economic development (Barteczko and Bocian 2000, 2004).

In the early 21st c., the research departments of the National Bank of Poland (NBP) were intensifying macromodelling activities aimed to construct quarterly

[4]The evolution of the W-8 and W-8 D models has been described in detail in the monographs by Welfe (1997, 2001, 2004b).

[5]The evolution of the WK models can be found in the following monographs: Welfe (1995, 1997) and Welfe et al. (2002, 2006).

models enabling the monitoring of inflationary processes. Attempts were also made to construct and use the CGE models (Kłos et al. 2005). To some extent, similar activities were also developing at the Ministry of Finance.

9.2.7.1 The Quarterly WK Model of the Polish Economy

The IEiS UŁ research project to construct a quarterly WK model was developing in the early 90s and in 1994 a medium-size model was operated on a regular basis. The model used data starting in 1989, representing the very beginning of the transition period. The early information from the recession period was abandoned then for its irregularity (Welfe 1995, 1997).

The WK models were demand oriented from the start. Following the assumption about market relationships being prevalent in the most recent years, the 2005 version of the models was made to account for typically market mechanisms. The models disaggregated the national economy into four sections, i.e. manufacturing and building industries, agriculture and services, and generate final demand, that is households' demand, public institutions' demand, investment demand, as well as exports and imports (by destination). The bridge equations linking the final demand components and the distinguished sections' demand for domestic output (value added) were specified using the pertinent I-O coefficients. The estimated value added determined employment and, assuming predetermined labour supply, the rates of unemployment. No potential output was generated yet.

The above models had an extended system of prices and deflators, of exchange rates and of wages and incomes. The latest versions included also the major components of financial flows and of the balance of payments (Welfe et al. 2002, 2006).

The model was updated every 3–4 years. The parameters of its equations were estimated with ECM. Starting with the 2005 version, the long and short-term relationships were distinguished. The models were carefully tested, a stochastic multiplier analysis was performed and the models' long-term properties were identified (Welfe et al. 2006).

All the successive versions of the models were systematically used for preparing short-term forecasts published in Polish and international economic journals (e.g. "Consensus Forecasts") and during numerous policy simulations. See also Welfe and Welfe (1986).

9.2.7.2 The Medium-Term W-8 Model

The medium-term annual W-8 model was constructed at the beginning of the transition period to enable analyses of the fiscal and monetary policy impacts (Welfe 1996; Welfe et al. 1996). To ensure that it meet its purpose, the blocks of equations assembled to explain financial flows and the money market were specified in detail. As a result, the early model had no sectoral decomposition and remained as a one-sectoral system. Its 2000 version was partially modified, though, by distinguishing

4 sections, i.e. manufacturing and building industries, agriculture and services. The links between final demand and value added were established using the I-O coefficients. The model's consistency at the macro-level was ensured owing to a top-down approach (Welfe 1999).

The W-8 model was medium size, with more than 300 equations. It covered all major blocks, i.e. explaining final demand, supply, prices and wages, revenues and expenditures of institutional sectors. Consumer demand depended on real disposable income, household financial wealth and nominal interest rates. The main determinants of investment demand were GDP and user costs. The Cobb-Douglas production function was used to generate potential output and its rate of utilization affecting prices. Prices additionally depended on unit costs. Real average wages were consistent with the Phillips curve. Employment was demand determined and given the labour force supply the rate of unemployment was obtained as a residual (Welf et al. 2003a, 2003b).

The W-8 model was systematically used to prepare medium-term forecasts and numerous policy simulations. It was also incorporated into the system of Project LINK world models.

Ecological issues were discussed in Plich (1998).

9.2.7.3 The Long-Term W-8D Models

The one-sectoral model W-8 was a stepping stone to the construction of a whole series of annual W-8D models intended for long-term analyses. i.e. covering 15–25 years. The models drew on the concept of empirical economic growth model developed by Welfe (2000). (See also Welfe 2001.)

The crucial innovation that was introduced starting with the model's first version was the endogenization of technical progress. The production function was extended by the explicit introduction of total factor productivity (TFP) that represented the total impact of the knowledge capital growth (Welfe 2004b). The TFP growth was explained with reference to the growth in the cumulated, domestic and foreign real expenditures on R&D and in human capital linked to investments.[6] The 2007 version of the model clearly distinguished the long-run relationships from the short-run adjustments (Welfe 2009a).

The W8-D models were used to produce numerous long-term forecasts extending to the years 2025 and 2030, as well as many scenario analyses (Welfe et al. 2003c; Florczak and Welfe 2004). The analyses demonstrated the impacts of varying rates of investment growth and of particular components of knowledge capital (Welfe et al. 2004; Welfe and Florczak 2009).

9.2.7.4 The IMPEC Model

The research activities developing in the transition period aimed to extend and utilize the large IMPEC model of the Polish economy, which was an integrated input-

[6] For details compare our monograph (Welfe 2009b).

output model enlarged by adding the SAM components. The research team led by Ł. Tomaszewicz kept updating the model on a regular basis and used it for numerous simulation analyses, performed within the international INFORUM system of models. The model has been recently added links to environmental conditions (Plich 2002). The links with the W-8 model of the Polish economy have been established.

9.2.7.5 The Macroeconometric Modelling Activities at the Central Bank of Poland (NBP)

Since the beginning of this century, the NBP's research departments that have been recently integrated into an Economic Institute have been involved in vigorous macroeconomic modelling activities. The quarterly models have been built, based on data samples starting in the years 1995/1997. The first in their range has been a small structural model of inflation with ca. 60 equations, which was built by the research team led by B. Kłos in the early years of this decade. Its demand has a Keynesian orientation in the short-run and supply follows a neoclassical orientation in the long-run. The model has been used to prepare short-term forecasts and to perform simulation analyses focusing on inflationary phenomena (Kłos 2002; Kłos et al. 2004).

Between 2002 and 2003, the research team headed by T. Fic constructed a new small quarterly model of Polish economy, which had a similar orientation as its predecessor and an enlarged price system. It was mostly used for forecasting purposes (Fic et al. 2003). In 2009 it was substituted by the NECMOD model provided with an extended system of price equations containing inflationary expectations and with equations explaining the labour market components (Budnik et al. 2008).

In the period 2006–2007 a quarterly SOE-PL model belonging to a new generation of models was built. Its construction was an attempt at creating a dynamic stochastic general equilibrium model (DSGE), which was made by a research team with G. Grabek et al. The model referred to an analogous model constructed at the RIKSBANK. It distinguished two components—a theoretical submodel and a submodel of dynamic adjustments. It was designed for experimental purposes and was used in policy simulations (Grabek et al. 2007).

The NBP operated also two large empirical models of general equilibrium (CGE models), which were built under international cooperation. They were mainly used for performing the scenario analyses (Tabeau et al. 1994; Laursen et al. 2004).

In the first years of the new century the Ministry of Finance made many attempts to construct quarterly models capable of supporting the preparation of annual government budgets and of monitoring their implementation. An outstanding representative of the models is the medium-size QPFSPM model, which was built by the research team led by M. Viren (Viren et al. 2003). It distinguished the private and public sectors and had an extended sector of financial flows. This forecasting tool was also used for carrying out simulations preceding the preparation of government budgets.

It might be interesting to know that the models of Polish economy have been the components of the macroeconometric models describing the European and world economies. For instance, the annual HERMIN model had a Polish submodel distinguishing 4 groups of sections, as well as the exposed and sheltered industries. It was mainly used for policy simulations, particularly as a tool analysing the likely impacts of the country's accession to the EU (Bradley and Zaleski 2003). In the later period, the Ministry of Regional Development also used it for policy simulations. Another submodel of Polish economy has been incorporated into the British NIGEM model of the world economy.

9.2.8 The Macroeconometric Models of Romania

9.2.8.1 The Annual Dobrescu Models

The first versions of the annual model of Romanian economy were developed at the Institute of Economic Research in Bucharest by the research team led by E. Dobrescu (National Institute for Economic Research 1992). They were fully supply-determined models that stressed restrictions in energy supply. The min condition was applied to two alternative GDP estimates, one generated from a two-factor production function and the other obtained by dividing the available energy stocks by the energy-output ratio assumed. The available imported energy was determined assuming exogenous balance of trade and estimated maximum volume of exports. The model had investment-generating equations and a system of deflators. It was treated as an experimental system.

Over the next several years, a new series of macroeconometric models of the transition economies was developed in Romania. The first of them appeared in 1996. The next versions were updated and re-estimated every year until 2000. The models were annual medium-sized macroeconometric models, where the production sector was disaggregated first into 5 and then into 4 sections. They had an explicit demand orientation. They generated final demand components determining output and indirectly employment.

The models' equations were specified in a somewhat atypical manner. All stochastic equations were linear and included lagged endogenous variables. The use of OLS did not necessarily lead to consistent parameter estimates. In many equations, the thus introduced inertia outnumbered the other explanatory variables. A striking example of the models' peculiarity was the way GDP was determined: it was derived not from an accounting identity, but from a regression equation. The explanatory variables included domestic final demand and exports, as well as money supply and the government budget deficit. After inertia were introduced, the autoregression coefficient increased to as much as 0.99. The specification of the consumer and investment functions were also far from standard. Demand depended mainly on real disposable incomes that were determined for the investment function as a residual from an identity balancing national income distribution in nominal terms.

The production function was not explicitly defined in the model. It was used to determine the labour productivity function, which was applied to generate labour

Table 9.2 Models of the Central and Eastern Europe in the period of transition to the market economies

Country Model Version Year	Institution Author	Frequency Sample	Equations total stochastic	Decomposition	Estimation method	Properties	Applications
BULGARIA							
1996	Economic Institute AN of Bulgaria, Sofia G. Minassjan	Annual 1990–		3 sectors	OLS		Simulation analyses, medium-term forecasts
CZECHOSLOVAKIA							
Short-term forecasting model 1991	Federal Statistical Office, Prague I. Šujan INFOSTAT Bratislava M. Olexa et al.	Quarterly	63 23		OLS	Demand determined, extended financial sector	Short-term forecasts
Annual 1991		Annual	40 20	–	OLS	Demand determined	Medium-term forecasts, simulation analyses
Annual with foreign trade 1992		Annual 1983–1991	106 31	4 commodity groups in foreign trade	OLS	Extended block of foreign trade equations	Experimental model

Table 9.2 (Continued)

Country Model Version Year	Institution Author	Frequency Sample	Equations total stochastic	Decomposition	Estimation method	Properties	Applications
CZECH REPUBLIC							
Quarterly 1999	Czech National Bank D. Vavra, J. Beneš INSEE S. Audne, J.L. Brillet, J. Beneš et al.	Quarterly 1994–1998			OLS TSLS	Extended block of foreign trade equations	Experimental model
Quarterly 2005	J. Beneš et al.	Quarterly			Calibration	DSGE model	Simulation analyses regarding monetary policy
CR-01 1994	Central Statistical Office of Czech Republic M. Šujanova, I. Šujan	Annual	47 12		OLS	Extended block of foreign trade	Forecasts, simulation analyses
HUNGARY							
NIGEM—Hungary 2000	National Bank of Hungary Z.M. Jakab, M.A. Kóvacs	Quarterly 1994 1995–1999	48		Long-term OLS SUR or DOLS, short-run ECM	Component of model NIGEM for 5 central banks	Forecasts

Table 9.2 (Continued)

Country Model Version Year	Institution Author	Frequency Sample	Equations total stochastic	Decomposition	Estimation method	Properties	Applications
MONTENEGRO							
2006	Universities of Maribor, Montenegro, Ljubljana F. Štiblar et al.	Quarterly			OLS	Demand oriented	Simulation analyses
POLAND							
WK Models	Institute of						
WK 91	Econometrics and Statistics UŁ	Quarterly 1989–1991		5 sections		Linear, demand oriented	Experimental model
WK 94	W. Welfe et al.	Quarterly 1989–1993	130 81	5 sections	Non-linear LS	Non-linear model, dynamic	Short-term forecasts, policy simulations
WK 98	A. Welfe et al.	Quarterly 1990–1996	204 103	4 sections	OLS	Demand determined	Short-term forecasts, policy simulations
WK 2000		Quarterly 1990–1996	256 113	4 sections	OLS ECM	Extended financial sector	Short-term forecasts, simulation analyses
WK 2005		Quarterly 1990–2003	4 sections	144 60	ECM	Short and long term relationships	Simulation analyses, stochastic simulations
W. Models	Institute of						
W5 Demand version	Econometrics and Statistics UŁ W. Welfe et al.	Annual 1960–1988	1076 436	11 sections, 5–7 commodity groups	OLS TSLS	Demand oriented, component of Project LINK	Medium term forecasts up to mid-90s, simulation analyses

Table 9.2 (Continued)

Country Model Version Year	Institution Author	Frequency Sample	Equations total stochastic	Decomposition	Estimation method	Properties	Applications
W8-96 1995–1996		Annual 1960–1993	158 51		Nonlinear LS	One-sectoral model with disequilibria	Medium term forecasts, simulation analyses
W8 2000 1999–2000		Annual 1960–1998	337 115	4 sectors, 5 commodity groups in foreign trade	Nonlinear LS	Disaggregation, endogenization of technical progress, extended financial sector	Medium-term forecasts, policy simulations
W8D 2000		Annual 1960–1988	211 79		Nonlinear LS	One-sectoral with extended production function	Forecasts and long-term simulation
W8D 2002		Annual 1960–2000	216 80	–	Nonlinear LS	Extended demand functions of house hold consumption and investment	Long-term forecasts, simulation analyses, up to the year 2025
W8D 2007–2008		Annual 1960–2005			Nonlinear LS ECM	Extended TFP function, distinguished long and short-term relationships	Long-term forecasts, simulation analyses up to the year 2030
W 10 1992–1994		Annual 1963–1992	Above 1200	18 industries, 3–7 commodity groups	Nonlinear LS TSLS	Continuation of W-5 models, demand and supply functions	Experimental, structural analyses
IMPEC	Instytut Ekonometrii i Statystyki UŁ Ł. Tomaszewicz	Annual		model I-O	OLS	Model integrated with I-O, extended for SAM	Simulation analyses, structural analyses

Table 9.2 (Continued)

Country Model Version Year	Institution Author	Frequency Sample	Equations total stochastic	Decomposition	Estimation method	Properties	Applications
MSMI Small Structural Model of Inflation 2000	NBP B. Kłos et al.	Quarterly 1995–2002	59–64 11	–	Long-term nonlinear, short-term QMN and OLS	Keynesian demand, long-term supply neoclassical	Short-term forecasts, simulation analyses stressing inflationary pressures
ECMOD 2002–2003	NBP T. Fic et al.	Quarterly 1995–1003	40 15		FM-OLS ECM	Neoclassical in the long-term, demand determined in the short-run; extended system of price equations	Short-term forecasts, simulation analyses
CGE 1994	NBP A. Tabeau et al., with participation of CPB, Hague	Annual 1990	45	4 groups, I-O, SAM	Calibration	Static and dynamic version	Simulation analyses
CGE 2004–2005	NBP, Min. of Finance, Min. of Economy, World Bank, P. Griffin et al.	Annual 2002		39 industries, I-O, SAM	Calibration		Simulation analyses
SOE-PL 2006	NBP G. Grabek et al.	Quarterly 1997–2003	25	–	Bayesian methods	DSGE model similar to Riksbank model of Sweden	Simulation analyses

Table 9.2 (Continued)

Country Model Version Year	Institution Author	Frequency Sample	Equations total stochastic	Decomposition	Estimation method	Properties	Applications
HERMIN 2001	The Economic & Social Research Institute Dublin J. Bradley Politechnika Wrocławska J. Zaleski	Annual 1994–2001		4 groups of sections	Calibration, OLS	Exposed and sheltered sections distinguished	Medium-term simulations regarding the impacts of the EU accession
QPFSPM 2002	Ministry of Finance M. Viren et al.	Quarterly 1995–2001	104 33	Private and public sectors	OLS	Demand neo-Keynesian, supply neoclassical, extended financial sector	Forecasts, policy simulations
ROMANIA							
Model of transition economy							
1991	The National Institute for Economic Research, Bucharest E. Dobrescu et al.	Annual			OLS	Supply determined with energy restrictions	Experimental version

Table 9.2 (Continued)

Country Model Version Year	Institution Author	Frequency Sample	Equations total stochastic	Decomposition	Estimation method	Properties	Applications
1999	The National Institute of Economic Research, Bucharest E. Dobrescu	Annual 1981–1998	134 16		ILS	Demand oriented, extended supply and financial sectors, participated in Project LINK	Scenario analyses, forecasts
2000		Annual 1981–1998	163		ILS		Scenario analyses for the pre-accession period
Model of a market economy 2005		Annual	182 62	6 sectors, I-O	OLS	Supply oriented, GDP generated from production function determines absorption	Forecasts, simulation analyses
HERMIN (HR-4) 1999	Institute of World Economy, Bucharest C. Ciupagea	Annual 1989–1997 1990–1997		4 sections	OLS Calibration	Demand determined	Forecasts, simulation analyses
Macroeconomic Model 2005	Academy of Economic Studies, Bucharest S. Stancu, C. Sava	Annual 12 observations	11 7		TSLS	Demand determined, highly aggregated	Experimental, forecasts

Table 9.2 (Continued)

Country Model Version Year	Institution Author	Frequency Sample	Equations total stochastic	Decomposition	Estimation method	Properties	Applications
RUSSIAN FEDERATION							
MACRO-LINK 1992	Central Bank, Moscow V. Stepanov	Annual	Above 100			Disequilibrium model, balancing demand and supply	Medium-term forecasts, simulations for planning process
CEMI-RAN 1991–1996	Central Economico-Mathematical Institute, Moscow M. Levinson	Annual		18 industries, I-O		Multisectoral model generated demand and supply of production	Simulation analyses
BEA/CEF 1999	Bureau of Economic Analyses, Moscow E. Gavrilenkov London Business School, London S.G.B. Havrey, J. Nixon	Quarterly	83		ECM	Demand determined	Simulation analyses
Model of the Russian Federation 1998	Ministry of Finance, University of Pretoria O. Basdevant	Quarterly	20		ECM	Disequilibrium model; final demand generated from demand functions and supply gap	Simulation analyses

Table 9.2 (Continued)

Country Model Version Year	Institution Author	Frequency Sample	Equations total stochastic	Decomposition	Estimation method	Properties	Applications
Medium-term model 2005	Centre of Macroeconomic Analyses and Short-term Forecasts, Moscow K. Mikhaylenko	Quarterly 1999–2003 2000–2003	350 100	10 industries	OLS	Econometric model combined with a balance model	Medium-term forecasts and policy simulations
SLOVAKIA							
SR-1S 1993	INFOSTAT Bratislava I. Šujan et al.	Half-year 1987–1992	49 17	–	OLS	Demand determined, SNA system used; Houthakker Taylor consumer demand function	Experimental model
SR-1Q 1994		Quarterly	41		OLS	Demand determined	Forecasts, simulation analyses
SR-1 1994	J. Haluška et al.	Annual	43		OLS	Demand determined, generated potential GDP	Forecasts
SR-02 1995		Annual 1987–1994	99 32	4 commodity groups in foreign trade	OLS	Disaggregation of foreign trade	Medium and long-term forecasts

Table 9.2 (Continued)

Country Model Version Year	Institution Author	Frequency Sample	Equations total stochastic	Decomposition	Estimation method	Properties	Applications
EMSE 1.0 1996	INFOSTAT, Bratislava M. Olexa et al.	Annual 1986–1995	61 16		OLS ECM	Mixed orientation, production generated gross-output	Experimental, structural analyses
EMSE 2.0 1997		Annual 1985–1996	82 25	4 commodity groups in foreign trade	OLS	Mixed orientation, global supply determined demand	Simulation analyses, forecasts
QEM-ECM-1 2001	J. Haluška, M. Olexa	Annual 1989–1997 1990–1997	80 25	5 income groups	ECM	Demand determined	Forecasts
Real Sector Model 2003	J. Haluška, C. Palčaková	Quarterly 1993–2002	43 14		ECM	Demand determined, distinguished long- and short-term relationships	Short-term forecasts
NBS 1998	National Bank of Slovakia M. Gavura, M. Tkač	Quarterly 1993–1999	56 32		OLS ECM	Demand determined, extended financial sector, long-run and short-run equations distinguished	Medium-term forecasts, simulation analyses of the impacts of monetary policy
ISWE	Institute of Economics SAV						
1993	M. Benčik, J. Pàlenik	Quarterly 1990–1994	28 12	–	OLS	Demand determined	Forecasts, policy simulations
1996		Quarterly 1992–1996	98 20	–	OLS	Production function added	Forecasts, policy simulations

Table 9.2 (Continued)

Country Model Version Year	Institution Author	Frequency Sample	Equations total stochastic	Decomposition	Estimation method	Properties	Applications
ISWE-ECM 2004		Quarterly			ECM	Demand determined	
SLOVENIA							
SLOPOL 1 2000	University of Klagenfurt Austria K. Weyerstrass, R. Neck	Quarterly 1994–1999	40 15		OLS	Keynesian with potential output dependent on labour force supply	Simulation analyses
3 2001		Quarterly 1994–2000	49 15		OLS	Neoclassical supply sector	Simulation analyses of impacts of fiscal and monetary policies using optimal control methods
4 2002		Quarterly 1994–2001	45 15		OLS	Potential output depends on labour force supply, reduced by natural unemployment	Simulation analyses of impacts of fiscal policy
6 2006		Quarterly 1995–2005	57 21		OLS ECM	Long- and short-term relationships distinguished	Simulation analyses of the impact of the EMU accession

Table 9.2 (Continued)

Country Model Version Year	Institution Author	Frequency Sample	Equations total stochastic	Decomposition	Estimation method	Properties	Applications
UKRAINE							
Annual model 1995	Kiev State University of Economics Gierassinenko, Shusticcov	Annual 1980–1994	12 10		TSLS	Demand determined, production function specified	Experimental model, forecasts
INTAS 1995–1996	Institute of Economic Forecasting NAS Ukraine V. Heyets, M. Skripni-chenko	Annual 1985–1995	33 11		OLS	Mixed orientation; total supply obtained from production function determined the components of final demand	Medium-term forecasts
2000		Annual 1985–2000	39 24		OLS	Extended financial flows sector	Forecasts, simulation analyses
Model VAR 1997	KIEV National University V.W. Parkho-menko	Quarterly 1994–1997	8 7		TSLS with IVV	VAR model	Experimental model

Table 9.2 (Continued)

Country Model Version Year	Institution Author	Frequency Sample	Equations total stochastic	Decomposition	Estimation method	Properties	Applications
Model of Ukraine development 2003	International Research Institute of Systems of Information and Technology O. Bakaev et al.						
Product version		Annual	110 75		OLS	Model with extended production sector	Forecasts, simulation analyses
Investment version		Annual	38 15		OLS	Aggregated model, mixed orientation	Forecasts, simulation analyses
MODELS OF CENTRAL AND EASTERN EUROPE							
LAM							
1 1993	Uniwersytet Gdański K. Strzała	Quarterly 1992–1997	24 12	Poland		Demand determined Long-term and short-term equations	Simulation analyses, forecasts
2 1993 3 1998	Leicester University W. Charemza			ČSRS, Estonia, Hungary, Lithuania, Latvia, Poland, Romania, Slovakia			

Table 9.2 (Continued)

Country Model Version Year	Institution Author	Frequency Sample	Equations total stochastic	Decomposition	Estimation method	Properties	Applications
NIGEM 2002	NIESR UK R. Barrell et al.	Quarterly	21 11	Czech Rep., Estonia, Hungary, Poland, Slovenia		Neo-Keynesian models being components of the NIGEM model of world economy; FDI major factor of growth	Simulation analyses i.a. for the pre-accession period

demand. The model had a large sector with equations accounting for financial flows and prices, but no transmission to the production sector was introduced (Dobrescu 1998).

The above versions of the model were frequently used in the simulations of the monetary policy impacts and in forecasting activities (Dobrescu 1999). The 2000 version that was systematically developed between 2001 and 2004 helped update Romania's pre-accession programmes (Dobrescu 2000, 2003).

The new annual macroeconometric model of Romanian economy built in 2005 had a sample starting in 1989 (the beginning of transition) and ending in 2004. It was argued therefore that unlike the previous models describing economies in transition whose data started with 1982 this model represented the market economy mechanisms (Dobrescu 2006). The new model was enlarged by linking it with an I-O submodel where the economy was divided into 6 sectors following data aggregation in the I-O table having 106 industries.

The structure of the new model changed considerably. Quite unexpectedly, it became supply determined. GDP was generated from the Cobb-Douglas production function with calibrated elasticity with respect to employment. An interesting novelty was the introduction of total factor productivity, which was explained in a rather nonconventional manner. Employment was dependent on output allowing for changes in the labour market. Final demand depended on disposable incomes being a component of value added (in nominal terms), on interest rates and, regarding investment demand, on FDI. As a result, the model did not have the consumption multiplier. It had the supply accelerator instead, because investments ensured an increase in fixed capital and output. The model had blocks of equations explaining prices and wages and financial flows. It was used to perform many policy simulations and medium-term forecasts.

Nearly parallel with the construction of the above models an attempt was made to build an empirical model of computable general equilibrium (CGE) for Romania. The model was based on the data from the I-O table (with 11 industries) and from national accounts covering the years 1989–1995. It was a static structure, which stressed the impacts of energy restrictions in the Romanian economy (Ciupagea et al. 1996). This research was discontinued because the structure of the Romanian economy changed too much and the authorities were insufficiently concerned.

9.2.8.2 The HERMIN Model

At the turn of the 20th c., C. Ciupagea constructed a subsystem for Romanian economy (2001) within the HERMIN system of macroeconometric annual models of the EU's "peripheral" countries (Bradley et al. 1995). The model had a modern structure. It was disaggregated into four sections. This increased the model's efficiency in describing economic processes taking place in a system moving towards a market economy and made it a better instrument for performing simulation analyses of the preconditions and impacts of accession to the EU.

The model was demand oriented. Following the convention assumed for the HERMIN models, three blocks of equations were distinguished in it: supply (enterprises), absorption and financial flows. In the enterprise block, the demand for products of the 4 distinguished industries was determined as a weighted average of the final domestic and foreign demand, the weights being obtained from the I-O tables. However, indicators of competitiveness representing the supply side were additionally added to the above equations. The demand for production factors was generated from the inverted CES production functions.

The model was used independently mainly for simulation analyses of the accession to the EU and for policy simulations performed within the HERMIN system of models.

9.2.8.3 The New Macroeconometric Models

More recently, new attempts at constructing models of Romanian economy were presented. Among the interesting proposals, there was an experimental, small annual model (Stancu and Sava 2007). It distinguished three blocks of equations explaining three markets: products, labour and financial markets, which were weakly dependent. The production block was demand oriented and contained a system of simultaneous equations generating final domestic and foreign demand. The financial market block generated money demand and state budget revenues, including bond issues. The small sample and collinearity being present despite the use of TSLS did not prevent getting economically unacceptable estimates of several parameters (e.g. negative propensities to consume or invest).

9.2.9 The Macroeconometric Models of the National Economy of the Russian Federation

In the early transition period attempts were made in the Russian Federation to construct annual macroeconometric models of the country's economy. Following the long-standing tradition, the models were multisectoral and stressed enterprise sector's activities under disequilibrium.

9.2.9.1 The MACROLINK Model

The first attempt was made by Y. Stepanov from the Central Bank of the Russian Federation. An experimental, medium-sized macroeconometric model MACROLINK was built. This was a disequilibrium model with equations explaining both demand and supply of material goods, useful for seeking balance between the two sides. It was designed for the planners to run simulation analyses and for developing medium-term forecasts (Stepanov 1995).

9.2.9.2 The CEMI-RUN Models

In the mid-90s, a large project called "Region" was launched at the Central Economic-Mathematical Institute (CEMI) in Moscow. One of its components was the large, multisectoral, annual macromodel CEMI-RUN (Levinson 1995, 1997). This model used an I-O submodel based on the 1994 data, which distinguished 18 manufacturing industries. CEMI-RUN generated demand as well as supply of particular groups of products, thus facilitating the search for equilibrium prices in the economic system. The model supported policy simulations.

9.2.9.3 The BEA/CEF Quarterly Model

The market economy development in the Russian Federation in the late 1990s stimulated the need to perform short-term analyses and forecasts, which called for the construction of quarterly macroeconometric models. An important stimulus came from the financial crisis that resulted in a deep devaluation of the rouble in 1998. An effective instrument was needed to analyse the consequences of the crisis. This situation led to the construction of the quarterly medium-sized model BEA/CEF of the Russian Federation's economy in 1999, but earlier on an experimental annual model was built. It was constructed by E.E. Gavrilenkov from the Bureau of Economic Analyses in Moscow in collaboration with the researchers from the London Business School (Gavrilenkov et al. 1999). The model was definitely demand determined. Consumer demand mainly depended on real disposable income. Investment demand was related to output, real interest rates and fixed capital. Final domestic demand with net exports determined GDP and indirectly employment. The model had equations explaining prices, wages and exchange rates. It was mainly applied to run policy simulations.

9.2.9.4 The Quarterly Disequilibrium Model

Towards the end of the 1990s, a group of experts working under the TACIS programme constructed a quarterly, one-sector model of the Russian Federation's economy at the Ministry of Finance (Basdevant 2000). They assumed that the Russian economy was still in disequilibrium, with prices being not capable of clearing the markets appropriately. This approach caused that a disequilibrium model was chosen. However, the model had to be specified differently than the previous models, because it did not explicitly generate demand and supply. It explained the effective transactions from the demand-side perspective using mixed equations representing demand (consumer, investment, foreign) adjusted for the functions of supply gaps.[7]

---

[7] It should be noted that in the demand-determined models supply gaps affect prices, thus clearing the market.

A serious difficulty that the modellers had to overcome was posed by the estimation of potential output that was needed to generate the supply gap. Potential output was obtained from the Cobb-Douglas production function with constant returns to scale, where fixed capital was the factor limiting production. The official data on fixed capital could not be used, because their level did not change considerably in the 1990s, although GDP declined by half over the decade. The "effective" level of fixed capital was estimated with the Kalman filters. This constructed model was primarily used to perform policy simulations.

9.2.9.5 The Annual Structural Model

In the first years of the 21st century, the economic situation in the Russian Federation stabilized, becoming increasingly similar to a typical market economy. In macroeconometric model building, this process was reflected by the tendency to construct models analogous to standard market economy models, but taking account of the likely imbalances.

The Centre of Macroeconomic Analyses and Short-term Forecasting in Moscow built multisectoral, annual model along these lines (Mikhaylenko 2004). Economic activities were decomposed in the model by distinguishing several sections and 10 manufacturing industries. Hence a large number of identities were introduced, which helped balance the results of economic activities. The model had a large number of stochastic equations, in excess of 100. The large number of exogenous variables in the model (around 100) somewhat impeded its operation.

The model was demand determined. However, the equations explaining final demand were specified in a definitely non-standard manner. Final demand almost exclusively depended on real incomes. The only determinant of household demand represented by retail sales was real disposable income. The demand from public institutions was determined by the federal budget's tax revenues. Investment demand was dependent on investment funds, its most important component being profits. The above incomes were generated in the production process (real wages were exogenous). As a result, the following feedback was defined in the model: incomes determined final demand, then output and finally incomes.

As shown by the above specification, transmission from the financial to the real sector (for instance via interest rates) was omitted; likewise expectations of economic agents.

A special property of the model was its mixed specifications of the equations in the production sector. Particular industries' demand for commodities was determined not only by the components of final demand (exports, retail sales, competitive imports) and intermediate demand, but also by the indicators of financial assets representing restrictions in supply.

The model's interrelated system of price equations was defined, including CPI, PPI, an exchange rate and a GDP deflator. In the PPI specification, energy (imports) prices played a crucial role, while labour costs were neglected. These specifications show that the second major feedback in the model ran through the balance of payments; it did not contain the inflationary loop.

The sector of financial flows including the equations explaining federal budget's incomes and expenditures and money flows was broadly represented in the model.

The main purpose of the model was medium-term forecasting and policy simulations.

9.2.10 The Macroeconometric Models of the Slovak National Economy

9.2.10.1 The SR Models

In Slovakia, the major force stimulating the development of macroeconometric modelling activities was the experienced research team first led by I. Šujan and then by M. Olexa from the research Institute INFOSTAT in Bratislava.

The first SR-1S model of the Slovak economy referring to the previous models was built in 1993. It was a small, semi-annual model run with data derived from the SNA system (recalculated for the 1980s from the MPS system). The model was definitely demand determined. GDP was generated as a sum of the final demand components. The equations explaining foreign trade were specified with some attention to detail in order to identify Slovakia's links with the Czech Republic. GDP determined employment and influenced the level of unemployment, taking into account labour supply. The model had a system of price equations and an equation explaining average wages that were dependent on the rates of inflation and unemployment. The model closed with the equations explaining state budget's revenues and expenditures. It was used for preparing forecasts and as a point of departure for the construction of its successive versions (Šujan et al. 1995).

In 1994 the model was substituted by the annual, one-sectoral model SR-01 of a similar size, which was enlarged later by decomposing foreign trade into ca. 100 equations (Haluška et al. 1995). Both models were demand oriented. Consumer demand was generated from the Houthakker-Taylor consumption function. Investment demand depended not only on GDP increase and the lagged value of investment, but also on investment funds corrected for real interest changes. The model generated also potential output from the Cobb-Douglas production function and its rate of utilization. The 1995 version distinguished equations explaining exports and imports for 4 groups of commodities in foreign trade, which qualified it for being a component of the Project LINK system of world models. The models were used to prepare medium-term forecasts and to run policy simulations (Haluška et al. 1995).

In the same period a small quarterly model of the Slovak economy was built for short-term forecasting and economic analyses.

9.2.10.2 The EMSE Models

In the second half of the 1990s the research team led by M. Olexa built a new series of macroeconometric models of the Slovak economy in transition, which were primarily intended as a forecasting tool (Olexa et al. 1997). The first version EMSE-1.0

was a multisectoral, annual model of medium size, whose structure differed considerably from its predecessors'. Firstly, it had a mixed orientation: output was generated from the production function and its balance with final demand was obtained through appropriate changes in inventories (in current prices). Secondly, production was assumed to equal total supply defined as the sum of GDP and imports. Following the W. Krelle's proposition, the model builders assumed that the imports were an important production factor. Hence the production function should explain total supply and the production should include imports. As a result, GDP was supply determined. Both GDP and imports were used for explaining changes in the consumer and investment demand and in employment. Therefore, the model had a supply accelerator, but not a consumption multiplier. It included a system of price equations and an equation explaining average wages that depended, i.a. on the rate of unemployment.

The next version EMSE 2.0 was larger, because foreign trade was decomposed into four commodity groups (Olexa et al. 1999). The parameters of the models equations were estimated with OLS.

9.2.10.3 The Quarterly QEM-ECM Macromodels

At the beginning of the 21st c., a new series of quarterly QEM-ECM models was developed at INFOSTAT. The parameters of the models' equations were estimated with ECM, hence the second component in their name. It was, however, more important that their authors resumed the "classical" specification of models meant for the market economies, stressing the need for conducting short-term analyses. The models were simultaneous and included many non-linearities.

The demand for domestic production (GDP) was determined from the known accounting identity, as a sum of final domestic demand and net exports. Consumer demand depended on real disposable income and consumer credits. Disposable income was mainly related to wage incomes, the latter being influenced by employment. Employment was dependent on GDP, which completed the feedback (the consumer multiplier). The main determinant of investment demand was GDP, assuming that the model had the accelerator. Although the model included a system of price equations, the most important GDP deflator was obtained by dividing nominal GDP by real GDP, which did not facilitate the analysis of inflation. Average wages were dependent on retail prices, labour productivity and the rate of unemployment. Unemployment was obtained as a residual from the difference between labour force supply and labour demand (Haluška and Palčákova 2004).

An attempt has been made very recently to utilize the information offered by economic cycle surveys in the macromodels (Haluška and Olexa 2005) and in the short-run estimates of the GDP growth rates (Haluška and Olexa 2007).

9.2.10.4 The Macroeconometric Models of the Institute of Economics

In the mid-90s macromodelling activities were conducted not only at INFOSTAT, but also at the Economic Institute of SAV in Bratislava, where the research team led by V. Pàlenik was active (Benĉik and Pàlenik 1993; Pàlenik and Benčik 1995; Benĉik 1996). The small, demand-determined quarterly model they constructed had several variants. Its 1996 version had a production function to deal with the supply side (labour productivity was, in fact, explained). It was used to produce regular forecasts of the development of the Slovak economy. In 2004, the model was reconstructed. The long-term relationships and short-term adjustments were distinguished and the ECM approach was used to estimate its parameters (Ďuraš et al. 2004).

9.2.10.5 The Macroeconometric Models of the Slovak National Bank

At the end of the 1990s, the Slovak Central Bank (NBS) expanded its macroeconomic modelling activities. A quarterly model of the Slovak economy was constructed to support regular forecasting activities and analyses of the monetary policy impacts (Gavura and Tkăč 2000). As well as having the real sector, the model was also provided with an extended financial sector including state budget's revenues and expenditures and many components of money supply to suit its purpose. It was a small, quarterly model. It distinguished the long term relationships, frequently calibrating the parameters at the unit level, and the short-term adjustments. The parameters of the stochastic equations were estimated with ECM.

The model was demand determined. Its equations were specified in line with the modern standards. The components of final demand determined GDP. GDP indirectly affected consumption and investment, thus introducing the relevant feedbacks into the model. Financial wealth was added in the consumer demand function as an additional explanatory variable. The investment demand function took account of the self-financing of firms, in addition to GDP. Employment was dependent not only on GDP, but also on real wages. Producer prices were cost determined. Real average wages were linked to labour productivity and the rate of unemployment. The financial sector was substantially extended with the state budget components and money circulation. The model was regularly used in forecasting and enabled many policy simulations.

9.2.11 The Macroeconometric Models of the Slovenian National Economy

At the end of the 90s, a quarterly model of Slovenian economy was built at the Klagenfurt University (Austria) by the research team led by R. Neck. The model was mainly intended for policy simulations (Weyerstrass et al. 2001). It had many versions. Versions 3 (2001) and 4 (2002) were used to analyse the likely impacts of fiscal and monetary policy using the optimal control methods (Haber et al. 2002). Version 6 was applied to study the potential effects of Slovenia's accession to the EU (Weyerstrass and Neck 2007).

The model and all its variants were one sectoral. They were demand oriented in the short-run, with a neoclassical enterprise sector. The potential output the model generated depended on labour supply reduced by structural unemployment. The most recent versions distinguished the long-term relationships and the short-term adjustments. The estimation procedures used the ECM approach.

9.2.12 The Macroeconometric Models of the Ukrainian National Economy

In the mid-90s, first attempts to construct the annual macroeconometric models of the Ukrainian economy were made in the research institutes of the Ukraine Academy of Science (UAS). The basic assumptions for their construction were formulated by L. Łukinov and O. Bakaev (Łukinov et al. 1997). The operational models were built later on.

9.2.12.1 The INTAS Model

The major research centre that promoted model building was the Institute of Economics and Forecasting, UAS. The research team led by V. Heyets and M. Skripnichenko constructed the small, one-sectoral, annual model INTAS of the Ukrainian economy, which was expanded in the next years. It was used in medium-term forecasting and policy simulations.

The INTAS model had a mixed orientation, thus resembling some other models built for economies in transition. The point of departure was global supply generated from the production function. In addition to the primary production factors (fixed capital and employment), imports (reduced by investment imports) were also used. GDP was obtained from an identity, by deducting imports from global supply. GDP was used to determine employment (allowing for fixed capital) and fixed capital increases. Consumer demand of both households and public institutions depended on GDP; investment demand was additionally linked to the real interest rates.

The exports equation had a mixed specification allowing for both demand and supply. Imports were decomposed by destination. Foreign trade balance was obtained as a residual. The model also presented the first estimates of state budget's revenues and expenditures. Prices were exogenous and were mostly used to calculate variables in nominal terms.

The uncommon specifications of the non-linear equations are worth noting. The reciprocals of the variables were explained through the reciprocals of the explanatory variables. It seems a rather strange way to introduce non-linearities, although it allows to use linear relationships. The equation parameters were estimated with OLS (Heyets and Skripnichenko 1997). The model was used to run policy simulations and to prepare the first medium-term macro-forecasts, which extended to the year 2005 (Heyets and Skripnichenko 1996).

The financial flows sector in the model was subsequently enlarged to enable the analyses of government revenues and expenditures (Heyets 1998). The next versions of the model supported policy simulations and medium-term forecasting.

In the late 1990s, several attempts were undertaken to construct minimodels that could be used to prepare the short-term forecasts of Ukrainian GDP. A VAR supported model was tried in Parkhomenko (1998).

9.2.12.2 The Annual Multisectoral Model

Between 2002 and 2003, research activities that aimed to construct a multisectoral, annual, medium-size model of the Ukrainian economy were undertaken under the leadership of O. Bakaev at the International Centre of Information Systems and Technology in Kiev. The model was to serve practical purposes. Its basic "product-oriented" version generated per capita output and consumption of several commodities (in agriculture also the sowing area and crops) from the pertinent regression equations. The "investment" version of the model was reduced to the major macrocharacteristics.

Presenting an unambiguous characterisation of the model is not easy. Generally, it seemed supply oriented. The rate of GDP growth in its basic version was obtained as an average of the growth rates of particular products. In the investment version, the rate of GDP growth depended on the exogenous share of investment in GDP. Final demand was not defined. Instead, retail sales were explained in the model. Investment demand (in the product-oriented version) depended on financial means. Employment was linked to GDP. Prices represented by the exogenous deflator of GDP were used to express GDP in nominal terms, to calculate GDP distribution and the primary incomes. The incomes were the source of budget's tax revenues (Bakaev et al. 2004).

This somewhat peculiar structure caused that the model was mainly used for policy simulation purposes.

9.2.13 The Multicountry Models of the Middle and East European Countries

In the period of the countries' transition to a market economy, macroeconometric models reflecting their transforming systems were developed. The models were either composed of identical submodels constructed for the particular countries, such as the LAM models, or of the submodels of larger systems explaining the peripheral European countries, e.g. the HERMIN models, or the world economy, for instance the NIGEM model.

9.2.13.1 The LAM Model

The quarterly, small model LAM (Long-Run Adjustment) was constructed by W. Charemza in the early 1990s. It was operated for many years at the Gdańsk

University's Centre of Macroeconomic and Financial Information and at the Leicester University. Its first version LAM 1 covered Poland and Czechoslovakia; the next one was extended to include the Czech Republic, Estonia, Hungary, Lithuania, Poland, Romania and Slovakia. The models included equations explaining the production and financial flows sectors. In the LAM 3 version of the model an equation explaining final demand was added (Charemza and Makarova 1998).

The final demand equation was neo-Keynesian—GDP was obtained as a net sum of the final demand components. The model generated also potential output and the rate of its utilization. Potential output was obtained from the Cobb-Douglas production function. The production factors did not include fixed capital; the energy use was introduced instead. The model generated prices and wages, as well as budget's revenues and expenditures and money demand.

Although the equations explaining foreign trade conformed to a standard specification, not a single attempt was made to distinguish the commodity flows within the Middle European countries. Hence the models did not represent an interlinked system. The model was regularly used in short-term forecasting and policy simulations.

9.2.13.2 The HERMIN Models

It the second half of the 1990s, the HERMIN model built for the peripheral European countries was extended to new countries that pursued EU membership. These research efforts were initiated and coordinated by J. Bradley. The models had similar structures, but they also took account of the special characteristics of the countries they described, as indicated by the models for Poland (Bradley and Zaleski 2003) and Romania (Ciupagea 2001). Because the models were not linked through the foreign trade flows, they were operated on a stand-alone basis, not as a system. They were mainly used to prepare simulation analyses of the likely impacts of the countries' accession to the EU.

9.2.13.3 The NIGEM Model

The initiative to extend the world economy model NIGEM by including the macroeconometric models of the five transition countries is worth noting. These were on behalf of respective five central banks models of the following countries: the Czech Republic, Estonia, Hungary, Poland and Slovakia (Barrel et al. 2002). The models were built to provide the central banks with instruments facilitating simulation analyses of the likely fiscal and monetary policy impacts, taking into account the conditions of the projected accession to EU and then to EMU.

Because only small data samples were available, the parameters of several long-term equations were calibrated relying on the NIGEM estimates for the other countries. The parameters of the short-run adjustment equations were estimated jointly with the sample covering all five countries. The country-specific effects were represented by constants treated separately for each country. The estimation process used the ECM approach.

The above models were structured in a way that did not greatly differ from the standard applied to the system of the NIGEM models. The approach stemmed from the belief that the countries in question had already adopted (more or less successfully) the market mechanisms. A more detailed description of the models can be found in the earlier presentation of the Hungarian model (see also Jakab and Kovács 2001).

The models of the five countries did not enter into an interrelated system, as they were not linked either by foreign trade financial flows. It must be stressed that the countries' economic growth was mainly driven by FDI and economic reforms, as illustrated by many policy simulations.

The models of the Middle and East European countries having different levels of disaggregation were present in the regional and world models, where they were frequently distinguished as separate submodels of the transition countries. In many such models, e.g. in the Project LINK system, they were linked to other models, mainly via foreign trade.

9.2.14 Final Remarks

In the countries moving towards a market economy the developing macroeconometric modelling activities had to overcome many serious difficulties. Many researcher centres believed pessimistically that the economic processes taking place in the transition countries were so much different from those typifying the previous regimes that the most reasonable thing was to wait for the data series on the transition period to become sufficiently long. For the quarterly models this was to happen in 1995 and for the annual models the new century seemed most appropriate.

A positive exception to the mainstream trend was the research centres in Bratislava (INFOSTAT) and Łódź (IEiS UŁ). They agreed that the data for the 1980s and earlier years could be useful, if a change in the regime was assumed to take place in the early 1990s. A verification process was applied to test the parameter stability hypotheses with respect to the functions of consumer and investment demand, production, average wages, prices, etc. The two research centres initiated and coordinated the work on constructing and implementing the new versions of the annual models for Czechoslovakia and then for Slovakia (INFOSTAT), and the W-5 and subsequently W-8 models for Poland (IEiS UŁ). At the same time, they also built and maintained quarterly models used in the short-term forecasting and simulation analyses of the fiscal and monetary policy impacts.

These developments encouraged many other transition countries to construct their own macroeconometric models in the second half of the 1990s. These were mainly the quarterly, demand-determined models that addressed economic agents' rising sensitivity to market signals. However, the macromodels of several countries

(Romania, the Russian Federation, and Ukraine) were initially mostly supply determined. Many mixed equations combining the demand and supply functions were specified. Such specifications were hardly acceptable, considering the economic content and problems of simulation analyses.

The practical significance of the quarterly models was enormous. They were regularly used to produce forecasts and numerous policy simulations. Their structure did not substantially differ from the standard models of the market economies. Only few countries maintained annual, large, multi-sectoral models to make medium-term forecasts and economic analyses. The long-term analyses were initiated by the Łódź centre. It also used the W-8D model to construct the development scenarios of knowledge-based economies.

The Czech and Polish central banks attempted to construct the quarterly DSGE models that following the hopes of their authors might replace the standard models in the future, which does not seem feasible, however.

Most macroeconometric models functioning in the transition countries were estimated with advanced techniques that allowed distinguishing the long-term relationships and the short-term adjustments.

References

Anders, H. D., et al. (1971). *Niektore probliemy aplikacie makroekonomickich modelov prognozovanii v planowani narodneho hospodarstvi NDR*. Material I miedzinarodneho symposia o aplikaci prognostickich modelov v socialistickom hospodarstve. Bratislava: VVS COSN.

Bakaev, O. O., Tkravczenko, T. T., & Mukminova, T. A. (2004). *Makroekonometriczne modeljuvanija ekonomicznogo rozvitku Ukrainy*. Kiev: Transport Ukrainy.

Barczak, A., Ciepielewska, B., Jakubczyc, T., & Pawłowski, Z. (1968). *Model ekonometryczny gospodarki Polski Ludowej (An econometric model of the economy of Poland)*. Warszawa: PWE.

Barrel, R., Holland, P., Pain, N., Kovacs, M. A., Jakob, Z., Smidkova, K., Sepp, U., & Cufer, U. (2002). An econometric macro-model of transition policy choices in the pre-accession period. In W. Welfe (Ed.), *Modelling economics in transition* (pp. 209–232). Łódź: Absolwent.

Barteczko, K., & Bocian, A. (2000). *Modele i symulacje makroekonomiczne (Models and macroeconomic simulations)*. Białystok: Wydawnictwo Uniwersytetu w Białymstoku.

Barteczko, K., & Bocian, A. (2004). *Modelowanie i symulacje makroekonomiczne (Modelling and macroeconomic simulations)*. Białystok: Wydawnictwo WSE w Białymstoku.

Barteczko, K., Bocian, A., & Sapocinski, T. (1985). *Studia i materiały: Vol. 17. Analiza wariantów zmian makro-struktury gospodarki Polski w okresie 1986–1995 (An analysis of the variants of the changes in the macrostructure of the Polish economy in the years 1986–1995)*. Warszawa: IGN.

Basdevant, O. (2000). An econometric model of the Russian Federation. *Economic Modelling, 17*, 305–336.

Benĉik, M. (1996). *Reihe Osteuropa. East European studies: Vol. 38. Revision of an econometric model for Slovakia*. Vienna: Institute for Advanced Studies.

Benĉik, M., & Pàlenik, V. (1993). *Makroeconomicky model SR I*. Bratislava: Institute of Economics SAV.

Beneš, J., Hlédik, T., Kumhof, M., & Vàvra, D. (2005). *An economy in transition and DSGE: what the Czech National Bank's new projections model needs* (Working Paper Series 12). Prague: Czech National Bank.

Bessarab, S., & Dadajan, V. (1991). *A medium-term model of the USSR economy* (Mimeo). Moscow: Central Economic and Mathematical Institute, USSR Academy of Sciences.

Bielkin, W. D., Buraczas, A. I., Devulis, G. S., & Satunovskij, L. M. (Eds.) (1978). *Mnogosektornyje regionalnyie modeli prognoza*. Vilnius: Mintis.

Bobińska, K. (1982). Bariery rozwoju w gospodarce centralnie planowanej (Barriers of growth in centralny planned economy). *Ekonomista, 3/4*, 377–399.

Bocian, A. (1984). *Modele makroekonomiczne w planowaniu gospodarki narodowej (Macroeconomic models in the planning of the national economy)*. Warszawa: PWN.

Bocian, A., & Burża, T. (1978). Badanie makroproporcji wzrostu w latach 1980–1985 przy wykorzystaniu modelu ekonometrycznego MEDIG (Investigating macroproportions of growth in the years 1980–1985, using MEDIG econometric model). *Gospodarka Planowa, 12*, 603–608.

Bodin, L. (1986). The Polish W-3 model. The OLS, TSLS and FP estimation. In W. Milo & M. Miszczyński (Eds.), *Macromodels and forecasts of the national economy. Proceedings of Models and Forecasts'83 and Macromodels'83* (pp. 7–34). Warszawa: PWN.

Bożyk, P., Maciejewski, W., & Piaszczyński, W. (1973). *Modele prognoz handlu międzynarodowego (Forecasting models of the international trade)*. Warszawa: PWN.

Brada, J. C., & King, A. S. (1980). A structural econometric model of Czechoslovakia: 1954–1975. *Empirical Economics, 4*, 199–219.

Bradley, J., & Zaleski, J. (2003). Modelling EU accession and structural funds impacts using the new Polish HERMIN model. Paper presented at the conference Modelling economies in transition, Cedzyna, 4–7.12.2002. In W. Welfe (Ed.), *Modelling economies in transition* (pp. 97–124). Łódź: Absolwent.

Bradley, J., Modesto, L., & Sosvilla-Rivero, S. (1995). HERMIN, a macroeconometric modelling framework for the EU periphery. *Economic Modelling 12*, 221–247.

Budnik, K., Greszta, M., Hulej, M., Kolasa, M., Murawski, K., Rot, M., Rybarczyk, B., & Tarnicka, M. (2008). *The new macroeconometric model of the Polish economy* (Mimeo). Warszawa: National Bank of Poland.

Charemza, W. (1989). *Large econometric models of an East European economy: a critique of the methodology* (Discussion Paper No. 121). Dept. of Economics, University of Leicester.

Charemza, W. (1991). Large econometric models of an East European economy: a critique of the methodology. *Economic Modelling, 8*, 45–62.

Charemza, W., & Gronicki, M. (1985). *Ekonometryczna analiza nierównowagi gospodarki Polski (Econometric analysis of disequilibria for the Polish economy)*. Warszawa: PWN.

Charemza, W., & Makarova, S. (1998). The LAM-3 model of East European economies: initial foundations and first results. In W. Welfe (Ed.), *Proceedings the 25th international conference MACROMODELS'98 and the 3rd conference of the international association AMFET—modelling economies in transition* (Vol. 1, pp. 75–116). Łódź: Absolwent.

Charemza, W., & Quandt, R. E. (1982). Models and estimation disequilibrium for centrally planned economies. *Review of Economic Studies*, 10–116.

Ciupagea, C. (2001). The Hermin Link model for the Romanian economy. In W. Welfe (Ed.), *MACROMODELS 2000* (pp. 9–53). Łódź: Absolwent.

Ciupagea, C., Voleu, A., Capros, P., & Georgakopulos, P. (1996). *The Romanian specific computable general equilibrium model* (ACE-Phare Working Paper, Mimeo).

Costa, A. M., & Mienszikow, S. M. (1979). Short-term forecasting models for the centrally planned economies of Eastern Europe. In J. A. Sawyer (Ed.), *Modelling the international transmission mechanism—applications and extensions of the Project Link system* (pp. 163–202). Amsterdam: North-Holland.

Courbis, R., & Welfe, W. (Eds.) (1999). *Central and Eastern Europe on its way to European Union. Simulation studies based on macromodels*. Frankfurt/Main: P. Lang.

Czerwiński, Z., & Welfe, W. (1982). Macroeconomic modelling based on econometric and simulation models for the Polish economy. In W. Hildenbrand (Ed.), *Advances in econometrics* (pp. 285–301). Cambridge: Cambridge University Press (Chap. 11).

Czerwiński, Z., Guzik, B., Jurek, W., Panek, E., Remka, H., & Śledziński, W. (1981). *Modelowanie i planowanie wzrostu gospodarki narodowej (Modelling and planning of economic growth)*. Warszawa: PWN.

Czerwiński, Z., Kiedrowski, R., Konopczyński, M., & Panek, E. (1998). KEMPO model as a tool for generating growth scenarios of the Polish economy by institutional sectors. In W. Welfe (Ed.), *Proceedings of the 25th international conference MACROMODELS'98 and the 3rd conference of the international association AMFET—modelling in transitions* (Vol. 1, pp. 53–74). Łódź: Absolwent.

Czyżewski, B., Dębski, W., & Welfe, W. (1976). Dynamic simulations based on revised W-1 model and the model W-3 of the Polish economy. *Prace IEiS UŁ, 26*. Łódź.

Davis, C. M. & Charemza, W. (Eds.) (1989). *Models of disequilibrium and shortage in centrally planned economies*. London: Chapman and Hall.

Dębski, W. (1987). *Ekonometryczne modele dochodów osobistych ludności Polski (Econometric models of personal incomes of the Polish citizens)*. Warszawa: PWE.

Dlouhy, V. (1984). Macroeconomic disequilibrium model of centralny planned economy: some remarks and alternative specifications. *Economico-Matematicky Obzor, 20*, 374–387.

Dobrescu, E. (1998). Macromodel of the Romanian economy in transition. In W. Welfe (Ed.), *Proceedings of 25th international conference MACROMODELS'98 and the 3rd conference of the international association AMFET* (Vol. 1, pp. 131–192). Łódź: Absolwent.

Dobrescu, E. (1999). *Macromodel of the Romanian transition economy, 1999 version* (Mimeo). Paper presented at Project LINK meeting, Athens. Bucharest: Centre for Economic Information and Documentation.

Dobrescu, E. (2000). *Macromodels of the Romanian transition economy* (3rd ed.). Bucharest: EXPERT.

Dobrescu, E. (2003). Macromodel estimation for the 2002 version of the Romanian "Pre Accession Economic Programme". In W. Welfe & A. Welfe (Eds.), *Macromodels'2002* (pp. 35–56). Łódź: Wydawnictwo UŁ.

Dobrescu, E. (2006). *Macromodels of the Romanian market economy*. Bucharest: Editura Economica.

Ďuraš, J., Kretan, V., Ondko, P., & Pàlenik, V. (2004). Koňstrukcia modolu ECM-ISWE 04 q 1. In *Ekomstat, zbornik vedecych prac* (pp. 56–68). Bratislava: Slovenska Statistica a demografickă spoločnost.

Fic, T., Gradzewicz, M., Kolasa, M., Kot, A., Murawski, K., & Rubaszek, M. (2003). *Kwartalny model makroekonometryczny gospodarki Polskiej ECMOD (Quarterly macroeconometric model of the Polish economy)* (Mimeo). Warszawa: NBP.

Florczak, W., & Welfe, W. (2004). The structure of the long-term W8-D-2002 econometric model of the Polish economy. In W. Welfe & P. Wdowiński (Eds.), *Modelling economies in transition* (pp. 19–46). Łódź: Absolwent.

Fomin, J., & Tomaszewski, W. (1971). *Socialist countries in the world trade model*. New York: UNCTAD.

Gajda, J. B. (1986). The system IES2 of minimodels of European centrally planned economies, a new version. In W. Milo & M. Miszczyński (Eds.), *Macromodels and forecasts of national economy, Proceedings of Models and Forecasts'83 and Macromodels'83* (pp. 113–140). Warszawa: PWN.

Garbicz, M., Kuszewski, T., Mazur, K., Maciąg, W., & Pawilno-Pacewicz, J. (1985). Prognostyczny model gospodarki narodowej SAPO (Forecasting SAPO model of the national economy). *Gospodarka Planowa, 1*, 34–40.

Garbicz, M., Kuszewski, T., Mazur, K., Maciąg, W., & Pawilno-Pacewicz, J. (1986). Średniookresowy model prognostyczny gospodarki narodowej SAPO. In *Prace i materiały ICiZ: Vol. 11. Modele makroporcji gospodarczych (Forecasting medium-term model of the national economy SAPO)* (pp. 115–192). Warszawa: SGPiS.

Gavrilenkov, E. E. (1990). *An experience of modelling the USSR economy* (Mimeo).

Gavrilenkov, E. E., Henry, S. G. B., & Nixon, J. (1999). *A quarterly model of the Russian economy: estimating the effects of a devaluation* (LBS Discussion Paper No. DP 08-99). London.

Gavura, M., & Tkáč, M. (2000). Econometric model NBS. In W. Welfe & P. Wdowiński (Eds.), *Macromodels'99* (pp. 95–127). Łódź: Absolwent.

Grabek, G., Kłos, B., & Utzig-Lenarczyk, G. (2007). SOE-PL model DSGE małej otwartej gospodarki estymowany na danych polskich. Metodologia, specyfikacja, wyniki estymacji i pierwsze zastosowania (SOE-PL model of DSGE type of a small open economy estimated using Polish data. Methodology, specification, estimation results and first applications). *Materiały i studia, 217*. Warszawa: NBP.

Grajek, M. (2000). Combined forecasts. Application for Poland. In W. Welfe & P. Wdowiński (Eds.), *Modelling economies in transition*, Rydzyna, 1–4.12.1999 (pp. 8–23). Łódź: Absolwent.

Green, D., & Higgins, C. (1977). *SOVMOD I: a macroeconomic model of the Soviet Union*. New York: Academic Press.

Greszta, M. (2002). Efficiency of Polish macroeconometric forecasts. In W. Welfe (Ed.), *Modelling economies in transition*, Krąg, 5–8.12.2001 (pp. 9–15). Łódź: Absolwent.

Haber, G., Neck, R., & Weyerstrass, K. (2002). Optimal fiscal and monetary polices for Slovenia under different exchange rate regims and targets. In W. Welfe (Ed.), *Macromodels'2002* (pp. 261–281). Łódź: Absolwent.

Halabuk, L., Kenessey, Z., Theiss, E., Kotasz, G., & Nyari, Z. (1966). *A magyar nepgazdasag M-1 statisztikai makromodellje*. Budapest: KSH.

Halabuk, L., Hulyak, K., Nyari, Z., & Kotasz, G. (1973). *The M-2 econometric model of the Hungarian national economy*. Budapest: Akademiai Kiado.

Hall, S. G. (1994). Modelling economies in transition. In W. Welfe & R. Kelm (Eds.), *Proceedings of MACROMODELS'94* (pp. 167–202). Łódź: Absolwent.

Haluška, J., & Olexa, M. (2005). An application of business tendency surveys in econometric models. In W. Welfe & P. Wdowiński (Eds.), *Modelling economies in transition 2004* (pp. 191–198). Łódź: Studio Poligraf. M. Color.

Haluška, J., & Olexa, M. (2007). Using business tendency survey. Results for flash estimates of GDP in Slovakia (an econometric approach). In W. Welfe & P. Wdowiński (Eds.), *Modelling economies in transition, 2006* (pp. 39–49). Łódź: GREEN.

Haluška, J., & Palčákova, H. (2004). Modelling real sector of the Slovak economy. In W. Welfe & P. Wdowiński (Eds.), *Modelling economies in transition* (pp. 11–18). Łódź: Absolwent.

Haluška, J., Orságová, J., Šujan, I., & Šujanova, M. (1995). *An experimental econometric model of the Slovak economy with disaggregated foreign trade* (Mimeo). Paper presented at the 22nd conference Macromodels'95, Warsaw.

Heyets, V., & Skripnichenko, M. (1996). *Macroeconomic model and forecast of development of Ukrainian economy for 1997–2005* (Mimeo). Kiev: Institute of Economies NASU.

Heyets, V., & Skripnichenko, M. (1997). Some data problems in econometric modelling of the Ukrainian economies. In W. Welfe & P. Wdowiński (Eds.), *Proceedings of the conference MACROMODELS'96 on integration and development* (Vol. 2, pp. 59–68). Łódź: Absolwent.

Heyets, V. (1998). Rozszirena ekonometriczna model finansovogo programuvanija ta vichidni położenija politik ekonomicznogo zrostanija v umovach finansowoj nestabilnosti. *Ekonomist, 5*, 12–21.

Hulyak, K. (1973). *Okonometriai fuzetek: Vol. 12. Model M-4; an econometric model incorporating input-output relationships*. Budapest: KSH.

Hulyak, K. (1986). An experimental disequilibrium macromodel for Hungary. In W. Milo & M. Miszczyński (Eds.), *Macromodels and forecasts of national economy, Proceedings of Models and Forecasts'83 and Macromodels'83* (pp. 247–256). Warszawa: PWN.

Jakab, Z. M., & Kovács, M. A. (2001). Hungary in the NIGEM world model. In W. Welfe (Ed.), *Modelling economies in transition* (pp. 233–258). Łódź: Absolwent.

Jaremienko, J. W. (Ed.) (1982). *Strukturnyje izmienienija v socjalisticzeskoj ekonomike*. Moskwa: Nauka.

Jaremienko, J. W. (Ed.) (1984). *Modelirovanije miezotraslievych vzaimodejstvij*. Moskwa: Nauka.

Jemielianov, A. S., & Kushnirskij, F. I. (1974). *Modelirovanije pokazatielej razvitija ekonomiki Sojuznoj Respubliki*. Moskwa: Ekonomika.

Juszczak, G., Kaźmierska, M. M., Welfe, A., & Welfe, W. (1997a). W5 model and the Polish economy in transition. In W. Welfe (Ed.), *Economies in transitions and the world economy. Models forecasts and scenarios* (pp. 253–294). Frankfurt/Main: P. Lang (Chap. 5).

Juszczak, G., Kaźmierska, M. M., Welfe, A., & Welfe, W. (1997b). The new annual W10 model of the Polish economy. Assumptions and first empirical results. In W. Welfe (Ed.), *Economies in transitions and the world economy. Models forecasts and scenarios* (pp. 295–341). Frankfurt/Main: P. Lang (Chap. 6).

Juszczak, W. (1987). The W7 S82 model of the Polish economy. *Prace IEiS UŁ, 58*. Łódź: Wydawnictwo UŁ.

Juszczak, W., & Welfe, W. (1983). A computational forecast for development of the Polish economy in the period 1983–1985. *Prace IEiS UŁ, 44*. Łódź.

Juszczak, W., Markowski, K., & Welfe, W. (1987). Analiza spójności narodowego planu społeczno-gospodarczego na lata 1986–1990 (An analysis of the consistency of the national social-economic plan for the years 1986–1990). *Prace IEiS UŁ, 56*. Łódź.

Kalecki, M. (1963). *Zarys teorii wzrostu gospodarki socjalistycznej (Outline of the theory of growth of a socialist economy)*. Warszawa: PWN.

Kalinin, D., Maciejewski, W., & Zajchowski, J. (1977). Kwartalny model gospodarki Polski 1967–1975 KP-3K (Quarterly model of the Polish economy 1967–1975 KP-3K). *Gospodarka Planowa, 3*, 128–135.

Kalisiak, Z., & Piaszczyński, W. (1972). *Makroekonomiczny model gospodarki narodowej z uwzględnieniem handlu zagranicznego (Macroeconomic model of the national economy, including foreign trade)*. Warszawa: IKiC HZ.

Klas, A., Šujan, I., & Kolek, J. (1979). Some applications of econometric models in Czechoslovakia. In *Probleme internationaler wirtschafts- und sozial-statistischer Vergleiche*. Heidelberg: Bund-Verlag.

Klein, L. R. (1991). Model building for a planned system. *Economic Modelling, 8*, 418–423.

Klein, L. R., & Welfe, W. (1982). *Wykłady z ekonometrii (Lectures in econometrics)*. Warszawa: PWE.

Klein, L. R., Welfe, A., & Welfe, W. (1999). *Principles of macroeconometric modeling*. Amsterdam: North-Holland.

Kłos, B. (2002). Mały strukturalny model inflacji wersja 3.5.6 (Small structural model of inflation, version 3.5.6). *Materiały i studia, 155*. Warszawa: NBP.

Kłos, B., Kokoszczyński, R., Łyziak, T., Przystupa, J., & Wróbel, E. (2004). Modele strukturalne w prognozowaniu inflacji w Narodowym Banku Polskim (Structural models in forecasting inflation at the Polish National Bank). *Materiały i studia, 180*. Warszawa: NBP.

Kłos, B., Kokoszczyński, R., Łyziak, T., Przystupa, J., & Wróbel, E. (2005). Structural econometric models in forecasting inflation at the National Bank of Poland. *Materiały i studia, 31*. Warszawa: NBP.

Kolek, J., & Šujan, I. (1978). *Ekonometriczskije modeli w socjalisticzeskich stranach*. Moskwa: Ekonomika.

Kornai, J. (1980). *Economics of shortage (Vols. A, B)*. Amsterdam: North-Holland.

Kotyński, J. (Ed.) (1990). *Kraje Europy Srodkowo-Wschodniej w handlu międzynarodowym. Modele i prognozy (Central and East European countries in the international trade. Models and forecasts)*. Warszawa: PWN.

Kudrycka, I. (1974). *Wielorównaniowy model gospodarki narodowej. Ogólna koncepcja i specyfikacja równań (Multiequation model of the national economy. General concept and specification of equations)*. Warszawa: ZBSE GUS.

Kudrycka, I. (1990). The quarterly forecasting model of the Polish economy. In W. Welfe & P. Tomczyk (Eds.), *Problems of building and estimation of large econometric models, part A*. Łódź: Wydawnictwo UŁ.

Kulikowski, R., et al. (1979). *Modelowanie rozwoju społeczno-gospodarczego kraju metodami analizy systemowej (Modelling of the socio-economic development of the country using system analysis methods)*. Wrocław: Ossolineum.

Kyn, O., Sekerka, B. L., & Hejl, A. (1967). Model for the planning of prices. In *Socialism, capitalism and economic growth, Essays presented to Maurice Dobb*. Cambridge: Cambridge University Press.

Laursen, T., Gradzewicz, M., Griffin, P., & Żółkiewski, Z. (2004). *An empirical recursive-dynamic general equilibrium model of Poland's economy* (Mimeo). Warszawa: World Bank and NBP.

Levinson, M. (1995). Simulation model of Russian economy as of 1994. In W. Welfe & M. Majsterek (Eds.), *Macromodels and forecasts* (pp. 23–28). Łódź: Absolwent.

Levinson, M. (1997). *Simulation model of national economy of Russia 1995–96* (Mimeo). Moscow: CEMI RAN.

Łapińska-Sobczak, N., Tomaszewicz, Ł., & Welfe, W. (1979). *Modelling of financial and monetary flows in Poland* (Mimeo). Paper presented to the Colloque international: models monetaires et financiers, A.E.A., Banca d'Italia, G.A.M.A. Rome, 7–10 February. Łódź.

Łukinov, I., Bakaev, O., & Bondarenko, G. (1997). Pro sistemu modelej regionalnogo prognuzovanija v umovach perechidtnoj ekonomiki. *Regionalna Ekonomika*, 5–22.

Maciejewski, W. (1976). *Zastosowania ekonometrycznych modeli rozwoju gospodarki narodowej (Applications of econometric models of economic development)*. Warszawa: PWE.

Maciejewski, W. (1980). *Ekonometria stosowana. Analiza porównawcza (Applied econometrics. A comparative analysis)*. Warszawa: PWE.

Maciejewski, W., & Zajchowski, J. (1974). Ekonometryczny model gospodarki narodowej KP-2 (Econometric model of the Polish economy KP-2). *Gospodarka Planowa*, *8*, 670–675.

Maciejewski, W., Opara, W., & Zajchowski, J. (1973). Ekonometryczny model gospodarki Polski KP-1 (Econometric model of the Polish economy KP-1). *Gospodarka Planowa*, *8*, 550–555.

Michalevskij, W. (1967). Sistema modelej dla rasczeta sbalansirowannogo srednesrocznogo plana. *Ekonomika i Matematiczeskije Metody*, *3*(5), 711–729.

Michalevskij, W. (1972). *Sistema modelej srednesrocznogo narodnochozjajstvennogo planirovanija*. Moskwa: Nauka.

Michalevskij, W., & Soloviev, J. P. (1966). Proizvodstwiennaja funkcija narodnogo choziajstva SSSR v 1951–1963 g. *Ekonomika i Matematiczeskije Metody*, *2*(6).

Mienszikow, S. M., & Costa, A. M. (1979). Models of centrally planned economies. In J. A. Sawyer (Ed.), *Application and extensions of the Project LINK system*. Amsterdam: North-Holland (Chap. VIII).

Mikhaylenko, K. (2004). Opyt postrojenija srednesrocznoj makrostrukturnoj modeli ekonomiki Rossii. In *Naucznyje trudy* (pp. 175–196). Moskwa: INP RAN.

Minassian, G. (1996). *Modelling Bulgarian economy: methodological and statistical issues* (Mimeo). Paper presented at the conference Macromodels'96, 3–6.12,1996. Łódź.

Minassian, G. (2012, in print). *Breaking the Bulgarian economic model: reasons and projections*. Saarbrücken: VDM Publishing.

National Institute for Economic Research (1992). *An econometric model for Romania* (Mimeo). Bucharest: National Institute for Economic Research.

Olexa, M., Haluška, J., Klein, T., & Sasinek, M. (1997). Econometric model of the Slovak economy for the period of transition. In W. Welfe & P. Wdowiński (Eds.), *Proceedings of the conference MACROMODELS'96 on integration and development* (Vol. 2, pp. 33–58). Łódź: Absolwent.

Olexa, M., Haluška, J., Klein, T., Orságovă, M., & Sasinek, M. (1999). Econometric model of the Slovak economy for the transition period—EMSE 2.0. In R. Courbis & W. Welfe (Eds.), *Central and Eastern Europe on its way to European Union. Simulation studies based on macromodels* (pp. 589–608). Frankfurt/Main: P. Lang.

Pajestka, J. (1961). *Zatrudnienie i inwestycje a wzrost gospodarczy (Employment, investment and economic growth)*. Warszawa: PWE.

Pàlenik, V., & Benčik, M. (1995). *Ekonometricky model ekonomiki Slovenska* (Economicky Ustav SAV, VP 27/7). Bratislava.

Parkhomenko, V. W. (1998). VAR-modelling of the Ukrainian economy, first results. In W. Welfe (Ed.), *Proceedings of the international conference MACROMODELS'98 and the 3rd conference of the international association AMFET—modelling economies in transition* (Vol. 1, pp. 117–130). Łódź: Absolwent.

Plich, M. (Ed.) (1998). *Model ekonomiczno-ekologiczny dla gospodarki Polski, cz. I. Problemy i metody (Economico-ecological model for the Polish economy, part I. Problems and methods*. Łódź: Broker Info.

Plich, M. (2002). *Budowa i zastosowanie wielosektorowych modeli ekonomiczno-ekologicznych (Construction and use of multisectoral economico-ecological models)*. Łódź: Wydawnictwo UŁ.

Portes, R., Quandt, R. E., Winter, D., & Yeo, S. (1987). Macroeconomic planning and disequilibrium: estimates for Poland, 1955–1980. *Econometrica*, *55*(1), 19–41.

Romański, J., & Welfe, W. (1981). Linear minimodel W-4L of the Polish national economy: alternative specifications and estimation techniques: fix-point compared with OLS and TSLS. *Prace IEiS UŁ*, *35*. Łódź: Wydawnictwo UŁ.

Romański, J., & Welfe, W. (1991). A macroeconometric disequilibrium model for Poland. In J. Gruber (Ed.), *Econometric decision models. New methods of modeling and applications* (pp. 591–609). Berlin: Springer.

Shapiro, H. T. (1977). Macroeconomic models of the Soviet Union and Eastern European economies: a tabular survey. *Econometrica*, *45*(8), 1747–1766.

Shapiro, H. T., & Halabuk, L. (1976). *Macro-econometric model building in socialist and non-socialist countries. A comparative study* (Working Paper, Mimeo). Dept. of Economics, University of Michigan.

Stancu, S., & Sava, G. (2007). Elaboration and analysis of a macroeconomic model of the Romanian economy. In W. Welfe & P. Wdowiński (Eds.), *Modelling economies in transition* (pp. 27–38). Łódź: GREEN.

Stepanov, Y. (1995). *Macroeconomic modelling of the development of the Russian economy during the transition period* (Mimeo). Moscow: Central Bank of Russia.

Štiblar, F., Oplotnik, Z., & Vukotič, V. (2006). *Montenegrin quarterly macroeconomic econometric model* (Prague Economic Papers No. 2).

Simon, A. (1978). *An econometric model of the Hungarian economy* (Mimeo). Budapest: Institute for Economic and Market Research.

Štrauch, D. (1990). Main features of the econometric model of the Czechoslovak economy version CEM-3.2. Paper presented to the 12th international conference Macromodels'85, Szczyrk-Bila, 10–13.12.1985. In W. Welfe & P. Tomczyk (Eds.), *Problems of building and estimation of large econometric models (part A)* (pp. 95–105). Łódź: Wydawnictwo UŁ.

Šujan, I. (1970). *The first tentative econometric model of the Czechoslovak economy* (Mimeo). Philadelphia: University of Pennsylvania.

Šujan, I. (1978a). Konstrukcia a aplikacia srednodobeho ekonometrickeho modelu CSRS. *Ekonomicky Casopis*, *26*(4), 373–388.

Šujan, I. (1978b). *A system of models for macroeconomic analyses and forecasts* (Mimeo). Paper presented to the conference on problems of building and estimation of large econometric models, Blazejewsko, 6–8.12.1978. Bratislava.

Šujan, I. (1986). Econometric models of Czechoslovakia. A survey. In W. Milo & M. Miszczyński (Eds.), *Macromodels of forecasts of national economy, Proceedings of Models and Forecasts'83 and Macromodels'83* (pp. 171–183). Warszawa: PWN.

Šujan, I. (1997). Economic outlook of the CSFR. Estimation of consequences of the Czecho-Slovak divorce. In W. Welfe (Ed.), *Economies in transition and the world economy* (pp. 475–484). Frankfurt/Main: P. Lang.

Šujan, I., & Kolek, J. (1974). *A quarterly forecasting model of the Czechoslovak economy* (Mimeo). Paper presented to the European meeting of the Econometric Society, Grenoble.

Šujan, I., & Olexa, M. (1984). Kvartalnaja ekonometriczeskaja model dla kratkosrocznogo analiza ekonomiki CSRS. *Ekonomika i Matematiczeskije Metody*, *20*(4), 646–655.

Šujan, I., & Štrauch, D. (1989). Agregatny nerovnovazny model ekonomiki CSRS. *Ekonomicko Matematicky Obzor*, *25*(1), 1–18.

Šujan, I., Gergelyi, K., & Kolek, J. (1974). *Prognosticky model ekonomiky CSRS*. Bratislava: Alfa.

Šujan, I., Haluška, J., Olexa, M., & Orságová, J. (1993). An experimental annual model of the CSFR economy with disaggregated foreign trade. Paper presented at MACROMODELS'92. In J. Gajda & W. Florczak (Eds.), *Large econometric models* (pp. 71–90). Łódź: Absolwent.

Šujan, I., Haluška, J., Olexa, M., & Orságová, J. (1995). A short-term forecasting model of the Czecho-Slovak economy for the transition period. Paper presented at the conference MACROMODELS'91, Warsaw, 4–6.12.1991. In W. Welfe & W. Florczak (Eds.), *Macromodels'91* (pp. 9–40). Łódź: Absolwent.

Šujanova, M., & Šujan, I. (1994). *The macroeconomic model of the Czech Republic CR-01* (Mimeo). Prague: The Statistical Office of the Czech Republic.

Szakolczai, G., Marjanek, K., & Sisak-Fekete, Z. (1985). *Adaptation of the Hungarian economy to change on the world markets. An analysis based on econometric model* (Mimeo). Paper presented to the ES 5th world congress, 17–24 August 1985. Cambridge.

Tabeau, A., Żółkiewski, Z., Frączek, R., & Szafrański, J. (1994). *Construction of an econometric model of the Polish economy* (Mimeo). The Hague: Ace programme, Final report.

Tomaszewicz, Ł. (1983). *Zintegrowane modele gospodarki narodowej (Integrated models of the national economy)*. Warszawa: PWE.

Tomaszewicz, Ł. (1985). Alokacja produkcji metodą input-output w warunkach nierównowagi (Production allocation using input-output models in disequilibrium conditions). *Acta Universitatis Lodzensis*. Łódź: UŁ.

Tomaszewicz, Ł., & Welfe, W. (1979). Econometric macromodels including the input-output analysis. Problems and practical solutions in the W-2 model of the Polish economy. In M. Janssen, L. F. Pace, & A. Straszak (Eds.), *Proceedings of the second international conference on dynamics and control of national economies*, Vienna 1979 (pp. 70–83). Amsterdam: North-Holland.

Vàvra, D., Beneš, L., Audric, S., Brillet, J. L., & Beynet, P. (1998). *A small model of the Czech economy* (Mimeo). Prague: CNB.

Viren, M., et al. (2003). *Quarterly public finance sector model of the Polish ministry of finance (QPFSPM)* (Mimeo). Warszawa: Ministerstwo Finansów.

Welfe, A. (1984). Analiza popytu w warunkach nierównowagi (Demand analysis in disequilibrium conditions). *Ekonomista, 5*, 1045–1064.

Welfe, A., & Welfe, W. (1986). Modele i prognozy rynku w warunkach nierównowagi (Market models and forecasts under disequilibria). *Ekonomista, 2*, 279–314.

Welfe, A., Karp, P., & Kelm, R. (2002). *Kwartalny makroekonomiczny model gospodarki Polski (Quarterly macroeconomic model of the Polish national economy)*. Łódź: Wydawnictwo UŁ.

Welfe, A., Karp, P., & Kębłowski, P. (2006). *Mechanizmy makroekonomiczne w gospodarce polskiej. Analiza ekonometryczna (Macroeconomic mechanisms in the Polish economy. An econometric analysis)*. Łódź: Wydawnictwo UŁ.

Welfe, W. (1973). A medium-term econometric model of the Polish economy. *Prace IEiS UŁ, 8*. Łódź.

Welfe, W. (1975). Forecasting econometric models of the Polish economy: their structure and use in planning and economic policy simulations. Paper presented to the III world congress of the ES, Toronto. *Prace IEiS UŁ, 23*. Łódź.

Welfe, W. (Ed.) (1979). *Ekonometryczne modele funkcjonowania gospodarki narodowej (Econometric models of functioning of the national economy)*, z. 36. Warszawa: Instytut Planowania.

Welfe, W. (1980). *New versions of the W-3 model of the Polish economy: structure-estimation-simulation* (Mimeo). Paper presented to the Vth international conference: Forecasting models in the socialist economy, Esztergom, 30.9–3.10.1980. Łódź: IEiS UŁ.

Welfe, W. (1982). Modele gospodarki socjalistycznej (Models of the Socialist economy). In L. R. Klein & W. Welfe (Eds.), *Wykłady z ekonometrii (Lectures in econometrics)*. Warszawa: PWE.

Welfe, W. (1983). Models of the socialist economy. In L. R. Klein (Ed.), *Lectures in econometrics* (pp. 197–227). Amsterdam: North-Holland (Appendix).

Welfe, W. (1985). Econometric macromodels of unbalanced growth. *Prace IEiS UŁ, 52*. Łódź.

Welfe, W. (1986). The econometric models of the Polish economy. In W. Milo & M. Miszczyński (Eds.), *Macromodels and forecasts of national economy, Proceedings of Models and Forecasts'83 and Macromodels'83* (pp. 185–213). Warszawa: PWN.

Welfe, W. (1987). New versions of W-models for Poland. Patterns of constrained growth and mixed government policies under economic reform. In W. Welfe & K. Markowski (Eds.), *Models and Forecasts'84* (pp. 101–111). Wrocław: Ossolineum.

Welfe, W. (1989). Macroeconomic disequilibrium models of Poland. In C. M. Davis & W. Charemza (Eds.), *Models of disequilibrium and shortage in centrally planned economies* (pp. 205–246). Cambridge: Chapman and Hall.

Welfe, W. (1991). Multisectoral econometric models of the centrally planned economies and the disequilibria. *Economics of Planning*, *24*, 203–226.

Welfe, W. (1992). *Ekonometryczne modele gospodarki narodowej Polski (Econometric models of the Polish national economy)*. Warszawa: PWE.

Welfe, W. (1993). Topics in macromodelling of East European countries in the period of transition. *Economics of Planning*, *26*, 105–126.

Welfe, W. (Ed.) (1995). *Kwartalny model gospodarki Polski. Struktura i własności (Quarterly model of the Polish economy. Structure and properties)*. Warszawa: PWN.

Welfe, W. (Ed.) (1996). *Średniookresowy ekonometryczny model gospodarki narodowej Polski w warunkach transformacji (Medium-term econometric model of the Polish economy for the transformation conditions)*. Łódź: Absolwent.

Welfe, W. (Ed.) (1997). *Economies in transition and the world economy—models, forecasts and scenarios*. Frankfurt/Main: P. Lang.

Welfe, W. (1999). The macroeconometric model W8 of the Polish economy. In R. Courbis & W. Welfe (Eds.), *Central and Eastern Europe on its way to European Union. Simulation studies based on macromodels* (pp. 507–582). Frankfurt/Main: P. Lang.

Welfe, W. (2000). Empiryczne modele wzrostu gospodarczego (Empirical models of economic growth). *Ekonomista*, *4*, 483–498.

Welfe, W. (Ed.) (2001). *Ekonometryczny model wzrostu gospodarczego (Econometric model of economic growth)*. Łódź: Wydawnictwo UŁ.

Welfe, W. (2004a). 30 years of macromodels. In W. Welfe & A. Welfe (Eds.), *Macromodels'2003, Proceedings of the 30th international conference*, Warszawa, 3–6.12.2003 (pp. V–XXVIII). Łódź: Wydawnictwo UŁ.

Welfe, W. (Ed.) (2004b). Długookresowy makroekonometryczny model W8D-2000 gospodarki polskiej (Long-term macroeconometric model W8D-2000 of the Polish economy). *Acta Universitatis Lodzensis, Folia oeconomica, 172*. Łódź: Wydawnictwo UŁ.

Welfe, W. (Ed.) (2007). *Gospodarka oparta na wiedzy (Knowledge based economy)*. Warszawa: PWE.

Welfe, W. (Ed.) (2009a). Makroekonometryczny model gospodarki opertej na wiedzy (Macroeconometric model of a knowledge economy). *Acta Universitatis Lodzensis, Folia oeconomica, 229*. Łódź: Wydawnictwo UŁ.

Welfe, W. (Ed.) (2009b). *Knowledge-based economies. Models and methods*. Frankfurt/Main: P. Lang.

Welfe, W., & Florczak, W. (2009). *Prognozy i scenariusze długookresowego rozwoju gospodarczego Polski (Long-term forecasts and scenarios of economic development of Poland)*. Łódź: Wydawnictwo UŁ.

Welfe, W., Gajda, J., & Żółtowska, E. (1991). *On the methodology of constructing large economic models of an East European economy (Poland). A comment* (Discussion Paper No. 142). Dept. of Economics, University of Leicester.

Welfe, W., Gajda, J., & Żółtowska, E. (1992). On the methodology of constructing large economic models of an East European economy (Poland). A comment. *Economic Modelling*, *9*, 137–145.

Welfe, W., Welfe, A., & Florczak, W. (1996). Symulacyjny makroekonometryczny model W8 gospodarki Polski (Macroeconometric simulation W8 model of the Polish economy). *Gospodarka Narodowa*, *12*, 32–46.

Welfe, W., Florczak, W., Sabanty, L., & Welfe, A. (2003a). The W8-2000 medium-term macroeconometric model of the Polish economy. In W. Welfe & A. Welfe (Eds.), *Proceedings of the 29th international conference MACROMODELS'2002* (pp. 11–34). Łódź: Wydawnictwo UŁ.

Welfe, W., Welfe, A., Florczak, W., & Sabanty, L. (2003b). Makroekonometryczny roczny model W8-2000 gospodarki narodowej Polski (Macroeconometric annual W8-2000 model of the Polish national economy). *Prace IEiS UŁ, 139*. Łódź: Wydawnictwo UŁ.

Welfe, W., Welfe, A., Florczak, W., & Sabanty, L. (2003c). Długookresowy makroekonometryczny model W8D gospodarki narodowej Polski (Long-term macroeconometric W8D model of the Polish national economy). *Prace IEiS UŁ, 141*. Łódź: Wydawnictwo UŁ.

Welfe, W., Florczak, W., & Welfe, A. (2004). *Scenariusze długookresowego rozwoju gospodarczego Polski (Scenarios of the long-term development of the Polish economy)*. Łódź: Wydawnictwo UŁ.

Weyerstrass, K., & Neck, R. (2007). SLOPOL 6: a macroeconometric model for Slovenia. *International Business and Economic Research Journal, 6*, 81–94.

Weyerstrass, K., Haber, G., & Neck, R. (2001). SLOPOL 1, a macroeconomic model for Slovenia. *International Advances in Economic Research, 7*, 20–37.

Woelfling, M. (1977). *Ein Oekonometrisches Modell der Volkswirtschaft der DDR*. Berlin: Akademie Verlag.

Chapter 10
The Macroeconometric Models of the European Union and the EURO Area

10.1 The Models of the European Union

10.1.1 Introduction

In the first years following the establishment of the European Union, forecasts and policy simulations for this expanding area were based on the world models. The models distinguished the macromodels of particular countries or their groups, for instance of the EU members. A relevant example is the Project LINK, MULTIMOD, INTRELINK, and QUEST models maintained by international organizations, as well as models run by central banks and research institutes. These models will be characterized in the last chapter.

Influenced by developments associated with PROJECT LINK, at the beginning of the 1970s the above situation contributed to the construction of the DESMOS model. The model covered 9 member countries of the European Economic Community (EEC). Waelbroeck and Dramais (1974) used it to analyse international transmission mechanisms, concluding that they were still weak.

In the mid-90s, many initiatives were undertaken to build macroeconometric models for countries being EU members along the lines used to construct the world models. No attempts were made, though, to link particular countries together. This group of models contains the small SLIM model (Douven and Plasmans 1986), the EUROMON model built at the Netherlands Central Bank in the mid-90s, the HERMIN model for the peripheral EU countries and the MAC SIM model for 6 EU countries.

The attempts to construct macroeconometric models for the European Union treated as a single economy were distinguished by their originality. They allowed for the growing number of EU members. The first of them date back to the second half of the 1980s, when A. Dramais proposed building the COMPACT macromodel for 10 countries. A new EU model was constructed outside this organization only in the year 2003. These models will be presented below. These developments are summarized in Table 10.1.

W. Welfe, *Macroeconometric Models*,
Advanced Studies in Theoretical and Applied Econometrics 47,
DOI 10.1007/978-3-642-34468-8_10,

10.1.2 The EUROMON Model

The macroeconometric model EUROMON (European Model of Netherlands Origin) was constructed at the Central Bank of Netherlands in the mid-90s to analyse the integration processes between the EU members. Covering 8 largest EU countries, i.e. Belgium, Denmark, France, Germany, Italy, Netherlands, Spain and United Kingdom, the model was open to include more members in the future. Since 1996 the model has regularly supported forecasting activities, as well as being used to run numerous policy simulations (de Bondt et al. 1997).

EUROMOD was a quarterly model. Its first version using the 1970–1990 data was subsequently extended to the year 1995. Because the models of particular countries were of medium size, the whole system totalled around 900 equations, ca. 250 of which were stochastic.

The country models were structured uniformly to ensure their comparability. The differences between particularly countries were expressed mainly by different values given to constants and elasticities. The equations in particular models were specified by referring to the solutions accepted for the MORKMON II model of the Netherlands economy run by the Central Bank (Fase et al. 1992).

The EUROMOD model had a Keynesian orientation. In the short-run, output and employment were determined by final demand. Prices and wages were rigid; prices adjusted to unit costs plus a mark-up with lags. Wages were formed in a bargaining process. The model included equations generating potential output and labour supply; the rate of capacity utilization was endogenized. This model had an advantage over the other model system in that it had an extended financial sector with equations explaining major components of banks' assets and liabilities.

The parameters of the model's equations were estimated using a two-step procedure. The parameters of the long-run equations were estimated in the first step and then the parameters of the short-term adjustments, using the ECM method.

10.1.3 The MAC SIM Model

At the end of the 1990s, the INSEE launched a research project involving six major EU countries, i.e. France, Germany, Italy, Netherlands, Sweden and United Kingdom. The project was led by J.L. Brillet and its aim was to provide an educational and analytical tool enabling the exploration of EU integration processes and in the future of countries' accession to the EMU (Brillet et al. 1999). The equations were specified in almost the same way for all countries, following the example set by the Micro DMS model built at the INSEE.

The model was demand determined, with strong emphasis laid on the likely disequilibria. In particular, the model builders took efforts that the utilization rate of potential output was generated. The rate affected investment demand, foreign trade and prices. Bilateral links between the countries were introduced into the model via foreign trade and exports and imports prices, which positively distinguished it

Table 10.1 Models of the European Union

Model Version Year	Institution Author	Frequency Sample	Equations total stochastic	Decomposition	Estimation method	Properties	Applications
DESMOS 1973	J. Waelbroeck, A. Dramais	Annual		9 EEC countries	OLS	Demand determined	Simulation analyses
EUROMON 1995	Central Bank of Netherlands G.J. de Bondt et al.	Quarterly 1970–1990 –1995	900 250	Distinguished 8 EU countries	Two-stage EMC	Models of particular countries medium sized; neo-Keynesian close to MORKMON II	Forecasts, simulation analyses of economic integration
MAC SIM 1999	INSEE J.L. Brillet et al.	Annual	671 66	6 EU[a] countries	ECM	Structure of country models similar to Micro DMS, model of mixed orientation	Simulation analyses of UE and EMU interactions for educational purposes
HERMIN 1995	Economic and Social Research Institute, Dublin J. Bradley et al.	Annual 1977–1990[a]	100–150	Covered peripheral EU countries[a]	OLS	Stressed supply factors of growth	Simulation analyses of the potential impacts of entering the UE
COMPACT	European Community	Annual	54 34				Simulation analyses
1986	A. Dramais	Annual 1960–1997	34 20	Aggregate of 10 countries	Two-stage EMC, using IIV	Prototype model for the EU as a single economy	
Aggregate Model for the EU 2001	University of Rome A. Bagnai, F. Carlucci			Aggregate of 12 countries		Excluded financial and monetary sector	Policy simulations

[a]Initially Ireland, Portugal and Spain, then extended

among the models constructed for groups of countries. Equation parameters were estimated using a two-step procedure; the ECM was used to estimate the equation parameters of the short-term dynamic adjustments.

10.1.4 The HERMIN Model

The system of the HERMIN models that J. Bradley developed for the peripheral EU countries was to serve special purposes, such as running simulations to study the conditions and effects of new countries' joining the EU (Bradley et al. 1995). The HERMIN models were medium size, with extended supply sectors. The CES production function played the major role in generating the dynamics of the production factors and in showing the effects of economic growth. The first model of the series was built for Ireland (Bradley and Fitz 1991); the next were the components of the international HERMIN project for Portugal, Spain and Greece (Bradley et al. 1995). Before all 10 new countries joined EU in 2004, successive models were constructed for several of the candidate countries, i.e. Hungary, Poland and Romania. These developments have been characterized in some detail in the previous chapter.

10.1.5 The GEM-E3 Model

Special CGE model was built in the late 1900s to provide details on the macroeconomies and their interactions with the environment and the energy system. It represented 14 EU member states either non-linked or linked throughout endogenous bilateral trade flows (Capros et al. 1997; Buscher et al. 2001).

10.1.6 The COMPACT Model

The comparative analyses of the economic characteristics of the countries being the first EU members led to the conclusion that the countries were similar enough to justify the construction of a macroeconometric model for the entire EU treated as a single economy. Many problems were common to all member states, such as fluctuating rates of inflation, the behaviour of the unemployment rate, etc. As a result, in the mid-80s the European Committee decided to construct a "prototype" model for the EU. The ensuing COMPACT macroeconometric model was built under the leadership of A. Dramais (Dramais 1986). This medium-sized model with 54 equations 34 of which were stochastic used annual data on 10 countries.

Output was determined in the model by final demand, but employment was generated from a disequilibrium condition applied to the Keynesian demand, neoclassical demand and labour supply. Investment depended on the ratio between the expected output and the available fixed capital. The expectations of production and inflation were determined through extrapolation. The COMPACT model contained a detailed specification of public sector' revenues and expenditures. It was systematically used as a tool for running the simulations of EU policy.

10.1.7 The Aggregated Model for the European Union

Several years elapsed before the research on an aggregated macroeconometric model for the European Union was resumed. It happened in the early 21st c. at the University of Rome (Bognai and Carlucci 2003). To some extent, the model was a continuation of the Dramais' model. It was a small, annual model with 34 equations, 24 of which were stochastic, which used the aggregated data for 12 EU countries. Its distinctive feature was the use of modern econometric techniques.

The model's structure assumed an open economy and macroeconomic equilibrium. Three sectors were distinguished (private, public and foreign) and four markets (production, domestic industry, foreign market and capital market). The model stressed the importance of analysing the monetary and fiscal policy impacts.

In the model, GDP was determined by demand. Consumer demand solely depended on real disposable income whose nominal value was residual; personal wealth as well as expectations occurred to be non-significant. In the long run investments were determined by output (accelerator) corrected for the user costs of capital.

As a result of cost minimization, employment was determined from the Cobb-Douglas production function. Wages were determined in the long-run by labour productivity and the rate of unemployment. Prices depended on labour costs and imports prices. Besides, the model had extended equations in the financial sector that explained, inter alia, interest rates as well as capital movements in the balance of payments.

The model was used many times to run policy simulations. The multipliers were calculated to show the impacts of declining public consumption and investments, deteriorating terms of trade and devaluation vis-à-vis the US dollar.

10.2 The Models of the EURO Area

10.2.1 Introduction

Following the decision establishing the European Monetary Union (EMU) at the end of the 20th c., international organisations made a series of attempts in the early years of the new century to construct macroeconometric models to describe the EURO area alone, or as the components of the world-wide models. The results of the efforts were discussed at international conferences, e.g. in Berlin in 2002, as well as becoming the subject of several comparative studies (Issing 2004). Let us mention first the activities of the Central European Bank that will be described more in detail below. Then a model was built at the Economic Research Institute in Halle (Dreger 2002). The OECD constructed a special model, in principle demand determined, including three groups of countries: the EURO area, USA and Japan and the rest of the world (Rae and Turner 2001). The INSEE also built a special model MZE. The forecasts and policy simulations generated by the world models began routinely to distinguish the EURO area. For summary description see Table 10.2.

10.2.2 The Models of the European Central Bank. The AWM Model

Before the European Monetary Union was finally established on 1 January 1999, attempts had been made to provide the European Central Bank (ECB) with efficient instruments enabling economic analyses of the EURO area. They were to illuminate the potential impacts of monetary policy for keeping inflation under control and its rates within the targeted limits. Within several years, a system of models was built. It included the extrapolation of the selected time series, a DSGE model for shock analyses and medium-sized macroeconometric models that played the major role in forecasting and policy simulations. An ambitious undertaking was the project to construct an annual Multi-country Model (MCM) to be composed of the annual country models belonging to the European System of Central Banks. The models were constructed sequentially, with the participation of the central banks of particular countries. Let us mention the models for Belgium (Jeanfils 2000), Ireland (Mc Guire and Ryan 2000) and Spain (Estrada and Willman 2002).

However, it was the aggregated, quarterly Area-Wide Model (AWM) constructed by the research team led by A. Dieppe and J. Henry just before the EMU was formed that played the main role in ECB's modelling activities (Henry 1999; Fagan et al. 2005). The model was improved over its lifetime and regularly supported the short- and medium-term forecasting and many policy simulations (Dieppe and Henry 2004; Fagan et al. 2005). It became the subject of an interesting comparative analysis where it was juxtaposed with three world models distinguishing the EURO Area: MULTIMOD, NIGEM, QUEST. The results were presented by K.F. Wallis during an international meeting held in Berlin in 2002 (Wallis 2004).

The successive versions of the model were largely alike. All were medium size, with only 15 stochastic equations. They stressed the specification of the supply sector. A neoclassical approach prevailed in the long-run, while in the short-run GDP was determined by demand.

In particular, output was generated from the Cobb-Douglas production function in the long-run. This function was used to determine the demand for production factors. A natural unemployment rate equal to NAIRU was assumed.

In the short-run, GDP was determined by final demand. Consumer demand depended on real disposable income and personal wealth. Investment activity was determined by the demand for fixed capital allowing for accelerator effects, on user investment costs (linked to interest rates) and on depreciation.

The model had several equations explaining prices and wages. The GDP deflator was dependent on labour costs, import prices and the demand gap. The rate of wages growth depended on labour productivity, the current and lagged rates of inflation and the unemployment rate's deviation from its natural value.

The block of equations for financial flows included the major components of the government budget. The balance of payments components were determined from the standardly constructed export and import functions for commodities and services flowing outside the EURO area.

As a result of ECB's modelling activities, a small macroeconometric model was additionally constructed in 2000. Its special property was that it used rational expectations. It distinguished inflationary expectations and lags in wage bargaining. These neo-Keynesian rigidities helped define the short-term adjustments in the model. The model was used in analysing the impacts of alternative monetary policies (Coenen and Wieland 2005).

The DSGE model for the EURO area that paved the road for other models of this class was constructed in a way resembling the ECB's approach to modelling (Smets and Wouters 2003).

10.2.3 The MZE Model

At the beginning of the 21st c., the INSEE built the small, quarterly model MZE for the EURO area (Beffy et al. 2003). It was one of the first models constructed for this area where it was treated as a single economic organism. The model had over 20 equations, 15 of which were stochastic. The parameters of the equations were estimated using the quarterly Eurostat data (partly adjusted) from the years 1992–2000. The model distinguished the long-term relationships with partly calibrated parameters and short-term adjustments with parameters estimated with ECM.

The model had a hybrid structure. Demand was neo-Keynesian, while supply was neoclassical in the long-run. Household consumer demand depended on real disposable income and real interest rate. Investment demand was linked to value added and real user costs of investment. The Cobb-Douglas production function with constant returns to scale was used to determine the demand for employment that additionally depended on real wages. It was also used to generate potential output in the long-run.

Wages were linked to the rate of inflation (in the long-run elasticities were calibrated assuming that their value was 1), labour productivity and the rate of unemployment. Labour supply was determined, likewise the rate of unemployment, including NAIRU. Prices depended on labour costs and in the short run on imports prices. The exchange rate was exogenous. The monetary sector was represented by an equation explaining the interest rate. The model was used for preparing forecasts and policy simulations.

Table 10.2 Models for the EURO area countries

Model Version Year	Institution Author	Frequency Sample	Equations total stochastic	Decomposition	Estimation method	Properties	Applications
MCM (Multicountry Model) 2000–2002	European Central Bank (ECB) J. Henry	Annual		Covered increasing country number[a]		Specific country structures	Forecasts and policy simulations (bottom-up) by ECB
AWM (Area Wide Model) 1997–1998	G. Fagan, J. Henry	Quarterly 1970–1997	89 15	Aggregate of 11 EURO area countries	ECM	Neoclassical consumption and supply, potential output, adaptive expectations	Forecasts and simulation analyses by ECB
2004	A. Dieppe, J. Henry, G. Fagan et al.	Quarterly 1970–2000	89 15		ECM		Forecasts and simulation analyses by ECB
2005		Quarterly 1970–	84 15		ECM		
EURO Area Model with Rational Expectations 2000	G. Coenen, V. Wieland	Quarterly 1974–1998	Several	Aggregate of France, Ger- many, Italy	VAR	Small model, rational inflationary expecta- tions, negotiated real wages, lags in adjustments	Model aimed at analyses of alternative strategies in monetary policy
EFN 2002	European Forecasting Network G. Dreger	Quarterly					Forecasts

Table 10.2 (Continued)

Model Version Year	Institution Author	Frequency Sample	Equations total stochastic	Decomposition	Estimation method	Properties	Applications
MZE 2003	INSEE P.O. Baffy et al.	Quarterly	Above 20 15	Aggregate	Two-step estimation: long-run, short-run ECM	Neo-Keynesian demand, potential output	Forecasts and policy simulations
2003	F. Smets, R. Wouters					Model DSGE	Simulation analyses
2005	C. Morana	Quarterly 1980–2004		Aggregate 12 EURO countries	Cointegration relationships F-VECM-X	Monetarist model	Simulation analyses
Structural model 2005	DIW Berlin, Universita Bocconi Milano C. Dreger, M. Marcellino	Quarterly 1991–2003	18 10	Aggregate 12 EURO countries	IV	Neoclassical supply, potential GDP, neo-Keynesian demand	Forecasts and policy simulation

[a]In the European System of Central Banks

10.2.4 The MORANA Model

Between 2005 and 2006, C. Morana constructed a small model covering the 12 EMU countries (Morana 2005). He aimed to provide the organization with an analytical tool for studying the sources and impacts of inflationary processes, as well as the effects of ECB's stabilization policy in this area. The model was structured to stress the monetary phenomena according to I. Fisher's monetarist theories. It was geared to seek long-term trends in the inflationary processes. The econometric techniques used in the model referred to cointegration theory, by applying the F-VECM-X method. The model was used for conducting simulation analyses of inflation stabilization processes.

10.2.5 A Structural Model for the EURO Economy

From 2004 to 2005 Ch. Dreger and M. Marcellino were working on a new quarterly model for the euro-area countries (Dreger and Marcellino 2007). They produced a small, structural model, where the EMU countries were treated as a single economic body pursuing common monetary policy, but not fiscal policy. This approach determined the model's structure. It was neo-Keynesian in the short run and neoclassical in the long run. Potential output and demand for production factors were determined from the Cobb-Douglas production function. Demand for labour was generated allowing for the NAIRU unemployment rate.

The model was a system of simultaneous equations with parameters estimated by means of the instrumental variables (IV) method. Because most variables were non-stationary, ECM was used. The model builders estimated the long-run and short-run parameters using the rather rarely used one-step method, which they believed was more resistant than the two-step procedure.

The model was used in preparing forecasts (for 2–4 quarters ahead) and policy simulations. They investigated the impacts of the following shocks: declining GDP growth rate in the USA, rising nominal interest rate in the euro area and the euro appreciation vis-à-vis the US dollar.

References

Beffy, P.-O., Bonnet, X., Paries, M. D., & Montfort, B. (2003). *MZE: a small macro-model for the Euro area* (Document de travail, G 2003/11). Paris: INSEE.

Bognai, A., & Carlucci, F. (2003). An aggregate model for the European Union. *Economic Modelling*, *20*, 623–649.

Bradley, J., & Fitz Gerald, J. (1991). The ESR I medium-term economic model. In J. Bradley, J. Fitz Gerald & D. McCoy (Eds.), *Medium-term review 1991–1996*. Dublin: The Economic and Social Research Institute.

Bradley, J., Modesto, L., & Sosvilla-Rivero, S. (1995). HERMIN, a macroeconometric modelling framework for the EU periphery. *Economic Modelling*, *12*(3), 221–247.

Brillet, J. L., et al. (1999). *The MacSim project. General economic description*. Paris: The MacSim team, INSEE, Conseil d'Analyse Economique.
Buscher, H. S., Buslei, H., Göggelmann, K., Koschel, H., Schmidt, T. F. N., Steiner, V., & Winker, P. (2001). Empirical macro-models under test. A comparative simulation study of the employment effects of a revenue neutral cut in social security contribution. *Economic Modelling*, *18*, 455–474.
Capros, P., Georgakopulos, P., Van Regemorter, D., Proost, S., Schmidt, T., & Conrad, K. (1997). The GEM-E3 general equilibrium model for the European Union. *Economic Financial Models*, *4*, 51–160 (spec. double issue).
Coenen, G., & Wieland, V. (2005). A small estimated Euro area model with rational expectations and nominal rigidities. *European Economic Review*, *49*, 1081–1104.
de Bondt, G. J., van Els, P. J. A., & Stokman, A. C. J. (1997). *EUROMON: a macroeconometric multi-country model for the EU* (DNB Staft Reports No. 17). Amsterdam: De Nederlandsche Bank.
Dieppe, A., & Henry, J. (2004). The Euro area viewed as a single economy: how does it respond to shocks? *Economic Modelling*, *21*, 833–875.
Douven, R. C., & Plasmans, J. E. J. (1986). SLIM: a small linear interdependent model of eight EU-member states, the USA and Japan. *Economic Modelling*, *13*, 185–233.
Dramais, A. (1986). COMPACT—a prototype macroeconomic model of the European community in the world economy. *European Economy*, *27*.
Dreger, C. (2002). *A macroeconometric model for the Euro economy*. Halle: Institute of Economic Research.
Dreger, C., & Marcellino, M. (2007). A macroeconometric model for the Euro economy. *Journal of Policy Modelling*, *29*, 1–13.
Estrada, A., & Willman, A. (2002). *The Spanish block of the ESCB multi-country model* (ECB Working Paper No. 149). Frankfurt.
Fagan, G., Henry, J., & Mestre, R. (2005). An area-wide model for the Euro-area. *Economic Modelling*, *22*, 39–59.
Fase, M. M. G., Kramer, P., & Boeschoten, W. C. (1992). MORKMON II. The Nederlandsche Bank's quarterly model of the Netherlands economy. *Economic Modelling*, *9*, 146–204.
Henry, J. (1999). Euro area-wide and country modelling at the start of the EMU. *Economic and Financial Modelling*, *6*, 103–148.
Issing, O. (2004). The role of macroeconomic projections within the monetary policy strategy of the ECB. *Economic Modelling*, *21*, 723–739.
Jeanfils, P. (2000). *A model with explicit expectations for Belgium* (Working Paper No. 4). Banque Nationale de Belgique.
Mc Guire, M., & Ryan, M. (2000). *Macroeconomic modelling developments the Central Bank* (Spring Bulletin). Dublin: Central Bank of Ireland.
Morana, C. (2005). A small scale macroeconometric model for the Euro-12 area. *Economic Modelling*, *23*, 391–426.
Rae, D., & Turner, D. (2001). *A small global forecasting model* (OECD Economics Department Working Papers No. 286). OECD Publishing.
Smets, F., & Wouters, R. (2003). An estimated dynamic stochastic general equilibrium model of the EURO area. *Journal of the European Economic Association*, *1*(5), 1123–1175.
Waelbroeck, J., & Dramais, A. (1974). DESMOS: a model for the coordination of economic policies in the EEC countries. In A. Ando, R. Herring & R. Marston (Eds.), *International aspects of stabilization policies*. Boston: Federal Reserve Bank of Boston.
Wallis, K. F. (2004). Comparing empirical models of the Euro economy. *Economic Modelling*, *21*, 735–758.

Chapter 11
The Macroeconometric Models of the Rest of the World Countries

11.1 Introduction

The models of the countries lying outside North America and Europe will be presented in a somewhat condensed form, the main reason being the difficulty of acquiring information on the development of macromodelling activities in those countries in the last 20 years. The details of the earlier developments can be found in the chapters of the monograph "A History of Macroeconometric" (Bodkin et al. 1991), whose authors discuss the course of macroeconometric modelling in Japan (K. Sato), India (K. Marwah) and the Latin American countries (A. Beltran-del-Rio). Considering that the interested reader may find there all substantial information on that period, only concise summaries will be presented in this monograph. The last two decades of model building in the countries mentioned, as well as in other Asian countries and more recently in the African countries will be described based on the available information. The crucial role of the "Economic Modelling" journal as a source of scientific information on this area has to be acknowledged.

The researchers who have particularly contributed to the development of macroeconometric activities in the above countries as both initiators and advisors were L.R. Klein and his collaborators, F.G. Adams in particular, as well as numerous Ph.D. students at the Pennsylvania University. One of their contributions brought an attempt to define the general rules for building the LDC models (Klein 1965). It had a positive influence on the construction of macromodels for Developing Countries by international centres and/or local research institutions (UNCTAD, Wharton Associates), which were frequently presented at the international Project LINK conferences. The attempts to establish the macromodel building rules for Developing Countries were continued at the IMF (Haque et al. 1990).

The development of macroeconometric activities in the rest of the world countries will be presented geographically—the presentation will start with Far East Asia to take the reader through Asia and Africa to Latin America. It seems natural to begin with Japan where modelling activities developed most intensively, seriously influencing the course of modelling in the other Far East Asian countries. The next to be outlined are the modelling activities relative to planning processes specific to

W. Welfe, *Macroeconometric Models*,
Advanced Studies in Theoretical and Applied Econometrics 47,
DOI 10.1007/978-3-642-34468-8_11,

China and different for India. The early models of the Arab countries will only be sketched out, unlike the more recent modelling activities in the African countries that will be presented more at length. The presentation will close with short information on the models for Latin American countries. The summary characteristic can be found in Table 11.1.

11.2 The Macroeconometric Models of the Japanese National Economy

11.2.1 The Early Models

In Japan, macroeconometric modelling started some 10 years later than in North America, but over the years it developed fast. The first models of the Japanese national economy were built at the research institutes in Tokyo and Osaka in the late 1950s.

At the Tokyo Centre for Economic Research (TCER) T. Uchida and T. Watanabe constructed a small, annual macromodel in 1959. It had several mutations, including a quarterly version. The TCER V version included a public finance submodel (Mori 1966).

In the same year, a special team of experts was formed at the Institute of Social and Economic Research, Osaka University (ISER). Its members were the well-known econometricians S. Ichimura, S. Koizumi, K. Sato, and Y. Shinkai having the support of L.R. Klein as an adviser. Their research led to the construction of a medium-size model containing 211 equations, 102 of which were stochastic, which covered all types of economic activity, including an extended sector for financial flows. However, the limited computational possibilities at that time prevented the model from becoming operational (Ichimura et al. 1964). A small model covering the years 1930–1936 and 1951–1958 was built instead. Its constants were modified to account for the structural changes (Klein and Shinkai 1963). Additionally, L.R. Klein constructed a small, long-term model, which was based on 5-years observations and covered the periods from 1878–1882 to 1933–1937 (Klein 1961). The next long-term model referring to the Klein-Shinkai model was built by H. Ueno. It spanned the periods 1920–1936 and 1952–1958 and its main purpose was to analyse structural changes in the manufacturing industry (Ueno 1963).

11.2.2 The Economic Planning Agency Models

Further development of macroeconometric modelling in Japan was facilitated by the successful application of macromodels to the planning of the Japanese economy. The central role was played by the Economic Planning Agency (EPA), whose Economic Research Institute established in 1958 was involved in the research on using macromodels for constructing the 7 plan for the years 1964–1968. A special committee was formed to analyse the structure and use of econometric models, which participated in the construction of a system of models. The system consisted of a long-term growth model and a medium-term 60-equation model based on half-year observations that drew on the TCER models (Tatemoto et al. 1967). In 1977, also based on half-year observations, a medium-term multisectoral model was built. It was linked to a large Input-Output model and was used to prepare medium-term, disaggregated economic plans.

In the mid-60s, a quarterly, medium-size model was built for the short-term forecasting purposes (Shishido et al. 1968). It was replaced by a quarterly model with more than 200 equations being a component of the EPA world model constructed in the years 1979–1981 (Sato 1991). All the above models were updated on a regular basis. They were first used to prepare the medium-term plans and then to make regular forecasts of the Japanese economy (Watanabe 1970).

Almost at the same time, macromodelling activities were developing at the Japan Economic Research Centre (JERC) linked with the daily "Japan Economic Journal". They resulted in the construction of a medium-term annual model used for producing 5-year forecasts (Sato 1981).

11.2.3 The Bank of Japan Quarterly Models

In the 1970s, the Bank of Japan established a short-term analytical framework by constructing a medium-size quarterly model with a fairly extended sector for financial flows. The subsequent quarterly model had over 170 equations. It had a Keynesian orientation. It was used for preparing short-term forecasts and simulations facilitating the analyses of the impacts of the monetary policy pursued by the Bank of Japan. It is worth adding that the Ministry of Finance used small quarterly models to monitor structural changes.

11.2.4 The Macroeconometric Models of Academic Institutions

In the second half of the 1970s macroeconomic modelling developed also in academic institutions. A small, annual model of the Japanese economy was distinguished in the system of models, which entered into the Tsukuba-Fais model of the world economy. It was used to prepare medium-term annual forecasts of the development of Japan's national economy (Shishido et al. 1980).

Another major stepping stone in the development of short-term macromodelling in Japan was the quarterly macromodel constructed at the Kyoto University in 1975 (Amano et al. 1976). It was systematically updated and enlarged (being added the financial and public sectors) under the leadership of C. Morigushi. It entered the Project LINK world model. The model remained a forecasting and policy simulations tool for many years.

The quarterly model of Japanese economy built at the Keio University was intended to support analyses of the interlinks between the real and financial sectors. It was used for both forecasting and policy simulation purposes (Hamada 1978). Another quarterly model that K. Mori built at the university endogenized the variables representing financial markets and financial flows.

11.2.5 The Macroeconometric Models of Commercial Institutions

The commercial institutions, especially large corporations, became involved in macroeconometric modelling intending to meet their special needs. Worth mentioning are the models built at Daiwa Sec. (Daiwa's quarterly model), Denken (a small annual model), Mitsubishi (MRI's annual model), Nomura Sec. (NRI's quarterly, medium-size model) and a commercial quarterly model NEEDS (Sato 1991).

The models predominantly had a Keynesian structure, with final demand determining output and employment. The supply sector was treated superficially. Prices mainly depended on labour costs and imports prices. The models switched to a neoclassical or monetarist approach with a considerable delay.

At the turn of the 1970s the first monetarist models were built in Japan: the EPA-Shimpo was a small model with 12 equations, while the ELSA model had only 7. Nominal national income in these models was determined by money supply and prices by demand pressures and inflationary expectations (Sato 1991).

In the next years, major research centres, both governmental and academic, vigorously continued macroeconometric modelling activities. In 1978, a macroeconometric annual model DEMIOS of the Japanese economy was built at the Economic Research Institute of North-East Asia (ERINA) in Niigata. It was a major component of the system of models for 8 north-eastern Asian countries NAMIOS described below.

DEMIOS was a large model with 81 sectors. It included detailed fiscal and monetary sectors. The total number of equations amounted to 4000. It represented a Leontieff-Keynesian framework. DEMIOS was expanded and updated from the very beginning. It was used in many policy simulations for Japan and the system represented by NAMIOS (Shishido et al. 2007).

Let us also mention the Economic and Social Research Institute (ESRI) where the Short-Run Macroeconometric Model of the Japanese Economy was built (ESRI 2008).

11.3 Macroeconometric Models for the Far East Asian Countries

11.3.1 The Systems of Models for Far East Asian Countries

In the late 1970s many research institutes in Japan made attempts to build systems of models for countries in the Far East, linked by commodity flows in international trade.

As a result, an Asian Link System was constructed at the Kyoto University. It included 10 ASEAN members (Indonesia, Malaysia, the Philippines, Singapore, and Thailand), Hong Kong, South Korea, Taiwan Province, as well as Japan, USA and the rest of the world. The countries were linked by an international trade matrix (Ichimura and Ezaki 1985).

The ELSA system built at the Institute of Developing Economies in Tokyo was somewhat similar to the Asian Link System in that it used the same set of countries. The difference was, though, that a monetarist orientation was built into the specifications of the Hong Kong, Japan and USA models (Institute for Developing Economies 1985).

In the late 1990s, the Keynesian-oriented model NAMIOS was built for seven north-eastern Asian countries. The model included I-O structures, which were interlinked via the international trade matrices (Shishido et al. 1999). The model was extended under the auspices of the National Institute of Research Advancement (NIRA) in Tokyo (Shishido et al. 2007).

Quite recently, in the years 2005–2006, T. Ozaki from the Kyoto University built an annual East Asian Link Model for China, Japan, S. Korea, and USA (2006), which included a bilateral trade linkage submodel. The model was designed with a view to evaluating the impacts of the recent fiscal stimulus packages (Ozaki 2006). The model had forward looking variables and its specification followed the neo-Keynesian approach (Ozaki 2011).

In the 1980s and afterwards, local research centres run by research institutes or central banks joined in the building of the macroeconometric models of particular countries. In many instances, their decisions to do so were inspired by their participation in the Project LINK conferences organized in their countries (Al-Din 1982). For the limited access to sources offering detailed descriptions of the macromodels further exposition will be confined to the models for Indonesia, Malaysia, the Philippines, South Korea and the Province Taiwan.

11.3.2 The Macroeconometric Models of the Indonesian National Economy

The first models of Indonesian economy were constructed at the National Planning Agency in the early 1980s, as part of a project aimed to build a system of models for the ESCAP countries. They were built into the Asian Link System of econometric models (Kobayashi et al. 1985). The Central Planning Agency and the Bank of Indonesia made a common decision to run their macromodelling activities jointly, as a result of which a short-term model for Indonesia was constructed in the mid-80s (Bank of Indonesia 1986).

In the early 1990s, a new, annual macroeconometric model of Indonesian economy was constructed at the Bank of Indonesia. It had more than 100 equations, 15 of which were stochastic. The parameters of its equations were estimated with OLS and the data used to this end covered the years from 1971 to 1989. This model was intended to enter the Project LINK world economy system.

The model was demand determined, but stressed the role of oil extraction and exports. Household consumer demand was determined by real disposable incomes and the real interest rate. Investment demand had a mixed specification, depending on both GDP in accordance with the accelerator rule and financial restrictions (real credits). The model was provided with an extended block of equations explaining public sector. Revenues included export-dependent income from the oil and gas manufacturing industries, direct taxes paid on disposable incomes, indirect taxes being a fraction of the national income, as well as other income categories. Public expenditures decomposed into expenditures on debt service and development (depending on the available sources of funding) and residual current expenditures were assumed to be equal to public revenues.

In the block of equations explaining foreign trade, crude oil and gas exports were distinguished and treated as a residual representation (potential supply). Given exogenous output, the domestic demand for oil and gas was met first and then deducted from output. The export of the other commodities was demand determined, allowing for changes in the relations between the domestic prices and the world and competitors' prices. Imports equations were decomposed into 4 SITC groups. The equations represented the ratios between imports and domestic non-agricultural output that depended on relative prices.

The model had a monetary sector with equations explaining money demand and credit supply. The model closed with a price equation, where the imports deflator was the only explanatory variable (Soekarni 1992).

A detailed description of a similarly structured annual model for Malaysia can be found in Semudram et al. (1989).

11.3.3 The Macroeconometric Models of the Philippine National Economy

In the 1990s, the Philippine government agencies built several, small-size, standard macromodels (Yap 2002). Quite recently, the Asian Development Bank (ADB) had a new quarterly model of the Philippine economy constructed, intended as a tool for preparing regular forecasts and scenario analyses (Cagas et al. 2006). The model used the 1990–2004 quarterly data and had 65 equations, 48 of which are stochastic. It distinguished the long-run relationships, mainly those generating supply according to the neoclassical assumptions, and the short-run relationships that described market mechanisms generating final demand, output and employment. Some of its equations had mixed specifications. The parameters of the equations were estimated with ECM. The use of dummies in the model was restricted to the isolation of the impacts of shocks.

The model distinguished three sectors: agriculture, manufacturing industries and others. Consumer demand was determined from a simplified demand function, where distributed national income net of tax and personal wealth were used as the explanatory variables. Investment demand depended on GDP and in the short-run on the interest rate and risks. The financial sector received much attention, especially the growing debt of Philippine economy.

In the model, the demand for imported commodities depended on final domestic demand, exports and relative prices. Exports were determined by the world imports of the Philippine-originating commodities.

Production was divided into sectors. Agricultural production depended on weather conditions and other sectors' demand. The production of the industrial and services sectors was determined by GDP in the short-run, but in the long-run it was generated from the homogeneous production function. Employment depended on the output of all three sectors, real wages and agriculture's share in GDP.

The model had equations explaining prices in all GDP components. The GDP deflator was a function of the components' deflators. In particular, CPI depended in the long-run on the value added deflators in industry and services and on the money supply-GDP ratio. In the short-run, it was also affected by the prices of imports. The model closed with a block of the money-market equations that contained the money demand and interest rate equations.

Having been successfully validated, the model is used now at the ADB for preparing forecasts and policy analyses.

11.3.4 The Macroeconometric Models of the South Korean National Economy

The Korea Development Institute has been involved in macroeconometric modelling for many years. In 1979, a quarterly model of the South Korean economy was constructed at the Institute. The model had many versions and was systematically used in economic forecasting and policy simulations. Its 1995 version had more than 100 equations, 45 of which were stochastic.

The model had a Keynesian orientation. GDP was demand determined. Personal consumption depended on real disposable income, real personal wealth and real interest rate. Investments were dependent on demand represented by consumption plus exports and user costs represented by the real interest rate, and on the available funding. The model had an extended supply sector. Potential output was generated from the Cobb-Douglas production function. Compared with effective output, it yielded an estimate of the capacity utilization rate that was used to explain the wholesale price fluctuations. Much space was given to a detailed specification of the equations explaining financial flows and the balance of payment components (Shim and Hong 1995).

11.3.5 The Macroeconometric Models of the Economy of Taiwan Province

A macroeconometric model for Taiwan Province was built at the beginning of the 1990s. As the model laid stress on the links between Taiwan's economy and the rest of the world, the equations in the foreign trade sector were specified in detail (Lo et al. 1992).

In the mid-90s, another model of the Taiwanese economy was constructed at the Chung-Hua Institute for Economic Research in Taipei. It was a quarterly, medium-size and demand-oriented model with 31 stochastic equations (Yu 1995). In the model, GDP was decomposed into three sections: agriculture, industry and services. Agricultural production depended on the demand for foodstuffs, allowing for inertia. Industrial output was dependent on domestic demand and exports. Production of services was treated as residual. The model had a production function to generate potential output. Except for fixed capital and employment—imports were included as an additional explanatory variable.

Consumer demand broken down into demand for foodstuffs and for other consumer goods was assumed to depend on real disposable income and personal financial wealth, allowing for inertia. Investment demand in the private sector followed the accelerator rule and depended on industrial output and real interest rate, with inertia being allowed for in this case too.

The paramount role of exports in the Taiwanese economy was addressed by a careful specification of the exports equations. Foreign demand for Taiwanese commodities was decomposed among major importing countries. Export to these countries was explained using their GDPs, relative prices and exchange rates, allowing for inertia. Taiwanese imports depended on the country's industrial output and the imports deflator, assuming respective inertia.

Labour supply was explained in the model as dependent on the population size and independent of the rate of unemployment being a function of industrial output changes.

The equation explaining nominal wages in industry was given a standard structure. The wholesale prices were determined by imports prices and labour costs. CPI depended on wholesale prices and a lagged money supply to GDP ratio. Exports prices depended mainly on imports prices.

The model had a block of equations that explained the money market components (money demand) and the public sector (budget tax revenues and others). The model was used in many simulation analyses dealing with the monetary and fiscal policy impacts.

A new annual macroeconometric model of the Thai national economy following its predecessor was constructed in the mid-80s. It was intended for the Project LINK world economy system (Arya 1986).

11.4 The Macroeconometric Models of the Australian and New Zealand Economies

In Australia, the computational general equilibrium (CGE) model ORANI that generated a detailed description of the country's national economy was to become the most renowned. Systematically updated and extended, the model helped produce many policy simulations. Its results influenced the development of the CGE models in many other countries (Dixon et al. 1984, 1986).

In New Zealand, macroeconometric modelling developed at the Reserve Bank of New Zealand as early as the 1960. The models constructed at the Bank were to serve analytical purposes, such as analysis of the monetary policy impacts and of the transmission of the Bank' decisions on the national economy. The models had standard structures and were built in many variants. The parameters of the XI version were estimated with non-linear LS and the number of the explanatory variables was reduced using the principal components method. The XII version was special in that it used the cointegration procedures (for the first time regarding the medium-size models): the parameters were estimated with the two-step Engle-Granger method. In the first step the parameters of the long-term relationships were estimated with OLS and in the second step ECM was applied to estimate the parameters of the short-run relationship. A broad spectrum of tests was applied (Brooks and Gibbs 1994).

The XII model had 105 equations, 43 of which were stochastic. The data were quarterly and covered the years 1965–1987. It was basically a one-sectoral model. However, consumption and foreign trade were disaggregated. Labour supply was also decomposed, which positively distinguished the model from similar quarterly models.

The model had a mixed orientation. Consumer demand was decomposed into demand for durables, non-durables and services. Demand depended on households' real disposable incomes as well as on their financial and real wealth. The impact of the interest rates occurred to be insignificant. The investment function was derived from dynamic optimization: investments depended on the marginal productivity of fixed capital, the interest rate and the capacity utilization rate. Inventory changes were residual, obtained by deducting production from final demand.

The long-run value added was generated from the CES production function with constant returns to scale. Time worked was used instead of employment and the level of fixed capital was modified by taking account of the potential utilization rate. The rate was determined based on the ratio between inventories and the total sales volume. Output determined time worked (paid) that also was dependent on real wages. The supply of the male and female labour force were explained separately; their shares in total labour supply depended on real wages (negatively), unemployment rates and social contributions.

The equations explaining foreign trade were extended. In particular, exports were linked with the world demand, commodity supply, and the relative exports prices. Imports were decomposed and made dependent in the long-run on the domestic demand and relative prices. The model was additionally provided with equations explaining changes in the particular balance of payments components and exchange rates being determined i.a. by the relationships between the Australian and New Zealand short-term interest rates.

The system of prices was based on producer prices in the long run. The equation explaining producer prices was built according to the 'cost plus a mark-up' principle. The costs consisted of unit labour costs, the exchange rate and world prices, assuming homogeneity restriction. The other prices were linked with producer prices. The impact of the demand gap occurred to be insignificant. Real wages depended on labour productivity, the active-to-total population ratio and on taxes.

The model had an extended block of equations explaining financial flows. The institutional sectors' revenues and expenditures were generated. As regards the money market, the equations were specified to explain money demand, credits for the private sector and interest rates determined by the exogenous short-term interest rate being the major instrument of monetary policy (Brooks and Gibbs 1994).

The XII model was regularly used in preparing short-term forecasts and numerous policy simulations.

11.5 The Macroeconometric Models of the Chinese National Economy

11.5.1 The Early Models

The development of macroeconometric modelling in China should be attributed to the activities of the Economic Forecasting Centre of the State Planning Committee, which was founded in the mid-1980s. The Centre participated in the construction of the 1981 I-O table that became part of national accounts in the MPS system. It supported the building of a multisectoral, medium-term, annual model of the Chinese economy (CMEM). The model had over 900 equations, 254 of which were stochastic. It was a supply-determined model containing a dynamic I-O submodel (Tong 1986). The model was involved in the preparation of the 7th 5-year plan covering the years 1986–1990. In more than 10 regions the satellite regional models were built.

At the other end, there is a small model of the Chinese economy whose construction was undertaken at the end of the 1980s. It explained GDP growth in 5 sections using the linear production functions based on the 1952–1984 annual data (Yu 1990).

11.5.2 The Models of the State Economic Information Centre

The models that were successively constructed in the 1990s to describe the properties of a centrally planned economy took into account the expanding elements of market relations. The first models of this type were built at the State Economic Information Centre in Beijing to support planning processes and particularly their implementation.

The annual model constructed in the mid-90s had 118 equations, 68 of which were stochastic. It was definitely supply oriented. GDP was obtained by transforming the gross output of 7 sections of the national economy. The sections' gross output was generated from the two-factor Cobb-Douglas production functions adjusted for the rate of factor utilization in order to take account of demand fluctuations. Nominal GDP was obtained as a sum of nominal consumption, investments, inventory increase and net exports. Consequently, the GDP deflator was determined from an identity.

In disaggregating the model, stress was laid on agricultural output and rural population engaged in agricultural production. This influenced the way the consumer demand function was specified. The specification of the investment function assumed that bank credits and other funding allocated by governmental institutions played the central role, thus accepting the government as the main factor stimulating the expansion of China's economic potential.

The commodity flows in the block of foreign trade equations were decomposed into 4 SITC groups assumed in Project LINK. The equations explaining the export of agricultural raw materials and oil represented supply, while equations explaining the export of manufactured goods showed the demand for the China-originating goods. The model was additionally provided with blocks of equations explaining financial flows, including money demand and price changes (Liang 1994).

The model's subsequent versions were incorporated into the world economy system Project LINK (Zhu and Liang 1999; Jiapei et al. 1995; Wu and Zhang 1995). The model was systematically updated and modified (Shen 1999, compare also Liang 2000).

11.5.3 The Demand-Oriented Models

The orientation of the Chinese models has evolved in the recent years following the development of a market economy in the country—they have become demand determined (Wang et al. 1999; Klein and Ichimura 2000). A relevant illustration of this tendency is the multiregional model of the Chinese economy that have received an extended specification of the final demand equations, while the supply sector was ignored (Gu and Chen 2005).

11.6 Macroeconometric Models of India, Pakistan and Sri Lanka

The evolution of macroeconometric modelling of the Indian national economy will be presented based on the works of Desai (1973) and Marwah (1991), who analysed this activity until the end of the 1980s.

Of essential importance for the early model building activities in India were the mathematical planning models (Mahalanobis 1953), which remained significant into the 1960s (Manne and Bergsman 1966).

11.6.1 The Early Macroeconometric Models for India

The first macroeconometric models of the country's economy were constructed outside India. They were the components of doctoral dissertations written mainly by authors studying at the University of Pennsylvania. The dissertation that N.V.A. Narashimanhan wrote in 1956 under the guidance of J. Tinbergen presented the first model of the Indian economy, certainly the first ever model of a developing country. This demand-driven system had 18 equations, 11 of which were stochastic, and the estimation procedure used the OLS method. The model data covered the years 1923–1948. This achievement went unnoticed for many years. New dissertations were prepared only in the years 1962–1963. These were N.K. Choudry's "An Econometric Model of India 1930–1955", K. Krishnamurty's "An Econometric Model of India 1948–1961" and K. Marwah's "An Econometric Model of Price Behaviour in India". All the models were demand oriented, but their sectors dealing with agriculture stressed the supply elements. The models were mainly used to perform structural analyses (Desai 1973).

11.6.2 The Macroeconometric Models of the Indian Economy in the Late 1960s and 1970s

At the end of the 1960s, K. Marwah presented her model 2. It had 48 equations, 39 of which were stochastic. The parameters were estimated with the annual 1939–1965 data, using OLS and TSLS. The model generated total supply from the production function that used fixed capital adjusted for the rate of its utilisation. Total demand was obtained by adding up domestic demand and net exports. The difference between total supply and demand determined inventory changes that affected the rate of capacity utilization and thus output in the next period. The model was disaggregated: output and prices were determined for 4 sections and in the foreign trade sector 7 SITC commodity groups were distinguished. The model was frequently used to perform simulation analyses, such as the analyses of the impacts of rupiah devaluation (Marwah 1969).

The model R. Agarwala presented in 1970 was rather special. It had 24 equations and it was mainly intended to explain the dynamics of agricultural and industrial production. The Cobb-Douglas production function made agricultural output depend on fixed capital and rainfalls, while industrial output was linked to fixed capital and employment. No equations were specified for final demand except those explaining investment demand that was used to determine fixed capital (Agarwala 1970).

Worth mentioning is the UNCTAD's model, which was built for the system of small models of developing countries. Being designed as tools for identifying the countries' financial needs, the models were structured in a special way and had a special meaning (United Nations 1968b). The Indian economy model had 32 equations, 17 of which were stochastic, and used the 1950/1951–1962/1963 data. The main role was given to the production function that was used to generate investment demand. This demand was contrasted with the total volume of savings—the difference was assumed to indicate the degree to which investment needs were fulfilled. On the other hand, the difference between export and import estimates was taken to indicate the level of necessary funding from abroad. This approach corresponded to Chenery's concept of two gaps (Chenery and Strout 1966).

At the end of the 1970s, P.K. Pani built a new annual macroeconometric model of the Indian economy. The model had 80 equations, 55 of which were stochastic. It was estimated with the 1950/1951–1969/1970 data, using the OLS and TSLS methods. Its structure did not significantly differ from its predecessors'. In the supply sector agriculture and the other (market) sections were distinguished. The final demand sector included consumption, investment and public institutions' expenditures. Investment demand was mainly determined by domestic and foreign savings. The model had extended blocks of equations that explained prices and the money market; it was the first model with a specified equation explaining money demand (Pani 1977).

11.6.3 Later Macroeconometric Models of India

In the 1980s, both academic and governmental research centres constructed a large number of new macromodels. Special credit for its contributions to macroeconometric modelling should be given to the Institute of Economic Growth in Delhi. The annual model constructed by B.B. Bhattacharya was a medium-size model with 121 equations, 55 of which were stochastic. The parameters of its equations were estimated with data covering 25 years (1951/1952–1975/1976), using the OLS procedure. The model was primarily developed to support studies into the public sector's impacts on the Indian economy. Accordingly, the block of the public sector and money market equations was substantially extended. Output and employment, as well as prices and wages were disaggregated. Agricultural commodity prices were affected by the relations between demand and supply, while the prices of the other goods depended on the unit costs and a mark-up. The equations explaining final demand and foreign trade had a standard specification. The model was mainly used to support policy simulations (Bhattacharya 1984). In 1984 results of research on inflation and growth were published (Krishnamurty et al. 1984).

In 1985, the Institute published a monograph including the descriptions of two similar, annual macroeconometric models of India's economy (Krishnamurty and

Table 11.1 Models of the rest of the world countries

Country Model Version Year	Institution Author	Frequency Sample	Equations total stochastic	Decomposition	Estimation method	Properties	Applications
JAPAN							
Medium term model 1960	EPA M. Tatemoto et al.	Half-annual	60	–	OLS	Continuation of TCER models	Forecasts, simulations
Multisectoral model 1977		Half-annual		–	OLS	Links with I-O model	Forecasts, policy simulations
Quarterly model	S. Shishido et al.						
1966		Quarterly		–	OLS	Demand determined	Short-term forecasts
1980		Quarterly	Above 200	–	OLS	Component of EPA world economy model	Forecasts for planning development, forecasts
BOJ	Bank of Japan	Quarterly	Above 170	–	OLS	Demand orientation, financial sector	Forecasts for the Bank of Japan
KYO	Kyoto University C. Morigushi	Quarterly		–	OLS	Extended public sector and financial sector	Forecasts, policy simulations
KEIO 1978	Keio University F. Hamada	Quarterly		–	OLS	Relations of real and financial sector stressed	Simulation analyses

Table 11.1 (Continued)

Country Model Version Year	Institution Author	Frequency Sample	Equations total stochastic	Decomposition	Estimation method	Properties	Applications
FAR EAST ASIA							
PHILIPPINES							
Model ADB 2005	Asian Development Bank M.A. Cagas et al.	Quarterly 1990–2004	65 45	Decomposition: 3 sectors	Equations long-term and short-term ECM	Neoclassical long-term relationships in production sector, mixed specification of short-term relationships	Forecasts, simulation analyses
INDONESIA							
Bank of Indonesia model 1990	Indonesia Bank M. Soekarni	Annual 1971–1989	100 15		OLS	Demand determined, extended public and monetary sector	Forecasts and simulation analyses stressing mining, manufacturing and exports of oil
SOUTH KOREA							
Model KDI	Korea Development Institute S. Shim, I. Hong						
1979		Quarterly		–	OLS	Keynesian orientation	Short-term forecasts, simulation analyses
1995		Quarterly	Above 100 45	–		Potential output generated from Cobb-Douglas production function	Short-term forecasts, policy simulation

Table 11.1 (Continued)

Country Model Version Year	Institution Author	Frequency Sample	Equations total stochastic	Decomposition	Estimation method	Properties	Applications
TAIWAN PROVINCE							
Quarterly model 1994	Chung-Hua Institute for Economic Research, Taipei T. Yu	Quarterly	31	3 sections	OLS	Demand determined, potential output	Simulation analyses
THAILAND							
1990	G.N. Arya	Quarterly	85 23	2 sections	OLS	Demand oriented	Simulation analyses
NEW ZEALAND							
Model of Central Bank 1970	Reserve Bank	Quarterly			OLS	Demand determined	Forecasts, policy simulations
XII 1993		1965–1987	105 43		Two-stage Engle-Granger procedure	Mixed orientation, distinguished long-term equations and short-term equations using ECM	Short-term forecasts, policy simulations
CHINA							
CMEM 1985	Economic Forecasting Centre W. Tong	Annual	900 254	Sectoral disaggregation using I-O	OLS	Supply determined	Simulation analyses

Table 11.1 (Continued)

Country Model Version Year	Institution Author	Frequency Sample	Equations total stochastic	Decomposition	Estimation method	Properties	Applications
Model SEFC 1993	State Economic Information Centre Y. Liang et al.	Annual	118 68	7 sections	OLS	Supply determined	Supporting planning process, simulation analyses
INDIA							
Model 2 1968	K. Marwah	Annual	48 39	4 sections	OLS TSLS	Mixed orientation, disequilibrium model, inventory change residual	Simulation analyses
Model UNCTAD 1968	UNCTAD	Annual 1950/51–1962/63	32 17			Mixed orientation	Simulation analyses
Model PANI 1976	P.K. Pani	Annual 1950/51–1969/70	80 55	2 sections	OLS TSLS	Mixed orientation	Simulation analyses
IEG models 1983	Institute of Economic Growth, Delhi B. Bhattacharya	Annual 1951/52–1975/76	121 55	Agriculture section distinguished	OLS	Extended public sector	Simulation analyses

Table 11.1 (Continued)

Country Model Version Year	Institution Author	Frequency Sample	Equations total stochastic	Decomposition	Estimation method	Properties	Applications
PAKISTAN							
1990	Pakistan Institute of Development Economies (PIDE) S.N.H. Naqui et al.	Annual	88 45	10 sections, 5 groups in imports	OLS	Mixed orientation, production functions in sections	Forecasts, policy simulations
SRI LANKA							
Authors model	W. Rankaduwa et al.	Annual 1960–1987	60 43	5 sections	OLS	Demand determined	Simulation analyses, forecasts
MEDITERRANEAN COUNTRIES							
ALGER							
1984	IEiS UŁ H.S. Ghali	Annual 1969–1979	21		OLS	Demand oriented, oil exports stressed	Simulation analyses
EGYPT (UAR)							
1989	IEiS UŁ R.A. Oteafy et al.	Annual 1970–1987		3 sections	OLS	Demand determined, I-O used	Policy simulations supporting planning process
1991	S. El-Sheikh	Quarterly 1961–1973	28		OLS	Mixed orientation	Simulation analyses

Table 11.1 (Continued)

Country Model Version Year	Institution Author	Frequency Sample	Equations total stochastic	Decomposition	Estimation method	Properties	Applications
IRAQ							
1982	IEiS UŁ N.A.Y. Al Din	Annual 1962–1978	80 46		OLS	Demand determined, industrialization issues stressed	Simulation analyses
1985	T.H. Najim	Annual 1960–1978	86 45		OLS	Extended price system	Simulation analyses
1984	H.S. Ghali	Annual 1960–1979	23			Oil exports equations stressed	Simulation analyses
LYBIA							
1984	IEiS UŁ H.S. Ghali	Annual 1969–1979	21		OLS	Demand oriented, oil exports stressed	Simulation analyses
KUWAIT							
1990	IEiS UŁ R. Amer et al.	Annual 1970–1984	66	3 sections	OLS	Demand determined	Simulation analyses
AFRICA							
GHANA							
1968	UNCTAD	Annual	29 21		OLS	Structure aimed at discovery of gaps; submodel of cocoa exports	Simulation analyses, forecasts

Table 11.1 (Continued)

Country Model Version Year	Institution Author	Frequency Sample	Equations total stochastic	Decomposition	Estimation method	Properties	Applications
1974	J.L.S. Abbey, C.S. Clark	Annual			OLS	Equations for financial sector stressed	Simulation analyses, forecasts
1990	E.E. Ghartey, U.L.G. Rao	Annual 1960–1979	45 31	3 sections	OLS	Supply determined, production function, population endogenous	Simulation analyses, forecasts
KENYA							
1964	C.W. Howe, H. Karani	Annual	23 20		OLS	Mixed structure, agricultural sector only	Simulation analyses
1968	UNCTAD	Annual	27 13		OLS	Structure aimed at discovering gaps	Simulation analyses, forecasts
1995	J. Elliot et al.	Annual	143 83		OLS	Extended financial sector	Forecasts, simulation analyses
2001	J.W. Musila, U.L.G. Rao	Annual 1970–1995	32 20	5 sections	Two stage procedure by Engle-Granger	Equations long-term and short-term estimated ECM	Forecasts, policy simulations
MALAWI							
1997	J.W. Musila	Annual 1967–1996	37 23		Two-stage procedure by Engle-Granger	Mixed structure; long-term and short-term equations estimated using ECM	Simulation analyses

Table 11.1 (Continued)

Country Model Version Year	Institution Author	Frequency Sample	Equations total stochastic	Decomposition	Estimation method	Properties	Applications
NIGERIA							
CEAR MAC IV 1985	CEAR, University IBADAN S. Olofin et al.	Annual 1960–1979	137 76	12 sections, disaggregation of agriculture and exports	OLS	Supply determined	Forecasts, policy simulations
1990	T.W. Oshikoya	Annual	72 35	5 sections, disaggregation of foreign trade	OLS	Demand determined, I-O submodel	Forecasts, policy simulations
SOUTH AFRICA							
1994	University of Pretoria G. De Wet	Annual	150 46	2 sections, 4 groups in foreign trade	OLS	Mixed orientation	Forecasts, policy simulations
TOGO							
SIGAPE 1995	Planning Ministry	Annual 1984–1994/95		Disaggregation, using I-O	Mainly calibration	Model used in planning, similar to the models RMSM of the World Bank	Simulation analyses

Table 11.1 (Continued)

Country Model Version Year	Institution Author	Frequency Sample	Equations total stochastic	Decomposition	Estimation method	Properties	Applications
LATIN AMERICA							
MEXICO							
CIEMEX 1 1969	Mexican Center of Economic Research, University of Pennsylvania A. Beltran del Rio et al.	Annual	50		OLS	Demand determined	Forecasts, policy simulations
11 1985		Quarterly	650 80		OLS	Extended monetary sector, subsystem I-O added	Short-term forecasts, policy simulations
VENEZUELA							
1976	WEFA, Philadelphia Macroeconom-ica P. Palma-Carillon	Annual			OLS	Demand determined	Forecasts, simulation analyses (oil extraction, debt)

Pandit 1985). The first model with 77 equations covered the years 1960/1961–1979/1980, while the other one, with 58 equations, was based on a longer data sample from the years 1950/1951–1977/1978.

In both models, the dominant role was given to the agricultural sector. The Krishnamurty's model explained its productivity, while the Pandit's model used the production function with explanatory variables such as the farming land area, the scale of irrigation and weather conditions.

The models had similarly specified price equations, but differed in how they explained the non-agricultural output and foreign trade. Exports and imports were endogenized only in the Pandit's model. Compared with the previous systems, neither of the two models had well-developed blocks for the public sector and the money market. The Krishnamurty's model was considered to be mainly concentrated on the economic growth analyses, while the main aim of the Pandit's model was the short- and medium-term analyses (Marwah 1991).

11.6.4 The Macroeconometric Model of Pakistan

The macroeconometric model of the national Pakistani economy was constructed in the early 90s at the Pakistan Institute of Development Economics (PIDE) in Islamabad. It was a medium-size model with 88 equations, 45 of which were stochastic. It emphasised the supply sector broken down into 10 sections. Each section was provided with a simple production function. The output generated by the functions determined employment and investment, as well as real disposable incomes. Inventory changes were used for balancing global supply and global demand. The model included detailed foreign trade equations and equations explaining financial flows and prices on a section-by-section basis. It was mainly used in preparing regular forecasts and policy simulations (Naqui et al. 1993).

11.6.5 Macroeconometric Models of the National Economy of Sri Lanka

The macroeconometric models of Sri Lankan economy were built at the academic centres in Canada. The first of the models was built in 1983 as part of a Ph.D. dissertation written by A. Karunasen. The model consisted of 89 equations, 35 of which were stochastic, and offered many (40) policy instruments. It was a supply-determined model decomposed into three sections: agriculture, industry and services, which were subdivided into industries and major agricultural products. The model explained budget revenues and expenditures, as well as the balance of payments components (see Rankaduwa et al. 1995). At the end of the 1980s, a fully supply-determined model of Sri Lankan economy was constructed, which emphasised the descriptions of the processes underlying the production of major agricultural products and their use (Perera 1989).

In the middle of the 1990s, a new, annual model of the country's economy was constructed in Canada. Following the tradition of the mainstream models, this model was demand determined (Rankaduwa et al. 1995). It had 60 equations, 43 of which were stochastic, and the data covered the years from 1960 to 1987. The equation parameters were estimated with OLS.

The consumer demand functions were specified separately for consumption of domestic and imported commodities. For the lack of the disposable income data, the distributed national income was used as the major explanatory variable. The demand for domestic commodities additionally depended on the real interest rate and financial wealth, and the demand for imported commodities depended on exports, which constrained the availability of imports. In both demand equations the inertia were found to be significant. A mixed specification of the private sector's demand for investments was used. Following the accelerator rule, investments depended on GDP increase and competitive investments in the public sector, but also on the import of investment goods and money supply. The GDP volume was obtained by adding up the demand of the public sector.

Production was decomposed into 5 sections. Value added in the sections was calculated as in the Egyptian model, i.e. using the bridge equations (based on the I-O data) linking with the particular final demand components, the major role being played by household consumption and exports.

The block of equations explaining foreign trade was expanded fairly well. Foreign demand for the major products (tea, caoutchouc, cocoa) was distinguished. Domestic demand for imported consumer, investment and intermediate goods was dependent on the GDP components, allowing for the impacts of the relative prices.

The model had equations explaining national budget's revenues (mainly tax revenues), expenditures and surplus and the components of the money market. The model closed with the system of price equations, where the major role was played by the GDP deflator determined by the exogenous unit costs.

The model was intended to be used as a forecasting and policy simulation tool.

11.7 The Macroeconometric Models of the North African and West Asian Countries

The modelling of economic development came later to the Middle East countries than to the developing countries in the other parts of the world. Small models of particular national economies in the region were constructed following the approaches that other developing countries adopted under the UNCTAD project (United Nations 1968a).

In the late 1970s, an attempt was made at the Institute of Econometric and Statistics (IEiS UŁ) in Łódź, Poland, to construct macroeconometric models for the Middle East countries. The models for Algeria, Egypt, Iran, Iraq, Kuwait and Libya were built within the doctoral dissertations supervised by W. Welfe. An important prerequisite for the dissertations to be written was the construction of databases for the above countries including the annual information on calendar information recalculated from the budget periods based on official sources.

11.7.1 The Structure of the Models for the Oil Extracting Countries

The models constructed for the oil and gas extraction and refining countries such as Algeria, Iraq and Libya had special properties. In most of the countries, crude oil extraction was nationalized and the revenues from its export were to fund the industrialization programmes. The programmes assumed the import of machinery and equipment that needed highly qualified labour force to be installed and maintained. The problems with their implementation seriously delayed the execution of the investment programs, which resulted in unplanned increases in foreign currency reserves (petrodollars). The above processes within import-led growth had to be appropriately accounted for in macroeconometric models (Welfe 1986).

The models separately treated the sections extracting and refining crude oil and gas and the other sections. As far as 'the other sections' are concerned, several models distinguished market sections, mainly the manufacturing industry built as a result of the industrialization process, and "traditional" sections (agriculture and services). The oil exports equations represented the world demand; the OPEC restrictions were exogenous. Production potential was not restricted.

The market sections' output was determined by both domestic demand and exports. However, growth was dependent on the import of investment and intermediate goods financed from the oil export revenues transferred by the state budget. This approach involved a special feedback: increasing imports augmented production potential and output, as well as boosting the export of manufactured products that provided more funds allocable to imports.

In the traditional sections, output was mainly determined by the supply of production factors, as well as being dependent on the weather conditions. Market clearing was achieved due to appropriate price adjustments. In the case of drastic shortages, complementary imports took place.

The models describing the North African and West Asian countries included all other sectors characteristic of the developing countries. They had standard consumption, investment, and employment demand functions, as well as the equations explaining foreign trade, prices and wages. They usually missed the blocks of equations explaining financial flows (Salman 1994).

11.7.2 The Macroeconometric Models of the Iraqi National Economy

The models of the Iraqi economy belonged to the class of macromodels outlined above. In 1982, an annual model was built using the 1962–1978 data. It had 80 equations, 46 of which were stochastic. The parameters of its equations were estimated with OLS (Al Din 1982). A new model was constructed in 1985 to analyse the process of industrialization. Its data sample was longer, spanning the period 1960–1978. This model had 86 equations (45 were stochastic). The equation parameters were estimated with OLS. The wage and price equations were more detailed compared with the previous model (Najim 1985).

A dissertation completed in 1984 concentrated on comparing the role of oil exports in the economic growth of Iraq with the situation observed in Algeria and Libya. The models presented in the dissertation had equation parameters estimated using the 1960–1979 sample for Iraq and the 1969–1979 sample for the other two countries; there were 23 and 21 stochastic equations, respectively. The models were smaller compared with their predecessors, mainly at the expense of the blocks explaining the production sector (Ghali 1984). All the models were used to generate interesting policy simulations (Mahmud 1985). Unfortunately, further work on updating and using them for forecasting purposes was discontinued.

11.7.3 The Macroeconometric Model of the Kuwaiti National Economy

Because of the different form of government in Kuwait, the macroeconometric model of the country's economy was structured somewhat differently. In addition to being one of the major crude oil extracting countries, Kuwait had a centralized, quasi feudal-capitalist management system resembling those in Saudi Arabia or in the United Arab Emirates. The extraction and refining of crude oil was within the domain of the public sector. The oil export revenues were used to enhance the state budget incomes (their portion was transferred to the Future Generation Fund). Having no industrialization programs, Kuwait's expenditures were mostly allocated to the development of infrastructure (desalting of the sea water) and to the social assistance for the native Kuwaitis citizens.

The model was built with the 1970–1984 data and had 66 equations. It was principally demand determined, but the export of crude oil was constrained due to the OPEC restrictions. Because information on disposable income was not available, consumer demand was assumed to depend on divided GDP. Investment demand was chiefly generated by the government, being mainly dependent on the budget development expenditures. Output was determined by final demand, including exports. It was decomposed into the extraction and refinement of crude oil (taking account of OPEC restrictions), manufacturing industry, agriculture (mostly gardening) and services. Output determined employment demand broken down into native workers and numerous immigrants (their number was residual and varied following the business cycle).

The model's large financial sector was primarily used to explain budget revenues and expenditures. The money market was also modelled. The model had a system of price equations and an industrial wage equation.

The model was mainly used in preparing policy simulations (Amer et al. 1990).

11.7.4 The Macroeconometric Models of the Egyptian and Iranian National Economies

Macroeconometric modelling in Egypt took a path that was characteristic of macro-modelling in the developing countries in the early years of their growth. In the 1970s and 1980s, the Egyptian models were regarded as tools facilitating the preparation of the medium-term plans for the country's development. The macroeconometric model built at the end of the 1980s was intended to support planning processes in the Egyptian economy in the years 1987–1997. It had to acknowledge the special properties that Egyptian economy showed in the 1970s and 1980s, such as growing openness, developing tourism and transportation services (Suez Canal), but first of all the developing export of cotton and corn.

The model was founded on a database compiled for the years 1970–1987/88. Economic activity was decomposed into three sections: agriculture, industry and the other sections. The model was demand determined, with domestic demand and exports functioning as output determinants. Household consumer demand depended on real disposable income, and public institutions' demand was linked with GDP, with inertia being allowed for in both cases. Investment demand, separately specified for each section, depended on value added and productivity. It was also decomposed into machinery and equipment, structures, etc., linking its volume with supply represented by industrial gross output. Potential output was calculated as a sum of sections' value added. Valued added was generated from the linear production functions. Regarding the production factors, fixed capital was obtained by adding up investments. Labour productivity was introduced instead of employment; this approach was justified considering the large surplus of labour supply.

The model had a detailed block of foreign trade equations. The export equations explained the supply of commodities, while the import equations accounted for domestic demand, allowing for relative prices. Wages were explained in the model, but prices remained exogenous. The model included equations explaining national budget revenues, expenditures and surplus.

It was frequently used to run policy simulations (Oteafy et al. 1990). The macroeconometric model of the Iranian national economy was constructed in the mid-80s along similar lines (Masoleh 1985).

A somewhat later attempt to build a macroeconometric model for the industrializing countries used the Egyptian economy as an example (El-Sheikh 1992). The same author constructed then a quarterly model of the Egyptian monetary sector, being the first model of that class designed for the developing countries (El-Sheikh 1994). The model used the quarterly data from 1961 (the socialist revolution) to 1973 (prior to the period of openness—Infitah) compiled by the author. No reliable quarterly data could be found for the next years. The data were used to estimate the parameters of the model's 28 equations. Only the financial sector of the Egyptian economy was described in the model. Its equations accounted for money demand, agricultural credits from banks, seasonal credits to cotton and rice traders, and other credits. The equations explaining changes in bank reserves and the components of intrabank balances were specified too; the interest rates were exogenous. The model was used to perform policy simulations.

11.7.5 Modelling the Maghreb Countries

Macromodelling activities in the other North African countries, including the Maghreb countries, were much less advanced than in the countries discussed. Worth mentioning are the efforts to design a model capable of explaining the development of the non-industrial sections in 4 countries: Algeria, Mauretania, Morocco and Tunisia. Their growth was made dependent on industrial output and imports; alternative estimation methods were used (Guisan and Exposito 2004).

11.8 Macroeconometric Models of the African Countries

Macroeconometric modelling in the African countries started to develop in the 1970s, after most of them became independent. First were the models constructed in Nigeria at the Ibadan University and in Kenya, and these will be presented below. Most small-size macromodels of standard structure were built for the African countries at the UNCTAD (United Nations 1968a). Towards the end of the 20th century, their role was taken over by UN-DESIPA and the Centre of Project LINK (Klein 1999).

A strong impulse stimulating the development of modelling activities in particular countries came from the African Research Network for Development Policy Analysis. It was founded in 2002 and associated 15 countries. The Pretoria University where the late G. De Wet established a research centre that initiated modelling activities in Central and South Africa also played an important role.

The macroeconometric models of the following countries were available to the author: Botswana, Ghana, Kenya, Malawi, Nigeria, South-Africa, Sudan and Togo. Their structures will be discussed below.

The first to be presented is a small macroeconometric model of Botswana's economy, which was built in the mid-90s. It was special in that it stressed the importance of education and total factor productivity in economic growth (Huff 1994).

11.8.1 The Macroeconometric Models of the Ghanaian National Economy

The first was a standard, small model that the UNCTAD built within a system of models for the developing countries (United Nations 1968a). The annual model which was constructed several years later contained the descriptions of both real and financial sectors and stressed the importance of Ghanaian exports, particularly of the cocoa (Abbey and Clark 1974).

At the end of the 1980s, an annual, medium-size macromodel was constructed. This model had 45 equations, 31 of which were stochastic, and used the 1960–1979 data. The parameters of its equations were estimated using OLS. Its authors were researchers residing in Canada and USA (Ghartey and Rao 1990). It had an improved structure compared with the previous models, which contributed to higher accuracy of forecasting.

The model was supply determined, with GDP generated from a simple Cobb-Douglas production function. Agricultural and industrial outputs were derived from separate production functions; the other sections' output was residual. Industrial employment depended on GDP, but the employment in agriculture was determined by exogenous fixed capital. The model was also provided with an equation explaining labour force migration from agriculture to industry. High population growth was explained in the model through equations explaining the dynamics of births and deaths.

Consumer demand was dependent on total real disposable income; investment increase was determined by fixed capital (replacement demand) and the interest rate on credits. Monthly wages depended on labour productivity and the GDP deflator. The model had detailed equations explaining financial flows (taxes and transfers) and money markets, including money demand.

The model found application as a tool for short-term forecasting and policy simulations.

11.8.2 The Macroeconometric Models of the Kenyan National Economy

The first model of Kenyan economy was constructed in the mid-60s (Howe and Karani 1965). It was a small model with 33 equations, 20 of which were stochastic. It dealt with the real sector of the national economy. The blocks of equations that were distinguished in the model explained output, investments, foreign trade and budget revenues. Prices were treated as exogenous. The UNCTAD model for Kenya was constructed almost concurrently. It had 27 equations, 13 of which were stochastic and 7 had calibrated parameters. Some of its blocks of equations resembled the first model's blocks, but it also contained equations generating savings and population dynamics (United Nations 1968a).

The model built twenty years later was a medium size, annual macroeconometric model with 143 equations, 53 of which were stochastic. It included both real sector and financial sector, as well as a system of price equations. The model was used for forecasting and policy simulation purposes (Elliot et al. 1986).

The next annual model was constructed in the early 21st c. by J.W. Musila and U.L.G. Rao, Canadian residents (Musila and Rao 2002). This small model with 32 equations (20 were stochastic) used the 1970–1995 data sample and distinguished the long-term relationships and the short-term adjustments. The parameters of the long-term equations were estimated using the cointegration techniques, while the short-term equations were estimated with ECM.

The model included the following sections: mining (with agriculture), manufacturing and building industries, public services and others. Value added in the sections was determined as a function of the appropriate components of final demand, the most important of them being consumption and exports. The impacts of the development programs were included.

Consumer demand depended on real disposable income, the interest rate and inflationary expectations. The main investment demand factors were real incomes and real interest rates. The public sector's demand was exogenous, but budget revenues were endogenized. Exports were depended on the world's real GDP and relative prices, while imports on Kenyan GDP and relative prices.

The model missed the production function on account of data scarcity. The GDP deflator depending on unit labour costs and surplus was central to the price system. The CPI was determined by the GDP deflator, imports prices and real money supply to reflect the impacts from the monetary sector. Money demand depended on GDP and nominal interest rates.

The model was used for running policy simulations and making development forecasts of Kenyan economy (Musila and Rao 2002).

11.8.3 The Macroeconometric Model of the Malawian National Economy

The macromodels of the national economy of Malawi were developed in the 1990s, mainly outside the country. A special supply determined model accentuating the role of foodstuff exports was constructed by Y. Van Frausum and D. Sahn at the Cornell University, as a part of the University Food Nutrition Policy Program (1993).

A new, annual macroeconometric model of Malawian economy was built nearly ten years later. It was a small model with 37 equations, 23 of which were stochastic. Their parameters were estimated with the 1967–1996 data sample and using the cointegration approach. The time series were mainly I(1), while the residuals of the cointegrating vectors were I(0). The parameters of the long-term equations were estimated first and then the parameters of the short-term equations using ECM; the appropriate tests were used. The structural changes were accounted for by means of dummy variables.

The model had a mixed structure. Output represented by value added was generated using a Cobb-Douglas production function with constant returns to scale. In the short-run, output depended on employment and price levels only. Final demand was represented by consumption, investment and net exports. Consumer demand depended solely on GDP (adjusted for direct taxes); other determinants (personal wealth, interest rates) occurred to be insignificant. Private investment was explained by specifying its supply that depended on real savings and the import of investment goods. Public investment, competing against private investment, was determined by lagged output, imports of investment goods and private investment.

The model distinguished the major export items, i.e. tea and tobacco as well as other commodities and services, that depended on the world's GDP and relative prices. Imports were dependent on Malawi's GDP, relative prices and the exchange rate (being a proxy for restrictions).

In the long-run, employment demand was determined by the domestic and foreign demand for production. In the short-run, it also depended on investments and the export of tea and tobacco.

The model generated budget revenues. Budget expenditures were exogenous and surplus flowed into the monetary sector. Within the sector, credits were generated for the private sector and the equation explaining money demand was specified. In the price equation system, the GDP deflator and CPI were dependent on unit costs plus a mark-up. The unit costs included lagged labour and imports costs, being homogeneous. Exports and imports prices were dependent on the world prices. In the wage equation nominal wages depended on lagged CPI and employment-to-population ratio (being a proxy for the rate of unemployment).

The model was carefully tested for its ex-post predictive accuracy and then used in policy simulations (Musila 2002).

11.8.4 The Macroeconometric Models of the Nigerian National Economy

Macroeconometric model building has long tradition in Nigeria. The first models of the country's economy (planning models and I-O models) were built in the 1960s. They were not operational, so they were discontinued. The operational macroeconometric models for Nigeria were constructed at the Centre for Econometric and Allied Research (CEAR), University of Ibadan, under the leadership of S. Olofin. The first versions of the MAC I, II, III models were experimental.

The MAC IV version being a medium-term model was developed in the mid-80s. It was a medium-size model with 137 equations, 76 of which were stochastic, and with annual data covering the years 1960–1979. The parameters of it equations were estimated with OLS.

The model was fully supply determined. Production in its 12 sections was generated from simple production functions (time trends were used in agriculture). Value added provided a point of departure for calculating real disposable income determining household consumption. Budget expenditures were an important variable determining investment. Budget revenues were largely dependent on the inflow of the receipts for exported crude oil, cocoa and minerals. The block containing the foreign trade equations was fairly detailed. Balance was achieved at the macro-scale through residual inventory changes.

The model included an extended price system. Because of data scarcity, its wage equations were limited to manufacturing industries, likewise those explaining employment.

The model was regularly used to prepare forecasts, as well as numerous policy simulations (Olofin et al. 1985).

A new macroeconometric model of the Nigerian economy was constructed in the early 1990s. It was fully demand determined. It had 72 equations, including the bridge equations that linked the final demand components with value added in particular industries through the I-O submodel. It was also provided with an extended financial sector (Oshikoya 1990).

11.8.5 The Macroeconometric Models of the South African National Economy

In the mid-90s, G. De Wet built an annual model of South African economy at the University of Pretoria. It was a medium-size model of mixed orientation that had 150 equations (46 were stochastic) and distinguished two sections: agriculture and other. The final demand components were separately explained. Consumption was decomposed into consumption of durables and semi-durables dependent on real disposable income and long-term interest rates, and consumption of non-durable goods and services determined by money supply, taking account of the relevant lags. Investment was dependent on GDP, with appropriate lags. GDP was obtained by dividing the nominal GDP represented by total final expenditures by the GDP deflator. Non-agricultural employment was assumed to depend on GDP.

The model's deflators and money market components were thoroughly specified. Special stress was laid on the foreign trade sector. Exports and imports were decomposed according to the SITC classification, allowing for the export of gold.

The model was used in forecasting and policy simulations (De Wet 1995). Because South Africa treated then monetary policy as a vital element of her economic policy, a small model of the country's monetary sector was constructed to support the analyses of the monetary policy impacts (De Wet et al. 1994).

Let us also mention the quarterly macroeconomic model built in 2005 by M. Koster, a student at the Economics Department, University of Johannesburg. This simultaneous linear equations model had 53 equations (23 were stochastic) and used the 1990–2004 sample. The model's structure was Keynesian. It was chiefly constructed for methodological purposes. Its equations were tested for linearity. Nonlinearity was discovered in 8 equations, mainly those explaining prices. To deal with this, the smooth transition autoregressive (STAR) estimation methods were applied (Koster 2005). The model was used in forecasting.

11.8.6 The Macromodel of the Togolese National Economy

In the mid-90s the Togolese Ministry of Planning built an annual model of the national economy intended as a planning tool. The model was called SIGAPE (Systeme Informatise de Gestion Applique a la Prevision Economique). Its role was to provide the authorities with forecasts extending to the year 2000, as well as running economic policy simulations (Kponsihoin and Philip 1996). The model used a data sample covering the years 1984–1994/1995.

The model was actually a planning tool whose structure followed the World Bank's RMSM model. Its key function was to supply information on the demand for imported goods and on foreign credits necessary to finance the country's development given the goals set, particularly the (exogenous) rate of GDP growth. The model was strongly disaggregated and used an I-O submodel. Its main building blocks were identities and equations with calibrated parameters. The model had blocks of equations that generated particular industries' value added from gross output, exports and imports, as well as blocks of equations explaining the incomes and expenditures of households and public institutions. This significantly distinguished the model from the macroeconometric models of Kenya or Malawi discussed above.

The macroeconometric model of Sudanese national economy was built in the mid-70s. The model was supply determined (Marzouk 1975). The small econometric model of the country's economy which was built nearly 10 years later showed the same orientation. It was made of 29 equations (8 were stochastic) and stressed the role of foreign trade in Sudan's economic growth (Zayid 1986).

11.9 The Macroeconometric Models of the Latin American Countries

The direct reason for macroeconometric activities to develop in the countries of South and Central America and in Mexico was the modelling activities that L.R. Klein and F.G. Adams conducted at the Pennsylvania University in Philadelphia. Many Ph.D. dissertations and research projects undertaken at the Wharton Econometric Forecasting Associates (WEFA) dealt with the modelling of the Latin American countries. The details of the modelling efforts conducted in the years 1965–1985 can be found in the exhaustive essay by Beltran-del-Rio (1991) that contains, for instance, a list of 187 macromodels built for the countries in that region.

The first model was constructed for the national economy of Brazil in 1969 as part of a Ph.D. dissertation. In the next years it was introduced into the system of the WEFA macromodels.

The macromodels of other countries were built on the inspiration from L.R. Klein, becoming more or less independent exercises.

11.9.1 The Macroeconometric Models of the Mexican National Economy

The macroeconometric model of the Mexican economy is very interesting for illustrative purposes. Built at the end of the 1960s by A. Beltran-del-Rio and H. Howell, it was consistently developed at the CIEMEX (Mexican Centre of Economic Research), Pennsylvania University. By 1985 it had been built in 11 versions and the number of its equations increased over its life cycle from 50 to over 650, including over 80 stochastic equations. The model was demand determined. In the mid-70s its monetary sector was extended and an I-O submodel was added. After the fuel sector and equations explaining the country's growing debt were incorporated into the model in the next decade, new opportunities for conducting vital analyses of the economic situation in Mexico became possible. This annual model was systematically used to make the quarterly forecasts of the Mexican economy (Beltran-del-Rio 1991).

11.9.2 The Macroeconometric Models of the Venezuelan National Economy

P. Palma Carillon constructed a macroeconometric annual model of the Venezuelan economy at the WEFA in 1976. He developed it in the next years within the special research organization METROECONOMICA, which was founded in 1978. The model was demand determined. In the later years it focused on describing the impacts of oil extraction and debt-related issues in the country. It was regularly used in preparing forecasts and numerous policy simulations.

In the 1970s the WEFA members constructed the macroeconometric models also for other Latin American countries, such as Argentina, Central America, Chile, Colombia, Costa Rica, Honduras, Panama and Peru. The models were mainly used in policy simulations. Some of them were discontinued. In the same decade, the government agencies and academic institutions in some countries, e.g. Chile, Colombia, Equator, Mexico and Porto Rico, also built macromodels.

In the 1980s macroeconometric modelling kept developing, however showing some tendency towards regionalization, as exemplified by the CIEMEX model for Mexico. New macromodels were built to meet the analytical needs of large corporations. Worth mentioning in this group are the macromodels constructed for Argentina, Brazil, Peru and Uruguay at SERFINA, a research institute owned by the Sao Paulo-based, grain-trading corporation Bunge&Born. Macromodels were also built at Vitro, a glass corporation in Mexico. The forecasting models maintained by the Junta del Acuerdo de Cartagena in Chile played an important role in predicting economic developments in the Andean countries.

UNCTAD also built and maintained small macroeconometric models for the developing countries in Latin America (United Nations 1968a).

References

Abbey, J. L. S., & Clark, C. S. (1974). A macroeconometric model of the Ghanaian economy 1956–1969. *Economic Bulletin of Ghana*, *4*, 3–32.

Agarwala, R. (1970). *An econometric model of India, 1948–1961*. London: Frank Cass.

Al-Din, N. A. K. (1982). *An econometric model of the Iraq economy*. Ph.D. Dissertation, University of Łódź, Łódź.

Amano, A., Ban, K., & Morigushi, C. (1976). *A quarterly forecasting econometric model of Japan: KYQ75* (Discussion Paper No. 095). Kyoto: Kyoto Institute of Economic Research, Kyoto University.

Amer, R., Sonbaty, A., Welfe, W., & Zatoń, W. (1990). *Econometric model of the national economy of Kuwait* (IEiS UŁ Discussion Papers No. 342). Łódź.

Arya, G. N. (1986). *An econometric model for Thailand under the LINK system* (Mimeo). Bangkok.

Bank of Indonesia (1986). *Central planning. A short term macromodel for Indonesia* (Mimeo). Bank of Indonesia.

Beltran-del-Rio, A. (1991). Macroeconometric model-building of Latin American countries, 1965–1985. In R. G. Bodkin et al. (Eds.), *A history of macroeconometric model-building* (pp. 389–448). Aldershot: E. Elgar.

Bhattacharya, B. B. (1984). *Public expenditure, inflation and growth. A macroeconometric analysis*. Delhi: Oxford University Press.

Bodkin, R. G., Klein, L. R., & Marwah, K. (1991). *A history of macroeconometric model-building*. Aldershot: E. Elgar.

Brooks, R., & Gibbs, D. (1994). A model of the New Zealand economy: Reserve Bank Model XII. *Economic Modelling*, *11*, 5–86.

Cagas, M. A., Ducanes, G., Magtibay-Ramos, N., Qin, D., & Quising, P. (2006). A small macroeconometric model of the Philippine economy. *Economic Modelling*, *23*, 45–55.

Chenery, H. B., & Strout, A. M. (1966). Foreign assistance and economic development. *American Economic Review*, *56*, 679–733.

De Wet, G. L. (1995). *A model of South Africa*. Pretoria: University of Pretoria.

De Wet, G. L., Jonkergouw, E., & Koekemoer, R. (1994). An econometric monetary policy model for South Africa. In W. Welfe & R. Kelm (Eds.), *Macromodels'94* (pp. 49–96). Łódź: Absolwent.

Desai, M. J. (1973). Macro econometric models for India: a survey. *Sankhya, Series B*, *3*, 169–206.

Dixon, P. D., Parmenter, B. R., & Powell, A. A. (1984). The role of miniatures in computable general equilibrium modelling: experience from ORANI. *Economic Modelling*, *1*, 421–428.

Dixon, P. D., Parmenter, B. R., & Rimmer, R. J. (1986). A summary description of ORANI. In T. N. Srinivasan & J. Whalley (Eds.), *General equilibrium trade policy modeling* (pp. 34–37).

El-Sheikh, S. (1992). Towards a macroeconometric policy model of a semi-industrial economy: the case of Egypt. *Economic Modelling*, *9*, 75–95.

El-Sheikh, S. (1994). A quarterly econometric policy model of an LDC monetary system: the case of Egypt. In H. M. Scobie (Ed.), *Economic progress and growth* (pp. 29–63). London: Chapman & Hall.

Elliot, J., Kwack, S. Y., & Tavlas, G. S. (1986). An econometric model of the Kenyan economy. *Economic Modelling*, *3*, 2–30.

ESRI (2008). *The ESRI short-run macro econometric model of the Japanese economy: basic structure, multipliers and economic policy analyses (2008 version)*. Tokyo: Economic and Social Research Institute, Cabinet Office of Japan.

Ghali, H. S. (1984). *The impact of oil revenues on national income and its distribution in Iraq 1960–1979 (with special reference to Libya and Algeria)*. Ph.D. Dissertation, An Econometric Study, University of Łódź, Łódź.

Ghartey, E. E., & Rao, U. L. G. (1990). A short run forecasting model of Ghana. *Economic Modelling*, *7*, 291–308.

Gu, Q., & Chen, K. (2005). A multiregional model of China and its application. *Economic Modelling*, *22*, 1020–1063.

Guisan, C., & Exposito, P. (2004). *Econometric models and causality relationships between manufacturing and non-manufacturing production in Morocco, Tunisia and other Northern African countries 1950–2000* (Working Paper Series Economic Development No. 78). Econometrics. Faculty of Economics, Santiago de Compostella University.

Hamada, F. (1978). A quarterly econometric model of Japan. *Keio Economic Studies*, *15*, 1–51.

Haque, N. U., Lahiri, K., & Montiel, P. J. (1990). A macroeconometric model for developing countries. *IMF Staff Papers*, *37*, 537–579.

Howe, C. W., & Karani, H. (1965). A projection model for the Kenyan economy: a study in development planning and comparative economic structures. *The East African Economic Review*, *1*, 21–31.

Huff, L. A. (1994). Predictive status of the Botswana economy into the next century. *The Botswana Technology Centre*, *16*, 29–31.

Ichimura, S. & Ezaki, M. (Eds.) (1985). *Econometric models of Asian Link*. Berlin: Springer.

Ichimura, S., Klein, L. R., Koizumi, S., Sato, K., & Shinkai, Y. (1964). A quarterly econometric model of Japan, 1952–1959. *Osaka Economic Papers*, *12*, 19–44.

Institute for Developing Economies (1985). *Econometric Link system for ASEAN* (Vols. 1, 2). Tokyo: IDE.

Jiapei, W., Youcai, L., & Yaxiong, Z. (1995). China's macroeconometric model for Project LINK. In M. Dutta (Ed.), *Economics, econometrics and the LINK. Essays in honor of Lawrence R. Klein* (pp. 543–563). Amsterdam: North Holland-Elsevier.

Klein, L. R. (1961). A model of Japanese economic growth, 1878–1937. *Econometrica*, *29*, 277–292.

Klein, L. R. (1965). What kind of macroeconometric model for developing economies? *Econometric Annual of the India Economic Journal*, *13*, 313–324.

Klein, L. R. (1999). Project LINK. In L. R. Klein, A. Welfe & W. Welfe (Eds.), *Principles of macroeconometric modeling* (pp. 127–151). Amsterdam: North-Holland.

Klein, L. R. & Ichimura, S. (Eds.) (2000). *Econometric modeling of China*. Singapore: World Scientific.

Klein, L. R., & Shinkai, Y. (1963). An econometric model of Japan, 1930–1959. *International Economic Review*, *4*, 1–28.

Kobayashi, K., Tampubobn, E., & Ezaki, M. (1985). Indonesia model. In *Econometric models of Asian Link*. Berlin: Springer.

Koster, M. (2005). *A macroeconomic model for South Africa: a non-linear economic modelling approach* (Mimeo). University of Johannesburg.

Kponsihoin, K., & Philip, J.-M. (1996). *Macro economic model SIGAPE, computerized management system applied to economical projections* (Mimeo). Republic Togolaise: Ministry of Planning.

Krishnamurty, K., & Pandit, V. (1985). *Macroeconomic modelling of the Indian economy: studies on inflation and growth*. Delhi: Hindustan.

Krishnamurty, K., Saibaba, P., & Kazmi, N. A. (1984). Inflation and growth: a model for India. *Indian Economic Review*, *19*, 17–111.

Liang, Y. (1994). *A long term econometric model of the Chinese economy* (Mimeo). Beijing: State Economic Information Centre.

Liang, Y. (2000). China's econometric model for project PAIR. In L. R. Klein & S. Ichimura (Eds.), *Econometric modeling of China* (pp. 201–248). Singapore: World Scientific.

Lo, J. C., Li, H.-Y., & Yu, T. (1992). A trade-oriented macroeconometric model of Taiwan. *Academia Economic Papers*, *20*, 625–677 (Post II).

Mahalanobis, P. C. (1953). Some observations on the process of growth of national income. *Sankhya*, *12*, 307–312.

Mahmud, I. K. (1985). *Foreign trade sector in Iraq. An econometric model*. Ph.D. Dissertation, University of Łódź, Łódź.

Manne, A. S., & Bergsman, J. (1966). An almost consistent intertemporal model for India's fourth and fifth plans. In I. Adelman & E. Thorbecke (Eds.), *The theory and design of economic development*. Baltimore: John Hopkins Press.

Marwah, K. (1969). An econometric model of India, estimating prices, their role and sources of change. *Indian Economic Review (New Series)*, *7*, 53–91.

Marwah, K. (1991). Macroeconometric modelling of South-East Asia: the case of India. In R. G. Bodkin et al. (Eds.), *A history of macroeconometric model-building* (pp. 356–388). Aldershot: E. Elgar.

Marzouk, M. S. (1975). An econometric model of Sudan: simulation experiments of growth prospects. *Journal of Development Economics*, *1*, 337–358.

Masoleh, A. S. S. (1985). *The econometric model for the Iran economy*. Ph.D. Dissertation, University of Łódź, Łódź.

Mori, K. (1966). Simulation analysis of fluctuations and growth of the Japanese economy: 1955–1960. In R. Komiya (Ed.), *Postwar economic growth in Japan* (pp. 189–205). Berkeley: University of California Press.

Musila, J. W. (2002). An econometric model of the Malawian economy. *Economic Modelling*, *19*, 295–330.

Musila, J. W., & Rao, U. L. G. (2002). A forecasting model of the Kenyan economy. *Economic Modelling*, *19*, 801–814.

Najim, T. H. (1985). *An econometric model for manufacturing industries sector in Iraq*. Ph.D. Dissertation, University of Łódź, Łódź.

Naqui, S. N., Khan, A. H., & Ahmed, A. M. (1993). *The P-LDE. Macroeconometric model of Pakistan's economy*. Islamabad: Pakistan Institute of Development Economics (PIDE).

Olofin, S., Iyaniwura, J., Adeniyi, J., & Olayide, S. O. (1985). An operational econometric model of the Nigerian economy. *Empirical Economics*, *10*, 231–262.

Oshikoya, T. W. (1990). *The Nigerian economy, a macroeconometric and input-output model*. New York: Praeger.

Oteafy, R. A., Welfe, W., & Zatoń, W. (1990). *Raport o stanie prac nad modelem ekonometrycznym Egiptu (Report on the econometric model of Egypt)* (IEiS Uł Discussion Papers No. 291). Łódź: IEiS UŁ.

Ozaki, T. (2006). An East Asian Link model and simulation analysis. *Journal of the Faculty of Economics, Kyoto Gakuen University*, *15*(3).

Ozaki, T. (2011). Econometric evaluation of the fiscal expansion and stimulus packages in three Asian countries and the United States. *The Journal of Econometric Study of Northeast Asia (JESNA)*, *7*, 17–50.

Pani, P. K. (1977). *A macroeconomic model of the Indian economy*. Delhi: Macmillan.

Perera, N. (1989). *An econometric model of the Sri Lankan economy* (Mimeo). Australia: Dpt. of Economics, University of Wallongong.

Rankaduwa, W., Rao, U. L. G., & Ogwang, T. (1995). A forecasting model of the Sri Lankan economy. *Economic Modelling*, *12*, 343–375.

Salman, A. K. (1994). A forecasting model of the OPEC countries. In W. Welfe & R. Kelm (Eds.), *Macromodels* (pp. 99–137). Łódź: Absolwent.

Sato, K. (1981). A survey of macroeconometric forecasting models of Japan: development and current state. *Japanese Economic Studies*, *9*, 3–60.

Sato, K. (1991). Econometric models of the Japanese economy. In R. G. Bodkin et al. (Eds.), *A history of macroeconometric model-building* (pp. 311–355). Aldershot: E. Elgar.

Semudram, M., Beng, G. W., & Chew, K. (1989). *Annual econometric model of the Malaysian economy* (Mimeo).

Shen, L. (1999). Yearly macroeconometric model of China. In H. Wang, B. Li & S. Li (Eds.), *China's applied macroeconometric model*. Beijing: China Finance Press.

Shim, S., & Hong, I. (1995). *The current KDI quarterly model of the Korean economy* (Mimeo). Korea Development Institute.

Shishido, S., Kohno, A., Nagaya, S., & Tanaka, S. (1968). Use of national accounts for a short-term econometric model, 1954–1966. *Review of Income and Wealth*, *14*, 267–309.

Shishido, S., Fujiwara, H., Kohno, A., Kurosawa, Y., Matsuura, S., & Wago, H. (1980). A model for the coordination of recovery policies in the OECD region. *Journal of Policy Modeling*, *2*, 35–55.

Shishido, S., et al. (1999). A multiregional econometric model for Northeast Asia (NAMIOS I): estimation and policy analysis. *The Journal of Econometric Study of Northeast Asia (JESNA)*, *1*, 13–31.

Shishido, S., Kawokami, A., Kurokawa, M., Morshuk, A., & Tomashiro, K. (2007). Policy simulations with an integrated model for Japan and Northeast Asia. *The Journal of Econometric Study of Northeast Asia (JESNA)*, *6*, 1–28.

Soekarni, M. (1992). *An econometric model for Indonesia under the LINK system, version 1.0* (Mimeo). Bank of Indonesia, Dpt. of Economics and Statistics.

Tatemoto, M., Uchida, T., & Watanabe, T. (1967). A stabilization model for the postwar Japanese economy, 1954–1962. *International Economic Review*, *8*, 13–44.

Tong, W. (1986). *China's economic forecast and project have made a gratifying step during "Sixth Five Year Plan" period* (Mimeo). Beijing: Economic Forecasting Centre, State Planning Commission.

United Nations (1968a). *Trade prospects and capital needs of developing countries*. New York: United Nations Conference on Trade and Development (UNCTAD), United Nations.

United Nations (1968b). *Trade projections for India, trade prospects and capital needs of developing countries* (pp. 298–317). New York: United Nations Conference on Trade and Development (UNCTAD), United Nations.

Ueno, H. (1963). A long-term model of the Japanese economy 1920–1958. *International Economic Review*, *3*, 171–193.

Van Frausum, Y., & Sahn, D. (1993). *An econometric model for Malawi: measuring the effects of external shocks and policies* (CFNPP Working Paper 13). Washington: Cornell University Food Nutrition Policy Program.

Wang, H., Li, B., & Li, S. (Eds.) (1999). *China's applied macroeconometric model*. Beijing: China Finance Press (in Chinese).

Watanabe, T. (1970). National planning and economic development: a critical review of the Japanese experience. *Economic Planning*, *10*, 21–51.

Welfe, W. (1986). *Topics in macromodelling of industrializing oil producing countries*. Paper presented at the AEA world congress on econometric modelling in semi-industrialized and developing economies, Istanbul, 10–12.12.1986. Łódź: IEiS UŁ.

Wu, J., & Zhang, Y. (1995). China's macroeconometric model for Project LINK. In M. Dutta (Ed.), *Economics, econometrics and the LINK, Essays in honor of Lawrence R. Klein* (pp. 543–563). Amsterdam: North-Holland.

Yap, J. T. (2002). *A perspective on macroeconomic and economy-wide quantitative models of the Philippines: 1990–2002* (Perspective Paper, Symposium Series). Manila: Philippine Institute for Development Studies.

Yu, Q.-L. (1990). An econometric model of China's national income. *Economic Modelling*, *7*, 258–262.

Yu, T. (1995). An analysis of the effects of economic policies on Taiwan's economic growth and stability. In M. Dutta (Ed.), *Economics, econometrics and the LINK, Essays in honor of Lawrence R. Klein* (pp. 519–541). Amsterdam: North-Holland.

Zayid, E. (1986). *The foreign sector and economic development: analysis and simulations of an econometric model of the Sudan*. Ph.D. Dissertation, Indiana University.

Zhu, B., & Liang, Y. (1999). Macroeconomic model of China: Project LINK. In H. Wang, B. Li & S. Li (Eds.), *China's applied macroeconometric model*. Beijing: China Finance Press.

Chapter 12
Macroeconometric Multicountry Models

12.1 Introduction

At the end of the 1960s, the Committee for Economic Stabilization and Growth at the Social Science Research Council in New York initiated a project for constructing a system of interlinked models of the world economy's industrialized countries. The system was intended as a tool for analysing the international transmission mechanism of economic activities and for forecasting its impacts (Hickman 1991). This initiative led to the formation of Project LINK, that is a system of the world models being special in that it linked the country models constructed by experts representing particular countries. The main methodological contribution of the project led by L.R. Klein was the construction of a world trade matrix, which allowed particular countries' exports and imports to be linked together. It was assumed that the volume of exports from a country was determined by other countries' demand (imports) and that the imports prices in a country equalled the weighted sum of exports prices in the world trade (Klein 1982).

These assumptions provided a basis for constructing the multicountry world models that were developed by international organizations (IMF, World Bank, OECD, UE), central banks (USA, Germany), research institutes (NIESR, CEPI, FUGI, Fair) and commercial organizations (Global Insight).

The special property of the macroeconometric models that the institutions built was (sometimes gradual) standardization of the specification of the equations for particular countries. The specification was extensive for large industrialized countries and modest for smaller countries or regions with mostly developing countries (Ball 1973).

The models were large, as the numbers of the equations approached several thousands in the 1980s, when efficient computer programmes were developed and potent personal computers became available (Hickman 1983).

Regarding their theoretical foundations, estimation methods and simulation procedures, the models evolved in a way which was typical of the models for major

W. Welfe, *Macroeconometric Models*,
Advanced Studies in Theoretical and Applied Econometrics 47,
DOI 10.1007/978-3-642-34468-8_12, © Springer-Verlag Berlin Heidelberg 2013

industrialized countries.[1] They drew on earlier experience (Waelbroeck and Dramais 1974; Dramais 1986).

Initially, a neo-Keynesian orientation prevailed among the models, but their supply sectors were expanding along the neoclassical lines. In the 1990s rational expectations were introduced into many models. In the early 21st c. the first attempts were made to build the DSGE models. The links between particular countries were initially limited to commodity flows, but then they were extended to include the migration of production factors and interest rate differentials (to account for capital flows), as well as the impacts of exchange rates and prices (Gana et al. 1979). The models' theoretical underpinning was derived from the extended Mundell-Fleming transmission mechanism concept (Whitley 1994).

The IMF team that constructed several versions of the annual model MULTIMOD and then the GEM model for policy simulations (Bayoumi et al. 2004) substantially contributed to these developments. Other quarterly and annual models were used to make regular world economy forecasts and international trade and policy simulations. Special credit should be given to the NIGEM model for its outstanding role in producing forecasts and economic policy simulations (Barrell et al. 2004).

The outlines of the models are presented in Table 12.1.

12.2 Project LINK

The macroeconometric modelling activities that were developing in the USA and Canada and then in Europe's largest industrialized countries and Japan created conditions for attempting to link the country models together into a consistent system. As a result, the international Project LINK was constructed in 1968, when the Committee on Economic Stability and Growth at the SSRC decided to call into being a project integrating the existing macroeconometric country models into a world system. The system was to be a tool supporting the studies into the properties and impacts of the international transmission mechanism.

The Committee's decision led to the establishment of the LINK Center in Philadelphia in 1969, with a Coordination Center headed by L.R. Klein supported by R.A. Gordon, B.G. Hickman and R.R. Rhomberg. The first years were spent on adjusting particular countries' models to make them comply with the unified foreign trade standards. Owing to this effort, the first forecast of international trade utilising a reduced form of intercountry linkages (MINI LINK) could be produced in 1970. The full model solution based on a specially constructed matrix of export shares in foreign trade was obtained already in 1971 (cf. Hickman 1991).

The construction of the matrix was crucial for the development of all other multicountry models. The point of departure was the observation that in most country

[1] An interesting comparative analysis of the models' properties in the first years of their development (until the early 1990s) has been presented by Whitley (1994).

models export was exogenous. Within a system, however, country's exports could be obtained as a weighted sum of other countries' imports, the weights being equal to the shares of the country's exports to particular importing countries. The country models generally assumed that the prices of the imported commodities were exogenous. Within a system, their estimates could be computed as the weighted sums of export prices charged by the countries exporting commodities to the importing country. The weights would be the relevant components of the matrix of export shares in foreign trade (Klein 1999).

In the early 70s, issues concerning the updating of the export share matrices were the object of lively discussions. The solutions proposed ranged from the application of the modified version of LES (Klein and Peeterssen 1993), the use of relative prices, given the elasticity of substitution for each exporter (Moriguchi 1973; Klein et al. 1975), to the use of relative prices, given identical substitution elasticity for all exporters (Hickman and Lau 1973). The future practice showed that the first approach was mainly applied.

From 1971–1972 onwards, the LINK model was systematically used for preparing world economy forecasts and in simulation analyses focusing on the major problems in the world economy. Prior to that, though, the system was tested for its properties with the interim and dynamic multipliers. In the 1980s, the financial sectors of the LINK models were extended, mainly by adding equations explaining capital flows and exchange rates. In the period following the liberalization of the exchange rates, the specification of the exchange rate equations prompted many discussions at the Project LINK meetings (Hickman and Klein 1985).

Project LINK systematically increased its membership, from 7–13 participants in the first half of the 1970s to more than 100 after 1987. The macromodels constructed by particular national centres also grew in number, substituting the UNCTAD minimodels (that mainly described developing countries and centrally planned economies) (Waelbroeck 1976; Sawyer 1979). Their number increased from 7 in 1960 to 79 in 1987 and to 80 in 1995.

Although the Project LINK models were generally demand oriented, some of them, mainly those of the centrally-planned and developing countries, were supply determined. There were both annual and quarterly models (the latter had to be aggregated in time) and they differed in terms of size, but all of them had unified foreign trade sectors. Both NSA exports and imports were decomposed into 4 SITC commodity groups: 0–1 (raw materials and agricultural products), 2–4 (other raw materials and half-finished products), 3 (fuels), 5–9 (consumer and investment goods).

New models being added, the system grew enormously. The initial number of 1,500 equations in the early 1970s increased to ca. 5,000 in 1975, around 20,000 in 1985 and to approximately 30,000 in 1998 (Klein 1999). An extension of available computer programs followed. An important step forward was the replacement of the mainframe with personal computers, for which effective simulation software was developed (Johnson and Klein 1979).

The system of models found manifold applications. As mentioned, it was systematically used for preparing biannual forecasts of the short- and medium-term development of the world economy. The forecasts were produced in collaboration with particular country partners, the solution being complex but efficient. The forecasts prepared by particular national centres were fed into the system at the LINK Center that used them to provide the members with a PRELINK forecast. This preliminary forecast generated from the system was discussed then with the partners during international meetings set up by the Center. The improved final version of the forecast POSTLINK was used by the UN Secretariat for drawing up UN Secretary General's Report on the global economic situation (Hickman and Ruffing 1995).

In the 1980, the LINK Center was moved to the Institute for Policy Analysis led by P. Pauly at the Toronto University in Canada and its computer system was transferred to the UN Department of Economic and Social Affairs in New York. With more direct links established between Project LINK and the United Nations, the role of the research undertaken under the Project grew considerably. The UN economic analyses of the world economy, its development and the likely changes were increasingly based on the results of LINK simulations. This was a very rich and multilateral cooperation. Its main areas were the impacts of the changes affecting capital markets and exchange rates, the effects of the oil shocks and of the changes in the economic policy pursued by the USA and other industrialized countries.[2] The activities were conducted in all the past periods, until the most recent analyses of the impacts of the global financial crisis and the 2008–2009 world recession.

Project LINK gave birth to many initiatives concerning the construction of new country models for developing countries in Asia and Latin America and more recently also in Africa. In the latter case, the activity has been supported since 2002 by the African Research Network for Development Policy Analysis having 15 country-members coordinated from the Pretoria University and DESA UN.

12.3 The World Trade Models

The work on developing international trade relationships within the Project LINK model was accompanied by attempts that were initiated to construct models capable of explaining the deeply disaggregated commodity flows, particularly of raw materials and fuels. In the mid-70s a COMLINK system was built. It was related to the system of models participating in the Project LINK and consisted of models accounting for 23 commodity groups.

The EITF[3] model of international trade built for the G-7 and East-Asian countries was also associated with Project LINK. It was a generalization of the NIRA-LINK model linking the national economies of Japan and the USA (Adams et al. 1996). It distinguished 35 industries in the models for Japan and the USA; a simplified model for 20 countries had 35 commodity groups operated in international trade.

[2] A broad description of the numerous analyses based on the LINK model can be found in: Sawyer (1979), Klein (1982, 1999), Hickman (1991).

[3] The abbreviated names of the cooperating institutions are: ERINA, International University of Japan (IUJ), University of Tsukuba, Foundation for Advancement for International Science (FAIS).

The particular country models were demand oriented, with expanded financial flows sectors and inter-industry relationships based on the I-O submodels. The EITF was used during numerous policy simulations.

The Michigan Model of Production and Trade built in the mid-1970s had special properties. It was a computable general equilibrium (CGE) model constructed by A. Deardorff and R. Stern from the Michigan University. Its purpose was to support research on employment in 29 industries in the major countries of the world. It initially covered 18 industrialized countries, but this number quickly rose to 34. The model was used to run numerous simulation analyses, for instance analyses of world trade liberalization.

In the late 1980s, its structure was enriched with new specifications derived from the new theory of international trade which took account of imperfect competition, growing returns to scale and product diversification. The model was reorganized to cover 12 countries and groups of countries and was given a new name: the Michigan Brown-Deardorff-Stern Model. Its numerous variants were used in many policy simulations, most of which concerned organizational changes in the world trade, such as the establishment of NAFTA (The Michigan Model of World Production and Trade 2007).

12.4 The OECD-INTERLINK Model

In the early 1970s, a model of international trade was built within the OECD Secretariat, which for many years served as a forecasting and policy simulation tool. Towards the end of the decade, the exogenous variables of particular countries were substituted in the model by respective equation systems, first mainly by identities and then stochastic equations were introduced. This gave birth to the system of half-annual INTERLINK models interlinked through commodity flows in foreign trade. They initially reproduced the existing country models, but sooner or later they were standardized (OECD 1989).

The INTERLINK model was used to prepare short- and medium-term forecasts and numerous policy simulations as soon as it was constructed (Helliwell et al. 1986). Numerous research contributions in the 1980s improved the system's specifications, which helped to stabilise its structure by the end of the 1980s. All variants of the models covered 7 large OECD countries, for which they had extended structures with 200–250 equations, up to 100 of which were stochastic, the 16 remaining OECD countries, for which they had a smaller number of equations (130–150, up to 50 of which were stochastic) and regions covering the remaining countries.

The models describing particular countries had similar structures—the differences mainly concerned the country-specific parameter estimates. They were demand oriented and the 7 large countries had neoclassical supply sectors (Richardson 1988).

The consumption functions were specified in an unusual manner. The consumption-to-income ratio was determined by the wealth-to-income ratio, prices and real interest rate. Investment in the enterprise sector of the large country models was generated from the production function to solve a system of equations explaining the demand for production factors. Investment in the small country models was explained traditionally, using the flexible accelerator rule. Investment in residential construction mainly depended on disposable income and real interest rates.

As regards the supply sector, the central role was given to a three-factor (including energy use) production function with constant elasticity of substitution. The function was used to determine the demand for production factors. Production supply and labour force supply were also determined within the supply sector. In the case of the smaller countries, a simplified specification was applied. The production function had only two factors. Employment directly depended on GDP and real labour costs.

The average wage equations were specified in a rather traditional way. An extended version of the Phillips curve with adaptive inflationary expectations was applied. The specification of the price equations was based on unit costs enlarged by a mark-up, assuming imperfect competition. The deflator of value added (omitting energy output) being crucial to a price system depended on unit costs, the terms of trade, competitors' prices and the capacity utilization rate. The long-term and short-term relationships were distinguished.

The country models had extended blocks of equations explaining financial flows, including budget revenues and expenditures, the functions of money demand (in large countries), as well as short- and long-term interest rates.

The models of particular countries were mainly linked through commodity flows and prices, which were the direct channels transmitting international shocks. They used the appropriate matrix of export shares. At the end of the 1990s, the matrix included 30 OECD countries and 6 groups of other countries/regions for 4 commodity groups (Le Fouler et al. 2001). In the more recent versions of the INTERLINK model the FDI flows were endogenized. The exchange rates providing the main financial linkage between the models were also endogenized. The expected exchange rates depended on the PPP ratios modified by the differences between country's short-term interest rates and the average interest rates in the other countries and by the foreign net assets accumulated.

The equations for regions including the non-OECD members and for the small OECD countries were specified similarly. Through the 1990s and afterwards, activities were continued to improve the specification of the models' equations, especially for the smaller OECD countries.

In the last versions of the model the equation parameters were estimated with ECM.

The above models were chiefly used by the OECD Secretariat in preparing short- and medium-term forecasts (Dalsgaard et al. 2001). To make this process more efficient, a new, quarterly Small Global Forecasting Model was constructed in the first years of the 21st century (Rae and Turner 2001). The forecasts it produced provided a starting point for updating the INTERLINK forecasts.

The model had 4 subdivisions: USA, the Euro area, Japan and the rest of the world (ROW) being of marginal importance. It was demand determined and consisted of several blocks of equations. Output was determined in a special way. The deviation of the effective GDP from its potential (exogenous) value was decomposed into final demand's deviation from its potential value and net exports. The first component depended on the interest rates and the budget deficit. Prices, among which CPI was the most important, were dependent on the unit cost components and demand pressure. The financial sector was represented by the short- and long-term interest rates that were treated as exogenous in forecasting and endogenized in simulation exercises.

The model was used in preparing numerous policy simulations, e.g. of the shocks affecting the world demand and of changes in the US monetary policy.

12.5 The Models of the World Bank

In 1973 the World Bank Minimum Standard Model was constructed to ensure consistency of the Bank's forecasts (Holsen 1973). It was in fact a planning model, because it was designed to help answer crucial questions for the developing countries about how large investments, imports and foreign credits were necessary to meet the target rates for GDP and export. Its later version, the Revised Minimum Standard Model (RMSM), was a point of departure for constructing the models of particular countries, capable of finding solutions to alternative planning tasks. For this reason, the investment and imports functions linked with the system of equations explaining the balance of payment components (mainly identities) were given the major role in the model. They did not include prices and wages (Addison 1989; Ventura 1991).

In the early 1990s, a new Bank-Gem macroeconometric model referring to the British GEM model was built at the Department of International Economic Analyses and Forecasts (IECAP). It was composed of around 150 country models, nearly 100 of them representing the developing countries. The models were linked via a system of relationships in the world trade and international finance. Because the specification and use of the developing country models was most important for the model builders, an Analytical Data Base (DAD) was constructed and efforts were intensified to adapt the MAXSIM computer system to personal computers (Petersen et al. 1991).

The standard developing country models had considerably different structure than the RMSM models did. The new models were classical, macroeconometric, jointly-interdependent systems with both real and financial sectors and with price systems. Together, they constituted a system whose components were linked through the commodity and financial flows. Consequently, the models' balance of payment components were endogenized.

The specification of the major stochastic equations distinguished between the long-term and short-term relationships. The parameters were estimated in two steps, the second step being the ECM method.

Although the general structure of the models was traditional, many equations were specified according to modern standards. Household consumer demand was represented by the demand of the credit-constrained households (thus being determined by disposable income) plus other households' demand, which depended on their joint personal wealth. In the short-run, consumption was increasing as a result of growing disposable income. Because personal wealth (including financial wealth) was introduced, consumption became indirectly dependent on the rate of inflation.

The investment demand functions were specified by taking fixed capital increase as a point of departure. An increase in fixed capital was determined by output (the accelerator rule), fixed capital increase in the public sector, real exchange rate affecting investment user costs and the level of debt being an investment risk proxy. In the imports equations, economic activity levels and relative prices functioned as the explanatory variables.

The supply sector was represented by a system of price equations generated by minimizing the cost functions. The producer price equations thus obtained depended on the unit costs of particular production factors. Labour costs were approximated by the consumption deflator. Imports prices and changes in the capital-output ratios were introduced. Other prices were dependent on producer and imports prices.

The country models had extended financial and monetary sectors, where special emphasis was laid on the balance of payment components.

The models were involved in the preparation of many policy simulations (Pedersen 1991a, 1994).

12.6 The Models of the International Monetary Fund (IMF)

12.6.1 The MULTIMOD Models

Toward the end of the 1980s, the annual model MULTIMOD was constructed at the IMF. Despite its neo-Keynesian orientation, many equations used rational expectations from the very beginning (Masson et al. 1988, 1990). It was enhanced by the INTERMOD 1.1 model where the role of the G7 countries was stressed (Helliwell et al. 1986). The most detailed version MULTIMOD MARK III was used in numerous policy simulations (Laxton et al. 1998). It is worth adding that the updated version of model MARK III B was used to perform interesting analyses of the development of the EURO area (Hunt and Laxton 2004). In the meantime MINIMOD was constructed (Haas and Masson 1986).

The main purpose for which the MULTIMOD model was constructed was policy simulations involving the major world countries. It distinguished the models for the 7 largest industrial countries and a block of equations accounting for the other industrialized countries, imposing a unified structure. Two additional blocks of equations were introduced for the indebted and debt-free countries (mainly oil producers).

The models' structure had neo-Keynesian elements, but the supply sector was not the only one where the neoclassical approach was broadly accentuated. In general, the model construction showed a prevalent tendency to specify the equations according to economic theory and to make a broad use of rational expectations.

This was reflected in the special structure of the models. A steady-state subsystem of equations explaining the long-term (equilibrium) relationships was distinguished, as well as a subsystem of dynamic adjustments moving the real processes towards the state of equilibrium. The parameters of the long-term equations were mostly calibrated, while the parameters of the short-term equations were estimated.

The industrialized country models distinguished five types of economic agents: households, enterprises, fiscal and monetary institutions, and foreign agents. In the world trade, raw materials, fuels and other commodities were singled out.

The household equations were specified for consumption, labour supply and personal wealth. The consumption function was comprised of two elements. One concerned households that, following the life cycle hypothesis, decided to spend a certain fraction of their personal and financial wealth on consumption. Personal wealth was represented by the discounted present value of the expected lifetime labour income, allowing for age distribution of the household members. The second element was households constrained by their current disposable income.

For the enterprise sector it was assumed that enterprises maximize their expected profits yielded by the production process described with the Cobb-Douglas production function. Fixed capital accumulation followed Tobin's "Q" concept, allowing for adjustment costs, and contributed to enterprises' investments.

The equations for the fiscal institutions explained expenditures on commodities and services, as well as transfers financed from taxes and loans. The models introduced long-term expenditure, transfers and the debt-to-GDP ratios. The short-term adjustments were assumed to meet their expected values.

The MULTIMOD model offered the possibility of introducing the monetary authorities' interventions involving nominal anchors such as money supply, nominal exchange rates, and the rate of inflation. Its models were also provided with equations adequately explaining prices and interest rates. In the industrialized countries, prices depended on exogenous oil prices, the prices of other raw materials, GDP deflators (except for oil prices) and exchange rates, allowing for a mark-up determined by the capacity utilization rate. The GDP deflator was derived from a reduced Phillips curve. The exchange rate equations were formulated following the uncovered interest rate parity concept.

The foreign trade equations had a standard specification. Exports and imports were dependent on the respective characteristics of economic activity (imports were determined using the I-O tables) and relative prices.

As mentioned, MULTIMOD was special in it broadly used expectations, particularly the rational ones, concerning households' future incomes, future profits and prices.

The MULTIMOD model served for many years, producing numerous policy simulations that complemented the analyses of the conditions and impacts of economic development in various countries, particularly those expecting IMF's financial support (Hunt and Laxton 2004).

12.6.2 The GEM Model

The new GEM model of the world economy was constructed at the IMF in the years 2003–2004. It was to incorporate the new trends in macromodelling, which involved the construction of a system belonging to the class of Dynamic Stochastic General Equilibrium (DSGE) models (see Lane 2001). Its structure proposed for the case with two-three countries has been discussed in the paper by Bayoumi et al. (2004).

The models built for particular countries rested on solid microeconomic foundations. The major parameters of the long-term equations were calibrated based on the results provided by microeconomic studies. The short-term equations explaining consumption, prices and wages assumed adjustment lags, which was typical of this class of models. Their parameters were estimated with the Bayesian methods.

The models included households, enterprises, and public institutions. The CES utility function was used, allowing for the appropriate specification of the consumption function (thus leisure impacts could be distinguished) and of the labour supply function. Personal disposable incomes included labour income and shares in profits. Consumer expenditures were split into purchases of domestic and imported commodities.

The production process was represented by the CES technology. The inputs of labour, fixed capital and sometimes of materials were used as the explanatory variables. The extended version of the model distinguished between the CES-based production of "intermediate" commodities and the production of "final" commodities that included also imported goods. This strange distinction was introduced across the DSGE models in order to differentiate commodity flows in the domestic trade according to their origin. There were also plans to make a distinction between the sheltered and open sectors. The equations for public institutions were explaining budget revenues and expenditures.

The model equations were dynamized mainly by the broad use of the cost adjustment concept, which involved the introduction of lags into equations explaining both nominal and real variables. Consumption and working time were dynamized using the concept of habit persistence, while costs adjustments were followed by lags in fixed capital increase, investments and imports.

The models were flexible enough to allow changes in their structures, for instance the making of extensions consisting in the introduction of the domestic trade sector or the addition of new countries. The models performed first policy simulations for 2–3 countries already in the early period of their use (Bayoumi et al. 2004).

Quite recently, the Global Integrated Monetary and Fiscal Model (GIMF) was constructed, as a tool supporting studies into the impacts of fiscal expansion (Kumhof and Laxton 2009).

12.7 The Models of European Community. The QUEST Models

In the early 1990s, the macroeconometric model QUEST of the world economy was built within the European Commission. It was initially operated by the Deutsches Institut fuer Wirschaftsforschung (DIW). The model was intended to support the analyses of EU's economic condition in relation to the economic situation in Japan, USA and the rest of the world.

In the middle of the same decade the model QUEST II was built, which was a new, extended and reconstructed version of the first model (Dramais et al. 1997). QUEST II was a quarterly model with approximately 2500 equations. It contained the models of 20 countries (14 being EU members) and of 6 regions. Each of the large country models had ca. 60 equations (Commission of the European Communities 1991).

This version was structured quite differently from its predecessor. QUEST II followed the concept of neoclassical Keynesian synthesis. The behavioural equations were founded on the microeconomic principles of intertemporary optimization of households' and enterprises' behaviour. The supply sector was modelled using the neoclassical production function. These solutions determined the specification of the long-run steady-state equations. The short-run equations were mostly specified using the neo-Keynesian approach—rigidities and lags in wage and price adjustments as well as cost adjustments in the investment demand equations were introduced (Roeger and in't Veld 1997).

The consumption functions were essentially built following the concept of life cycle hypothesis. It was, however, assumed that the consumption expenditures of around 30 % of households were determined by their current real disposable incomes. The other households' consumption depended on the expected lifetime incomes and real personal financial wealth.

The investment demand functions were derived from the production functions. In the CES production function fixed capital and energy functioned as the production factors. The impacts of these variables were combined with employment and the effects of its quality through the Cobb-Douglas production function, which was used to determine potential output and the rate of its utilization. The demand for labour and energy depended on output and, respectively, on real wages and relative prices. The long-term elasticities were calibrated assuming that their values were 1, while the short-term elasticities were estimated.

Financial flows were modelled in a traditional manner. Budget revenues were obtained by adding up the tax incomes and those coming from other sources. Budget expenditures were decomposed into the purchases of commodities and services, remunerations, transfers and debt service.

Producer prices depended on lagged prices, unit labour costs and cyclical markup changes represented by variations in the rates of capacity utilization and employment. CPI was determined by producer prices, imports prices and VAT. The real average wages were assumed to result from a bargaining process between the employees' and employers' representatives. Thus they depended on labour productivity, unemployment rates (modified for the changes in unemployment benefits), the estimated value of leisure (dependent on the level of personal wealth) and on the average number of hours offered by the employees.

The particular country models were linked through commodity, service and capital flows constituting the balance of payments components. An important role was given to the equations explaining exchange rates that were mainly determined by the interest rate differentials.

Table 12.1 Multicountry models

Model Version Year	Institution Author	Frequency Sample	Equations total stochastic	Decomposition	Estimation method	Properties	Applications
PROJECT LINK 1968–1971	Social Science Research Council, Committee on Economic Stability and Growth	Annual and quarterly	1500	Specific country models[a], common structure of exports and imports, 4 SITC commodity groups distinguished	Different	Models of different orientation with endogenized imports and export prices linked through the international matrix of export shares, hence endogenization in the system of exports and import prices	Forecasts presented every half year, simulation analyses of functioning world economy and regarding international transmission mechanisms
1974	University of Pennsylvania L.R. Klein et al.	Annual and quarterly	3000		Different		Forecasts and simulation analyses
1980	University of Toronto, DIESA UN P. Pauly et al.	Annual and quarterly	5000		Different	Endogenization of capital flows and exchange rates	Forecasts, policy simulations
1989		Annual and quarterly	20000		Different		Forecasts, policy simulation
NIRA-LINK 1990	University of Pennsylvania, NIRA Tokyo F.G. Adams, B. Gagnes, S. Shishido	Annual	ca. 500	Initially Japan and USA, sectoral disaggregation	OLS	Demand determined, satellite model w.r.t. LINK system	Simulation analyses

Table 12.1 (Continued)

Model Version Year	Institution Author	Frequency Sample	Equations total stochastic	Decomposition	Estimation method	Properties	Applications
EITF 1995	ERINA, IUJ	Annual		G7, 9 countries of East Asia and 20 other countries and regions, 35 sectors	OLS	Demand oriented I-O model	Simulation analyses
MICHIGAN MODEL of World Production and Trade 1970	University of Michigan A. Deardorff, R. Stern	Annual		Initially 18 industrialized countries up to 34 countries, 27 sectors		CGE model	Simulation analyses
Michigan BDS model	University of Michigan D.R. Brown, A. Deardorff, R. Stern	Annual		Different versions composed of models of 39 countries, 29 groups of commodities		CGE model	Simulation analyses
INTERLINK 1979	OECD	Half-annual			OLS, first differences	Models demand determined, country specific structures	Forecasts, policy simulations

Table 12.1 (Continued)

Model Version Year	Institution Author	Frequency Sample	Equations total stochastic	Decomposition	Estimation method	Properties	Applications
1991	P. Richardson et al.	Half-annual	Large countries: 230–260 ca. 100 Small countries: 130–150 ca. 50	23 OECD countries/regions	ECM	Extension of supply and financial flows sectors, unification of the models structure	Forecasts, especially of foreign trade, policy simulations
Small Global Forecasting Model 2000	OECD D. Rae, D. Turner	Quarterly 1963–2000 1977–2000	15 14	USA, Japan, Euro area, others			Forecasts preceding the main forecasts based on INTERLINK
RMSM-X 1973	WORLD BANK D. Addison	Annual	ca. 430		Mainly calibrated	Planning model, assuming the GDP rates of growth that determined the investment shares, imports and foreign credits requirements	Simulations, forecasts
BANK GEM	K.N. Pedersen et al.	Annual 1960–1989 1965–1989	260[b] 22[b]	150 countries, of which 100 developing countries grouped in blocks of equations	ECM	Mainly neoclassical specification, extended block of balance of payments	Medium-term forecasts, simulations analyses

Table 12.1 (Continued)

Model Version Year	Institution Author	Frequency Sample	Equations total stochastic	Decomposition	Estimation method	Properties	Applications
MULTIMOD MARK II 1988–1989	IMF P. Masson et al.	Annual		Models of industrialized countries and others	Calibration	Two segments distinguished: steady-state and dynamic adjustments	Simulations analyses
MULTIMOD 1998	IMF D. Laxton et al.	Annual		Models of industrialized countries and others	Calibration	Neoclassical: optimization of activities of households and firms, rational expectations	Simulations analyses
MULTIMOD MARK III	IMF B. Hunt, D. Laxton	Annual		Models of 7 highly industrialized countries, aggregation of remaining industrialized and developing countries, linked through commodity flows	Calibration		Numerous policy simulations, forecasts
MARK III B 2004		Annual 1981–2001			Calibration		Policy simulations

Table 12.1 (Continued)

Model Version Year	Institution Author	Frequency Sample	Equations total stochastic	Decomposition	Estimation method	Properties	Applications
GEM 2003–2004	IMF T. Bayoumi et al.	Quarterly		Gradually extended	Calibration, Bayesian methods	DSGE model, microeconomic foundations: maximization of the activities of economic agents, integration of demand and supply (CES production function) and of international capital markets	Policy simulations, mainly regarding the impacts of shocks in the world prices, exchange rates and competitive conditions
QUEST 1991	European Commission	Annual		Models of the EU countries, USA, Japan	OLS	Neo-Keynesian adoptive expectations	Simulations analyses
QUEST II 1996	A. Dramais, W. Roger, J. in't Veld	Annual 1975–1995		Models of EU countries, USA, Japan and 10 regions interlinked through commodity and financial flows 62 coun-tries/regions		Neoclassical, dynamic optimization, rational expectations	Simulation analyses mainly w.r.t. the EU

Table 12.1 (Continued)

Model Version Year	Institution Author	Frequency Sample	Equations total stochastic	Decomposition	Estimation method	Properties	Applications
FUGI Global Model	Soka University A. Onishi	Annual	37000		OLS	Mixed	Simulations analyses
7.0 M 80 1991		Annual 1971/72–1990	40000	80 countries/regions	OLS	Developed countries demand determined, developing countries and CPE supply determined	Forecasts, simulation analyses
M 180			150000	180 countries/regions			
MIMOSA 1989	CEPII OFCE, Paris H. Delessy, H. Sterdyniak et al.	Annual		18 countries/regions, 6 of which major industrialized; each country: 6 sections, I-O, 6 groups in foreign trade	OLS	Neo-Keynesian, supply represented by production functions, foreign trade-links of commodity groups	Medium-term forecasts, simulation analyses
1995		Annual	5000 400–500[c] 150[c]		OLS, NLS, 3 stage LS		Forecasts, simulations analyses

Table 12.1 (Continued)

Model Version Year	Institution Author	Frequency Sample	Equations total stochastic	Decomposition	Estimation method	Properties	Applications
MCM 1979	FED, USA G. Stevens	Quarterly		G-5 and ROW	OLS	Countries linked via commodity flows and price systems	Quarterly forecasts, simulation analyses (impacts of shocks outside the USA)
1991–1992		Quarterly		G-7 and 5 regions	ECM	Demand determined, neoclassical growth	Simulations analyses
FRB/GLOBAL 1996	FED, USA	Quarterly	1400 160[d] 60[d]	G-7 and 5 blocks	ECM, rational expectations	Extended equations explaining exchange rates, multilateral structure of international trade introduced	Policy simulations, forecasts
SIGMA 2005	C.J. Erceg et al.	Quarterly		6 country blocks	Calibration	Neo-Keynesian in the short-run, neoclassical in the long-run, expectations alternatively adaptive or rational, dynamic adjustments, DSGE model	Simulation analyses, forecasts
GEM	National Institute of Economic and Social Research, London R. Barrell et al.	Quarterly	750	Models of OECD countries and 6 regions	OLS	Demand determined, generated prices and links between countries	Medium-term forecasts, simulation analyses

Table 12.1 (Continued)

Model Version Year	Institution Author	Frequency Sample	Equations total stochastic	Decomposition	Estimation method	Properties	Applications
NIGEM 2002		Quarterly	1000 200	Models of OECD countries, 9 other countries, 6 regions	Two-stage ECM, rational expectations procedure by Fair-Taylor	Neo-Keynesian, expectations *for ward-looking*, lagged nominal adjustments, CES productions function, financial flows sector	Quarterly forecasts, numerous policy simulation
OEF 1990	Oxford Economic Forecasting M. Burridge et al.	Quarterly	780	Europe (11), Japan, USA, 4 regions	VECM	Extended supply sector, costs adjustments, expectations based on learning process *versus* rational expectations	Policy simulations, forecasts
PRIAMO 1990	PROMETEIA Bologna C. D'Adda et al.	Annual		7 large countries OECD, 5 remaining OECD countries, 8 regions	OLS	Demand determined	Simulation analyses, including the likely results of joining the EU

Table 12.1 (Continued)

Model Version Year	Institution Author	Frequency Sample	Equations total stochastic	Decomposition	Estimation method	Properties	Applications
MC 2004	R.C. Fair	Quarterly for 15 countries, annual for the remaining countries, started in different years, the most early in 1960 up to 2000–2001	488 362 37[e] 15[e]	USA and 38 countries	TSLS	Demand oriented, adaptive expectations, financial sector	Numerous policy simulations, forecasts

[a]Initially 7 industrialized countries, 13 countries in 1973, 20 countries and 4 regions in 1980, 79 countries in 1981, 80 countries in 1995

[b]For particular countries

[c]It applied to the G7 countries; the remaining industrialized countries had a total 120 equations, 45 of which were stochastic, OPEC and ROW ca. 40 in total, 15 of which were stochastic

[d]For particular countries

[e]For the next 19 countries the shares in the world trade were only estimated

The system of the QUEST models was used in numerous policy simulations, mainly those accentuating potential monetary and fiscal policy impacts, especially in the context of budget expenditure expansion (Roeger and in't Veld 2004).

12.8 The Models of the US Federal Reserve Bank

12.8.1 The MCM Quarterly Models

The Federal Reserve Bank (FRB) was the first central bank to notice the necessity to analyse a country's economic situation in the context of its links with the major partners in the world economy. The quarterly model of the world economy MCM (Multi-Country Model) was constructed at the FRB already in 1976. It covered Canada, Germany, Japan, the United Kingdom, the USA and the rest of the world (ROW) and had around 1,000 equations in total. The model was demand determined, but its supply sector showed a neoclassical orientation. It was systematically used to carry out policy simulations since 1979.

In the 1980s, the model's equations underwent a major modification, particularly its investment function. The equations explaining capital flows were replaced with the exchange rate equations. Following the changes in the monetary policy, the detailed specification of the banking sector was abandoned, but the equations explaining interest rates were introduced instead. The estimation of the equation parameters switched to ECM.

At the beginning of the 1990s, the MCM model incorporated the models of the G-7 countries and Mexico, as well as of 4 regions that were composed of other OECD countries, new industrialized countries, OPEC and ROW. The bilateral links were substituted by multilateral ones (Stevens et al. 1984).

The research projects launched at the end of the 1980s aimed to build a new model where the neoclassical concepts would be used more broadly, assuming interactive decision making and more extensive application of rational expectations in the specified equations. The project resulted, first of all, in the construction of a small quarterly model MX3 that only covered 3 countries, i.e. Japan, USA, West-Germany, and the ROW. The model had 32 equations, 11 of which were stochastic. Their characteristic property was that they distinguished the long-term relationships and clearly utilised rational expectations; one-period lags were introduced (Edison et al. 1989; Gagnon 1991).

12.8.2 The FRB/WORLD Model

In the next years, the large quarterly model was deeply reconstructed. It was also decided then that the US model and the world model should be combined into a single system. Thus a new macroeconometric world model FRB/WORLD was built.

Its main component was the model of the US economy FRB/MCM presented in Chap. 3. The new world model had only around 250 stochastic equations, 40 of which concerned the US economy (Levin et al. 1999).

The country models distinguished long-term relationships and short-term adjustments. Output was determined by demand. Supply was generated from the Cobb-Douglas production functions, allowing for energy use. The demand for production factors was derived from the production functions. The adaptive and/or rational expectations were broadly introduced, primarily into the equations explaining interest rates and exchange rates, but also into those explaining wages and prices. Long lags were built into the short-term equations accounting for wages and prices.

In the G-7 models, the parameters of the consumption and investment functions were estimated. The models for the other countries had a simplified standard structure; the parameters of their equations were calibrated (Brayton et al. 1997).

An attempt has been made recently to construct a new DSGE-type macromodel. This new model called SIGMA is a quarterly model with 7 blocks of countries. It is mostly Keynesian in the short-run, but in the long-run the neoclassical orientation predominates. The adaptive and rational expectations are alternately used in the model. The parameters of the long-run equations have been calibrated (Erceg et al. 2006).

12.9 The MEMMOD Model of the Deutsche Bundesbank

At the end of the past century, the quarterly multicountry model of the world economy called MEMMOD was constructed at the Deutsche Bank (2000). It covered 9 countries (including G-7) and country groups with the remaining EU and OECD countries, and the rest of the world. It presented a mixed orientation. In the country models final demand (consumption, investment) determined output and indirectly employment and imports. On the other hand, the production functions were used to generate potential output and its utilization rate affecting prices. The models had specified equations explaining financial flows. The MEMMOD model was used in preparing short-term forecasts and policy simulations.

12.10 The Models of the National Institute of Economic and Social Research (NIESR): GEM and NIGEM

12.10.1 The GEM Model

In the mid-1980s, the UK Treasury set up a research project to construct a world multicountry model. This gave birth to the world models built at NIESR and in Oxford. NIESR called its model GEM (Global Econometric Model) and maintained it together with London Business School (LBS). GEM was a quarterly model containing large models for the G-7 countries, small models for the 3 remaining OECD countries and models for 6 regions that included between 4 and 5 stochastic equations. Altogether, the model had around 750 equations.

The country models were, in principle, demand oriented and had rational expectations that were mainly used to determine exchange rates. Consumer demand was dependent on real disposable income, personal wealth and interest rates. Investment demand was derived under the profit maximization rule, assuming imperfect competition. Employment depended on output and real wages. Prices were determined by unit costs plus a mark-up affected by the capacity utilization rate. The models contained equations explaining money demand and exchange rates. The links between the country models were mainly expressed through commodity flows. The GEM model was systematically used in preparing the quarterly forecasts of the world economy and numerous policy simulations (Whitley 1994).

12.10.2 The NIGEM Model

At the turn of centuries, the decision to merge together the quarterly models of the UK and the world economy was made. As a result, the new multicountry model of the world economy NIGEM (National Institute Global Econometric Model) was constructed at NIESR. The model covered all OECD countries and additionally 9 countries (including China and Russia) and 6 regions. All country and country-group models included at least equations explaining domestic demand, exports and imports, prices and the balance of payment components (NIESR 2002).

The models of the OECD countries included the neo-Keynesian concepts with forward looking expectations and lags in the nominal variables, making the adjustment processes sluggish. They had blocks of equations determining consumption, incomes and personal wealth of households, the production process (production functions), and blocks explaining prices and wages, budget revenues and expenditures, financial markets and international trade. Rational expectations were entered into the equations explaining consumption, wages and exchange rates. The equations discriminated between long-term relationships and short-term adjustments (Barrell et al. 2004).

The central role in the models' structure was given to the production functions, i.e. the CES functions with constant returns to scale and technical progress linked to labour input. They were used to derive the functions representing demand for production factors and estimate the capacity utilization rates. Producer prices were dependent on the unit costs, including import prices, and on a mark-up. The CPI was determined by producer prices, import prices and unit labour costs. Real wages were assumed to result from wage bargaining, so their determinants were labour productivity and unemployment rate and in the short-run also the difference between the expected and effective rates of inflation.

In the model, consumption was dependent on real disposable income and real personal wealth being comprised of financial assets and possibly of residential real estates. There was a large block of equations concerning financial flows and financial assets. The equations explaining exchange rates were built assuming that the changes in the rates were caused by changing relations between interest rates. The international links were represented by the commodity and service flows, the impacts of prices and exchange rates and by capital flows.

The NIGEM model was, and still is, widely used in forecasting the world economy and in running numerous, interesting policy simulations (NIESR 2002).

12.11 The Multicountry Model MIMOSA

The annual model MIMOSA (Macroeconomic Integrated Model for Simulation and Analysis) was constructed as a result of cooperation between two French research institutions, Centre d'Etudes Prospectives et d'Informations Internationales (CEPII) and Observatoire Français des Conjonctures Economiques (OFCE). Its authors were a special research team MIMOSA, which was formed in 1986. Since 1989 the model was used for producing medium-term forecasts and analyses of the world economy developments.

It was the number of countries and regions that made MIMOSA one of the largest multicountry models of the world economy, rather than frequently applied disaggregation of economic activities and decomposition of foreign trade. The model represented a neo-Keynesian orientation. Its models of the 6 largest countries had detailed, extended structures, while the other country models had a simplified structure (Mimosa Modelling Group 1990).

The model was deeply restructured in 1994, which involved the addition of new countries. Besides the 6 major countries (France, Italy, Japan, the UK, the USA and West Germany), the model distinguished 5 groups of EU and OECD countries and 7 other regions. In the industrialized countries 4 sections were distinguished, i.e. agriculture and foodstuffs industry, fuel industry, manufacturing industry, and other industries, linked through the I-O relationships; foreign trade was decomposed into 4 commodity groups. The model thus produced was large, with ca. 5000 equations. Particular country models had a unified structure; those describing the 6 major countries had 400–500 equations, while the other country models had considerably smaller numbers of equations. The equation parameters in the country models were estimated using OLS, NLS and the three-stage LS, mainly with the 1965–1992 sample (Delessy et al. 1996).

The country models had a neo-Keynesian orientation, with fairly well extended specifications for the large countries. Consumer demand (per capita) was dependent on lagged consumption to allow for inertia, current and lagged real disposable income (per capita). It also depended on real interest rates and the rate of inflation that substituted financial wealth which was not available. Consumer demand for particular commodity groups or sections was additionally determined by relative prices, like the investment demand for residential real estates. Labour supply was determined using the exogenous coefficients of the economically active population decomposed by gender and age (5 groups). In the long-run, their dynamics was described by a logistic function, but in the short-run their changes depended linearly on the rate of unemployment.

The supply sector was modelled assuming intertemporal profit maximization of enterprises. Potential industrial output was estimated using the survey data and its increase was dependent on cumulative investment. Investment demand depended on the cumulated increase in output (the accelerator), the effects of labour substitution by fixed capital and on cumulative real profits (i.e. profitability). Employment was determined by output, labour costs-to-capital costs ratio, the exogenous growth rate of technical progress (the trend) and also by changes in the average working time per employee.

In the price system, value-added deflators played a decisive role. They mostly depended on the unit labour costs modified by mark-up variations resulting from the changes in the capacity utilization rates or the debt service changes. The wage equations were specified using an extended Phillips curve approach: the rate of wage increase depended on the lagged rate of inflation and the unemployment rate, allowing for the labour productivity growth rate. With the above equations the rate of NAWRU (Non-Accelerating Wage Rate of Unemployment) was estimated.

The models were additionally provided with simple descriptions of the processes generating budget revenues (tax systems) and expenditures, including social transfers. The characteristics of the financial flows were parsimonious. Only the interest rate changes were explained.

As mentioned, the equations for the distinguished regions had a simpler specification. The country models were linked through commodity flows in 4 commodity groups and 2 groups of services. Country imports were dependent on domestic demand, its competitiveness and the rate of capacity utilization. The exporting countries allocated commodity flows following the components of the international exports shares matrix, which was updated on an annual basis (as opposed to the initial version of the model). This allowed including new country models into the system.

The MIMOSA multicountry system was frequently used in international economic analyses, e.g. to investigate the effects of fiscal and monetary expansion in Europe and/or the world economy (Delessy et al. 1996).

12.12 The Model FUGI

In the early 1960s, A. Onishi from the SOKA University in Japan constructed the annual system of macromodels GEM (Global Economic Model) for the purpose of making regular forecasts. Initially covering 15 Asian countries, the system was extended in the next decade to the rest of the world. Its version prepared at the end of the 1970s accounted for 62 countries/regions and the total number of its equations was around 37,000. This model was used into the 1980s, i.a. at the DIESAP of the UN to prepare long-term forecasts and policy simulations of the world economy.

In 1991, it was substituted by the more detailed annual multicountry FUGI Global Model 7.0. This model was built in two versions: a "small" one covering

80 countries and country groups and a "large" one containing 180 countries and country groups. The number of the equations in the large version exceeded 150,000. This necessitated the development of a special computer programming system. The equation parameters were estimated using OLS, the sample starting in the first half of the 1970s and ending in 1990.

The system was built of several subsystems, i.a. an economic subsystem and subsystems for the ecological conditions, freedom and security and human rights. The economic subsystem distinguished developed and developing countries, as well as centrally planned economies. The models of the developed countries were assumed to be demand oriented (potential GDP > effective GDP) and had more detailed structures. The models of the developing and CPE countries were supply determined (potential GDP < effective GDP demand). Hence, all the models had blocks generating global supply and demand. The specifications of the equations were universal, but special properties of the large countries were taken into account.

Potential output was calculated from the labour productivity functions that were dependent on the capital-labour ratios, real R&D expenditure per employee, the ratios of 5-year investment totals to fixed capital, and energy constraints. The models explained unemployment rates and indirectly employment.

Country's demand for GDP was obtained by adding up domestic final demand and net exports. Consumer demand was dependent on either domestic GDP or real disposable income and short-term real interest rates. For some developing countries consumption was residual. Investment demand was mostly the function of the available funds and the interest rates on long-term credits. Exports were mainly derived from the transaction matrices in international trade.

The country models explained the major components of distributed national income, i.e. operational surplus and wage funds. The real operational surplus was dependent on GDP (reduced by wage funds), interest rates and the terms of trade, and the nominal average wages depended on CPI, labour productivity, unemployment rate and the operational surplus' share in GDP.

The models included extended systems of prices and deflators. The equations explaining wholesale prices reflected the impacts of money supply and the unit cost changes. Import prices were endogenized by means of the transaction matrix in international trade, allowing for predetermined export prices.

The models contained blocks of equations explaining money demand, as well as interest rate systems. Public finances were represented by the equations explaining particular state budgets' revenues and expenditures. An attempt was also made to explain changes in the major balance of payments components and in the exchange rates (Onishi 1993).

The FUGI system was used in numerous policy simulations of the world economy developments, particularly in the context of the links between the Japanese and US economies, and in forecasting.

12.13 The Model of Oxford Economic Forecasting

At the beginning of the 1990s, a forecasting system incorporating the quarterly multicountry model of the world economy OEF (Oxford Economic Forecasting) was constructed in the United Kingdom. The OEF model was a medium size model with 780 equations, which included 15 countries and 4 regions (Burridge et al. 1991).

The country models were demand determined. Household consumer demand was dependent on real disposable income, personal wealth and interest rates. The production functions were used to derive equations explaining employment. Real average wages were determined by lagged productivity of labour and unemployment rates. Prices resulted from unit costs plus a mark-up that depended on the rate of potential capacity utilization. Exchange rates were dependent on the interest rate differences and the current account surplus. Interest rates were mostly exogenous. The equations explaining money demand in the major countries were specified.

The OEF model was systematically used in forecasting and in the economic analyses of the world economy changes.

12.14 The Multicountry Model PRIAMO

In the early 1990s, the small, annual multicountry model of the world economy PRIAMO was constructed at the research institution PROMETEIA in Bologna, Italy. It was mainly designed for simulation analyses exploring the world economy changes having effects on the Italian economy, which could be performed in tandem with policy simulations based on the macroeconometric model of Italy maintained at PROMETEIA.

The multicountry model included 7 large OECD countries (France, Germany, Italy, Japan, Spain, the United Kingdom, the USA), 5 remaining OECD countries and groups of countries, and 8 regions covering the rest of the world. The country models were demand determined, their structure being characteristic of the prevailing mainstream models. The structure was the most detailed for the large 7 OECD countries, moderately detailed for the other OECD countries, and simplified for the rest of the world.

In the large country models consumer demand was dependent on real disposable income, real personal wealth and real interest rates. Investment demand was determined, following the accelerator rule, by smoothed GDP and user costs (primarily short-term real interest rates). Domestic final demand plus net exports determined demand for GDP. On the other hand, the models had equations that generated potential output using the Cobb-Douglas production functions. The emerging capacity utilization rates were used for determining prices that depended also on unit labour costs and import prices.

The models of the particular countries were linked through transaction matrices in international trade (D'Adda et al. 1997).

The PRIAMO model was used in numerous policy simulations. It is worth mentioning that it supported the analyses of the potential impacts of Central European countries' entering into the EU (D'Adda et al. 1997).

12.15 The Multicountry MC Model by R.C. Fair

The MC model of the world economy was constructed in the early 21st c., as a result of the research efforts of just one author, R.C. Fair. It is known for its originality. The model was built within a macroeconomic framework that the author developed for his previous models of the USA. The model's equations were specified using solid microeconomic foundations: economic agents making their decisions solved the maximization problems under imperfect competition; the forecasting errors entailed disequilibria in commodity markets (inventory fluctuations) and in labour markets (unemployment); expectations were not rational (Fair 1974, 1984, 1994). In building his MC model R.S. Fair broadly drew on the rich experience he had accumulated while constructing the previous versions of his models of the US economy, which are presented in Chap. 3.

The MC model was composed of the macromodels for 39 countries. The US economy model presented in Chap. 3 was the largest (31 stochastic equations). The models for the other countries had a unified structure and were smaller, each having 15 stochastic equations. The total number of the stochastic equations was 362 (Fair 2004). Depending on their availability, either quarterly or annual data were used, the first year of the sample being basically 1960.

Unlike the US model, the other country models were specified like the structural models of a neo-Keynesian orientation. Lagged endogenous variables were introduced into the majority of the equations as the explanatory variables, so that the short-term and long-term impacts could be distinguished from each other. The downside of this specification was that the autoregression coefficient estimates were frequently extremely high and the estimates of the other parameters were barely significant. The estimation process was based on the Cowles Commission methodology: TSLS was applied, assuming that the deviations from the deterministic trend were stationary.

Consumption demand (per capita) was dependent on GDP (per capita), financial assets-to-potential output ratio, interest rate and lagged consumption (per capita). Investment demand was determined by its lagged volume, GDP (the accelerator rule) and interest rate. Global demand (sales) was obtained from an identity and demand for GDP was generated from a stochastic equation, allowing for inventory changes. Employment increase was obtained in a rather complex manner. The starting point was a lagged ratio of effective-to-potential employment, which was calculated by dividing GDP by exogenous labour productivity. It was modified by the current and lagged increases in GDP. The labour force supply was obtained using the coefficients of the economically active populations (by gender), being mainly dependent on real wages. The rate of unemployment was residual.

Prices were represented by the GDP deflator. They were dependent on the deflator's lagged value, import prices and the rate of capacity utilization (represented by the deviations from the GDP trend). Nominal wages were determined by the GDP deflator, labour productivity and unemployment rate. Export prices in relation to the world prices were dependent on the ratio between domestic prices and world prices. Exchange rates, estimated separately for the US and DM, were dependent on the ratios between domestic prices and respective US and DM deflators and on the differences in interest rates. In the country models, interest rates were endogenized. The short-term interest rates were dependent on the growth rates of GDP deflators, the rate of capacity utilization and the US and German interest rates. The long-term interest rates were determined by their lagged values. The models contained also the equations explaining money supply.

Import (per capita) was dependent on its lagged value, domestic demand (per capita) and relative prices. Export was generated in the system using the exports shares matrix of international trade. The quarterly shares were available. They depended on the ratios between particular countries' export prices to the world prices. The equations were used to forecast the components of the aforementioned matrices.

The MC model was used in many interesting analyses. On the methodological plane, R.C. Fair investigated the impacts of the nominal versus real effects and tested for the NAIRU appearance. He also tested the correctness of using rational expectations, finding no empirical justification for this decision. He propagated the use of the bootstrap methods in the estimation process. On the economic plane, he investigated, for instance, stabilization issues with respect to the introduction of the Euro (Fair 2004).

12.16 Other Multicountry Models

For the scarcity of information on the other multicountry models, their characteristics will only be sketched. The same concerns the computational general equilibrium models and small models of rather historical significance (cf. Whitley 1994).

In the last years of the previous century, the ATLAS model was constructed at the Ministry of Finance in France. It was a large quarterly model that included 10 regions and had 1450 equations, 540 of which were stochastic. It was used in forecasting and policy simulations. The quarterly multicountry EPA model (Economic Planning Agency, Tokyo) was of similar size. The EPA model covered 9 countries and 6 regions, having ca. 1,200 equations in total (EPA 1999). In Japan, S. Shishido constructed the large multicountry model of the world economy TSUKUBA-FAIS. It had the same goals and structure as Project LINK, but all its country models were constructed by the common standards developed at the Tsukuba University (Shishido 1980). Multicountry models including respectively 23 countries and 6 regions and 3 countries and Europe as a single region were developed also at the Wharton Associates and at the Data Resources Inc. (DRI); at the beginning of the 21st c. they were merged into one system maintained by Global Insight.

In the 1990s, B. Guerts and H. Timmer built the multicountry computational general equilibrium model World Scan at the Office of Economic Policy Analyses, CBS, in the Hague (CPB Netherlands Bureau for Economic Policy Analysis 1992). Its core version built at the end of the previous century was used for constructing its special versions intended for the long-term analyses of the world economy (globalization, ecological threats, energetic raw materials, etc.). It was a large

CGE model with 12 regions and 7 sections that distinguished between the high- and low-qualified labour force (CPB Netherlands Bureau for Economic Policy Analysis 1999).

In the United Kingdom, a multicountry, annual model with rational expectations was developed in the mid-80s, in the tradition of the Liverpool models. It included 9 major OECD countries and 3 blocks having in total 153 equations. The model had a neoclassical orientation and addressed market equilibria. Household demand was dependent on the expected real incomes and real financial assets. Prices were determined by unit costs and a flexible mark-up (Minford et al. 1986).

In France at GAMA a multicountry model GAIA was constructed. It is a quarterly model used in short and medium-term forecasting.

The MSG multicountry model that W. McKibben and J.D. Sachs developed in the beginning of the 1990s gained historical meaning. It was an annual CGE model with rational expectations (McKibben and Sachs 1991). Its new version was built at the end of the same decade (McKibben 1999).

At the beginning of the 21st c. a small G-3 model intended for policy analyses was constructed at the Centre for International Macroeconomics in Oxford. It covered Japan, USA and the Euro Area treated as a single economy (Chamberlin et al. 2003). Its structure was complex. Stress was given to the specification of the equations explaining the supply sector, especially to the specification of the real wage equation representing the wage bargaining process. The model included both rational expectations and expectations based on the learning process. The latter expectations, preferred by the authors, were followed by very long dynamic adjustments. The parameters of the country models' equations were estimated using the cointegration techniques to allow for the non-stationary time series.

References

Adams, F. G., Ganges, B., Shishido, S., & Movchauk, A. (1996). *Disaggregate modeling of the US-Japan-world economy* (Mimeo). Paper prepared for the meeting of Project Link, Lausanne.

Addison, D. (1989). *The World Bank revised minimum standard model. Concepts and issues* (Policy, Planning and Research Working Papers). Washington: The World Bank.

Ball, R. J. (1973). *The international linkage of national economics models*. Amsterdam: North-Holland.

Barrell, R., Becker, B., Byrne, J., Gottschalk, S., Hurst, I., & van Welsum, D. (2004). Macroeconomic policy in Europe: experiments with monetary responses and fiscal impulses. *Economic Modelling*, *21*, 877–931.

Bayoumi, T., with assistance from Laxton, D., et al. (2004). *GEM: a new international macroeconomic model* (Occasional Paper No. 239). Washington: IMF.

Brayton, F., Levin, A. T., Tryon, R. W., & Williams, J. C. (1997). *The evolution of macro models at the Federal Reserve Board* (Finance and Economics Discussion Series No. 1997-29, Mimeo). FRB.

Burridge, M., Dhar, S., Mayes, D., Meen, G., Neal, E., Tyrell, N., & Walker, J. (1991). Oxford economic forecasting's system of models. *Economic Modelling*, *8*, 227–413.

Chamberlin, G., Henry, S. G. B., & Satchi, M. (2003). A model of the G-3. *Economic Modelling*, *20*, 1083–1095.

Commission of the European Communities (CEC) (1991). Quest: a macroeconomic model for the countries of the European Community as part of the world economy. In *European economy* (Vol. 47, pp. 51–236). Brussels: EEC.

CPB Netherlands Bureau for Economic Policy Analysis (1992). *Scanning the future*. The Hague: CPB Netherlands Bureau for Economic Policy Analysis.

CPB Netherlands Bureau for Economic Policy Analysis (1999). *World scan, the core version*. The Hague: CPB Netherlands Bureau for Economic Policy Analysis.

D'Adda, C., Poli, G., & Vincenzi, L. (1997). PRIAMO: Prometea international annual model. In W. Welfe (Ed.), *Economies in transition and the world economy. Models forecasts and scenarios* (pp. 425–455). Frankfurt/Main: P. Lang.

Dalsgaard, T., Andre, C., & Richardson, P. (2001). *Standard shocks in the OECD INTERLINK model* (Economics Department Working Papers 306). OECD Publishing.

Delessy, H., Sterdyniak, H., et al. (1996). *The new version of MIMOSA. A macroeconomic model of the world economy* (version 1995, Mimosa Team No. 96-01). CEPII/OFCE.

Deutsche Bank (2000). *Macro-econometric multi-country model: MEMMOD* (Mimeo). Frankfurt/Main: Deutsche Bank.

Dramais, A. (1986). COMPACT. A prototype macroeconomic model of the European community of the world economy. *European Economy, 27*.

Dramais, A., Roeger, W., & in't Veld, J. (1997). *Quest II. A multi-country business cycle and growth model* (Mimeo). Paper presented at the Project Link meeting, Malaysia. European Commission.

Edison, H., Marquez, J., & Tryon, R. W. (1989). The structure and properties of the Federal Reserve Board multicountry model. *Economic Modelling*, *4*, 115–315.

Economic Planning Agency (1999). *EPA world econometric model—fifth version*. Tokyo: Economic Planning Agency.

Erceg, C. J., Guerrieri, L., & Gust, C. (2006). *SIGMA: a new open economy model for policy analysis* (Mimeo, International Finance Discussion Papers No. 835). FRB.

Fair, R. C. (1974). *A model of macroeconomic activity, Vol. I: The international model*. Cambridge: Ballinger.

Fair, R. C. (1984). *Specification, estimation and analysis of macroeconometric models*. Cambridge: Harvard University Press.

Fair, R. C. (1994). *Testing macroeconometric models*. Cambridge: Harvard University Press.

Fair, R. C. (2004). *Estimating how the macroeconomy works*. Cambridge: Harvard University Press.

Gagnon, J. (1991). A forward-looking multi-country model for policy analysis: MX-3. *Economic and Financial Computing*, *1*, 311–361.

Gana, J. L., Hickman, B. G., Lau, L. J., & Jacobson, L. R. (1979). Alternative approaches to the linkage of national econometric models. In J. A. Sawyer (Ed.), *Modelling the international transmission mechanism*. Amsterdam: North-Holland.

Haas, R., & Masson, P. (1986). *MINIMOD: specification and simulation results* (IMF Staff Papers No. 33, pp. 722–767).

Helliwell, J., Sturm, P., Garrett, P., & Salou, G. (1986). *The supply side in OECD's, macroeconomic model* (OECD Economic Studies No. 6). OECD Publishers.

Hickman, B. G. (Ed.) (1983). *Global international economic models*. Amsterdam: North-Holland.

Hickman, B. G. (1991). Project LINK and multi-country modelling. In R. G. Bodkin, L. R. Klein & K. Marwah (Eds.), *A history of macroeconometric model-building* (pp. 482–506). Aldershot: E. Elgar.

Hickman, B. G., & Klein, L. R. (1985). Recent developments in project LINK. *Social Science Research Council Items*, *39*, 7–11.

Hickman, B. G., & Lau, L. J. (1973). Elasticities of substitution and export demand in a world trade model. *European Economic Review*, *4*, 347–380.

Hickman, B. G., & Ruffing, K. G. (1995). Project LINK: past, present and future. In L. R. Klein & F. Lo (Eds.). Tokyo: UNU Press.

Holsen, J. G. (1973). *The minimum standard model for GPP. System and global economic projections*. Washington: Country Economic Projection Department, The World Bank.
Hunt, B., & Laxton, D. (2004). Some simulation properties of the major Euro area economies in MULTIMOD. *Economic Modelling*, *21*, 759–783.
Johnson, K., & Klein, L. R. (1979). Error analysis of the LINK model. In J. A. Sawyer (Ed.), *Modelling the international transmission mechanism*. Amsterdam: North-Holland.
Klein, L. R. (1982). Model LINK. In L. R. Klein (Ed.), *Wykłady z ekonometrii (Lectures in econometrics)* (pp. 261–286). Warszawa: PWE.
Klein, L. R. (1999). Project LINK. In L. R. Klein, A. Welfe & W. Welfe (Eds.), *Principles of macroeconometric modeling* (pp. 27–151). Amsterdam: North-Holland.
Klein, L. R., & Peeterssen, A. (1993). Forecasting world trade within Project LINK. In R. J. Ball (Ed.), *The international linkage of national economic models*. Amsterdam: North-Holland.
Klein, L. R., Moriguchi, C., & van Peetersen, A. (1975). The LINK model framework of trade with applications to 1972–73. In P. Kenen (Ed.), *International trade and finance*. New York: Cambridge University Press.
Kumhof, M., & Laxton, D. (2009). *The global integrated monetary and fiscal model (GIMF)*. In IMF macro-linkage, oil prices and deflation workshop, 6–9 January 2009.
Lane, P. R. (2001). The new open economy macroeconomics: a survey. *Journal of International Economics*, *34*, 235–266.
Laxton, D., Isard, P., Faruqee, H., Prasad, E., & Turlelboom, B. (1998). *MULTIMOD MARK III: the core dynamic and steady state models* (IMF Occasional Papers No. 164).
Levin, A. T., Rogers, J. H., & Tryon, R. W. (1999). *A guide to FRB/GLOBAL* (Mimeo, International Finance Discussion Papers No. 588). Washington: FRS.
Le Fouler, L., Snyker, W., & Turner, D. (2001). *Trade linkages and trade matrices in the OECD INTERLINK model* (OECD Economics Dpt. Working Papers No. 310). OECD Publishing.
Masson, P., Symansky, S. A., Haas, R., & Dooley, M. (1988). *MULTIMOD: a multi-region econometric model* (Staff Studies for the World Economic Outlook). Washington: IMF.
Masson, P., Symansky, S. A., & Meredith, G. (1990). *MULTIMOD MARK II: a revised and extended model* (IMF Occasional Paper No. 71). Washington: IMF.
McKibben, W. (1999). Solving large scale models under alternative policy closures: the MSG 2 multicountry model. In A. Hughes-Hallet & P. McAdam (Eds.), *Analyses in macroeconomic modelling*. Boston: Kluwer Academic.
McKibben, W., & Sachs, J. D. (1991). *Global linkages: macroeconomic interdependence and cooperation in the world economy*. Washington: Brookings Institution.
Mimosa Modelling Group (1990). *MIMOSA, a model of the world economy* (OFCE Working Paper No. 90-02).
Minford, A. P. L., Agenor, P., & Nowell, E. (1986). A new classical econometric model of the world economy. *Economic Modelling*, *3*, 154–174.
Moriguchi, C. (1973). Forecasting and simulating the world economy. *American Economic Review. Papers and Proceedings*, *63*, 402–409.
NIESR (2002). *NIGEM. The world model manual*. London: NIESR.
NIESR (2005). *Overview of NIGEM*. London: NIESR.
OECD (1989). *The OECD INTERLINK system* (Mimeo).
Onishi, A. (1993). FUGI global model 7.0. A new frontier science of global economic modelling. *Economic and Financial Computing*, *3*, 8–67.
Pedersen, K. N. (1991a). *A first version of a prototype bank GEM/STO model* (Mimeo). Washington: International Economic Analysis and Prospects Division, The World Bank.
Pedersen, K. N. (1991b). *A macroeconomic developing country model: first step of a World Bank global economic model, ICECAP* (Mimeo). Washington: The World Bank.
Pedersen, K. N. (1994). The prototypical developing country model of the World Bank global economic model. In H. M. Scobie (Ed.), *Economic progress and growth* (pp. 1–28). London: Chapman and Hall.

Petersen, C. E., Pedersen, K. N., Riordan, E. J., Lynn, R. A., & Bradley, T. (1991). *BANK-GEM: a World Bank global economic model* (Mimeo). Paper presented at the Project LINK conference, Moscow. SECAP, The World Bank.

Rae, D., & Turner, D. (2001). *A small global forecasting model* (OECD Economic Department Working Papers No. 286). OECD Publishing.

Richardson, P. (1988). *The structure and simulation properties of OECD's INTERLINK model* (OECD Economic Studies No. 10). OECD Publishing.

Roeger, W., & in't Veld, J. (1997). *QUEST II, a multicountry business cycle and growth model* (Economic Papers No. 123). European Commission.

Roeger, W., & in't Veld, J. (2004). Some selected simulation experiments with the European Commission's QUEST model. *Economic Modelling, 21*, 785–832.

Sawyer, J. A. (Ed.) (1979). *Modelling the international transmission mechanism*. Amsterdam: North-Holland.

Shishido, S. (1980). A model for coordination of recovery polices in the OECD region. *Journal of Policy Modelling, 2*, 33–55.

Stevens, G., Berner, R., Clark, P., Hernandez-Cata, E., Howe, H., & Kwack, S. (1984). *The U.S. economy in an independent world: a multicountry model* (Mimeo). Washington: Board of Governors of the FRS.

The Michigan Model of World Production and Trade (2007). Description. http://fordschool.umich.edu/rsie/model/description.html.

Ventura, J. (1991). *RMSM-X: an exposition* (Mimeo). Washington: Country Economics Dpt., The World Bank.

Waelbroeck, J. (Ed.) (1976). *The models of project LINK*. Amsterdam: North-Holland.

Waelbroeck, J., & Dramais, A. (1974). DESMOS: a model for the coordination of economic policies in the EEC countries. In A. Ando, R. Herring & R. Marston (Eds.), *International aspects of stabilization policies*. Boston: FRB of Boston.

Whitley, J. (1994). *A course in macroeconometric modelling and forecasting*. New York: Harvester-Wheatsheaf.

Chapter 13
The Epilogue

Within macroeconometric modelling that has been developing worldwide for more than 60 years now almost every country has built and maintains its own national macromodel. The industrialized countries have several or even more country models. Macroeconometric models have become an indispensable instrument for forecasting and programming economic development, as well as for making analyses of the economic policy impacts at the country and international levels. This tendency seems to be permanent and the models' applications are likely to grow in number with progressing globalization.

The large variety of macroeconometric models that exist today has several sources. Firstly, they differ in terms of scopes and sizes that depend on their purpose. Secondly, they have different structures and specifications that result from their theoretical underpinning. Thirdly, they use different estimation methods, which range from the traditional ones based on the Cowles Commission methodology to the methods founded on the cointegration analysis of time series.

Their sizes are very diverse. Unlike the early 1980s, this is no longer the subject of disputes (Kmenta and Ramsey 1981). The model size depends on the purpose it is to fulfil. The computational power has ceased to be a factor that restricts it. The annual models are in principle large, especially if the I-O submodels have been included. Among the quarterly models the medium-size models prevail. The monthly models that have been built in the recent years have small numbers of equations, though.

The development of the macroeconometric modelling activities has been neither smooth nor regular, especially during the last 10 years. In the first 25 years after macromodels started to be constructed academic centres played the major role in developing the macroeconometric modelling rules and in shaping the models' structures. The American centres were in the lead, among which L.R. Klein and the University of Pennsylvania played the central role, followed by the UK university centres. Those centres originated the "mainstream", demand-oriented models; initially annual and then quarterly, the models systematically grew in size from several hundred to several thousand equations along with the substantial disaggregation of economic activities.

W. Welfe, *Macroeconometric Models*,
Advanced Studies in Theoretical and Applied Econometrics 47,
DOI 10.1007/978-3-642-34468-8_13,

The well-known Lucas critique provoked broad discussions on the role of expectations. The rational expectations conception drew many adherents and enriched the specifications of many macroeconometric models. Nonetheless, many research institutions continued to use the concept of adaptive expectations or, alternatively, made attempts to apply the idea that economic agents learn systematically. It has been broadly realized that the knowledge of the theoretical mechanisms governing economic adjustments remains the exclusive domain of experts in the large corporations (Fair 2004). In many countries the information on economic agents' expectations gathered from regular firm and household surveys is also used.

Over the last 20 years international organizations, research institutes and central banks in particular countries have constructed numerous multicountry models of the world economy, where the numbers of the countries range from several to several hundreds. The models have detailed structures for large industrialized countries, but the structures for the other countries are less developed. The large models have in total many thousand equations. The models are operational and mainly used to run policy simulations. In many developed countries the world economy models have "absorbed" their country models. This tendency has been more and more common.

The macroeconometric mainstream models had well-developed structures. Within the real sector the models distinguished the demand sector, on one hand, and the slowly developing supply sector, on the other, as well as the financial flows sector. This framework has recently typified many macromodels of developed countries and the newly built models of developing countries. It is maintained for many reasons. It facilitates sectoral disaggregation, establishment of links with the I-O type models, extension of the supply sector to include the endogenized impacts of technological progress (allowing for the properties of a knowledge-based economy) and, last but not least, the ecological conditions.

In this class of macroeconometric models built in the recent years the long-term (equilibrium) relationships and the short-term adjustments are regularly distinguished. The models use estimation methods developed by R.F. Engle and C.N.J. Granger and apply ECM to estimate the short-term parameters.

Over the last 20 years certain twilight of these macromodelling activities has been observed. New theoretical approaches developed in the last two decades have shifted the centre of gravity to the microeconomic foundations of macromodelling. The economic agents are assumed to optimize their behaviour, i.e. to maximize the utility and profits under imperfect competition. An important role has been given to the concept of lifetime expected and/or permanent income of households. The neoclassical production function has been used to determine the demand for production factors and in generating producer prices and wages.

This process has been followed by a change in the general structure of macroeconometric models. Within the real sector, household subsectors (consumer demand, labour force supply, etc.) and enterprises (output, demand for production factors, prices and wages) have been distinguished. The economic agents are assumed to make basically rational decisions. This has led to the introduction of the consumption function concept assuming lifetime expected incomes and to its verification in many models of industrialized, stable countries. Producers have been assumed to maximize profits under imperfect competition—the solution of this optimization problem determined the specification of the production function and of the equations explaining the demand for production factors, as well as producer prices and wages.

This theoretical underpinning has been primarily used to specify the long-term (equilibrium) relationships being frequently used as the basis for calibrating the equations' parameters. In the specification of the short-term relationships' parameters, the effects of expectations and inertia have been introduced as a rule, mainly using ECM.

The above tendencies can be seen in many models constructed in the last years by both research centres and public institutions. The academic institutions have rarely contributed to this development.

A new theoretical impulse came at the turn of centuries with the development of the "new economy" (Lane 2001). According to the new approach, the economic agents are optimizing their activities under monopolistic competition; the adjustments are rigid (neo-Keynesian); this chiefly applies to price and wage adjustments, but also to cost adjustments.

The above ideas provided a basis for the construction of a new class of models based on the Kydland-Prescott business cycle theory. These were the dynamic, stochastic, general equilibrium models (DSGE). They differed from the computational general equilibrium (CGE) models mainly in that they assumed imperfect competition and rigidities in time adjustments.

The DSGE models have been and are being developed mainly at the central banks' research centres, because the banks are interested in their possible applications to investigating the monetary and fiscal policy impacts. The structure of the models has become special, somewhat different from that characterising the mainstream models. In the production sector the flows of the domestic and imported commodities are sharply distinguished. Hence the "production of intermediate goods" is distinguished, i.e. domestic output is explained through the production functions and the "production of final goods" covering the flows of domestic and imported commodities to final users (that may be in the domain of domestic trade). Domestic production is frequently decomposed into sheltered industries and those open to foreign competition.

In the models, the demand equations following the neo-Keynesian approach induce nominal rigidities. The consumption functions typically include lags (the Brown effect), likewise the investment functions, allowing for the impacts of adjustment costs.

The specification process founded on theoretical assumptions has usually led to parameter calibration in the long-term equations, less frequently in the short-term equations. The latter have been mostly estimated with ECM and more recently with the Bayesian methods. This has opened new prospects for the applications of this class of estimation methods.

The development of macromodelling activities associated with the use of the DSGE models awoke the hopes for the renaissance of the activities in the academic centres. J. Verbruggen from the CPB in the Netherlands, the oldest macromodelling institution, has stated: "The DSGE models are a blooming business..." (Verbruggen 2008, p. 1). However, sceptical opinions can also be frequently heard. It is indicated that the DSGE models fail to cover all interesting economic phenomena and are weakly linked with the observations. These shortcomings tremendously impede the attempts at using the DSGE models to fulfil many functions of the large mainstream models (Fair 2009).

New tendencies developing within estimation theory that draw on methods ranging from the time series analysis to the cointegration analysis of the economic relationships have evoked the revival of modelling activities in the academic centres. In the applications they were limited to the construction and analysis of the small equation systems describing relatively isolated economic relationships. They have been mainly applied to systems presenting the dynamics of inflation. The limited computational potential has restricted the systems to a few equations.

Therefore, the attempts to use this methodology in macroeconometric modelling have been rare and have been restricted to the analyses of monetary markets. Let us mention the VECM models for the USA (Anderson et al. 2002), the SVAR models for the United Kingdom (Garratt et al. 2006) and a model for Norway (Bårdsen et al. 2005). The British model has been compared with the simultaneous model COMPACT for the United Kingdom, but this exercise has not provided any definite conclusions on the effects of the analysed shocks (Jacobs and Wallis 2005). The authors of the models have expressed their hopes that appropriate submodels of other economic sectors may be built in the future and integrated into one large system representing the national economy as a whole. This may point to the likely path of future development of macroeconometric modelling, being a "theoretical" alternative to the DSGE models.

References

Anderson, R. G., Hoffman, D. L., & Rasche, R. M. (2002). A vector error-correction forecasting model of the US economy. *Journal of Macroeconomics*, *24*, 569–598.

Bårdsen, G., Eitrheim, Ø., Jansen, E. S., & Nymoen, R. (2005). *The econometrics of macroeconomic modelling*. Oxford: Oxford University Press.

Fair, R. C. (2004). *Estimating how the macroeconomy works*. Cambridge: Harvard University Press.

Fair, R. C. (2009). *Has macro progressed?* (Cowles Foundation Discussion Paper No 1728). Cowles Foundation for Research in Economics at Yale University.

Garratt, A., Lee, K., Pesaran, M. H., & Shin, Y. (2006). *Global and national macroeconometric modelling. A long-run structural approach*. Oxford: Oxford University Press.

Jacobs, J. P. A., & Wallis, K. F. (2005). Comparing SVARs and SEMs; two models of the UK economy. *Journal of Applied Econometrics*, *20*, 209–228.

Kmenta, J., & Ramsey, J. B. (Eds.) (1981). *Large-scale macroeconometric models. Theory and practice*. Amsterdam: North-Holland.

Lane, P. R. (2001). The new open economy macroeconomics: a survey. *Journal of International Economics*, *54*, 235–266.

Verbruggen, J. (2008). Models: from "out" to "in"! *CPB Newsletter*, April 1, p. 1.

Part II
The Structure of Macroeconometric Models

Chapter 14
Introduction

The second part of this monograph is an attempt to present the structures of macroeconometric models run by major macromodelling centres. It will be described how equations of macroeconometric models are specified, including consumer demand and investment demand functions, production functions and equations explaining prices, wages and financial flows. In each case the underlying economic theories will be summarized, whose detailed presentations are available in the mathematical economy monographs.

The above description will be accompanied by a presentation of the results of empirical research based on macroeconometric models. The results, most frequently having the form of the estimates of respective elasticities, will be presented for the major countries in the world to show the degree to which the macro-characteristics of economic agents' behaviour were close to each other and stable. This explains why we refer to the macromodels of particular countries to describe the specifications of equations, but in characterising the estimation results we use the results obtained from the multicountry world models.

The construction and use of macroeconometric models is connected with many other important issues which are not addressed in this book, such as the methods for estimating equation parameters, especially those involving cointegration analysis, or numerous tests that are broadly used by the authors of contemporary macroeconometric models. Also, the descriptions of numerous model applications, including the results of multiplier analyses, policy simulations and forecasts, will not be provided. These issues require special discussion and more space, which can be found in many monographs and contributions to the world literature.[1]

It is not feasible to define what a "typical" structure of a macroeconometric model is, as several types of models were developed in the past, having different theoretical underpinnings and various arrangements of the specified equations. They were predominantly the demand determined Keynesian models, called the main stream

[1]Cf. Hendry (1995), Juselius (2006), Majsterek (2008), Welfe (2009); the models' applications were described e.g. in Klein et al. (1999) and Fair (2004).

W. Welfe, *Macroeconometric Models*,
Advanced Studies in Theoretical and Applied Econometrics 47,
DOI 10.1007/978-3-642-34468-8_14,

models in the previous years. They distinguished the sectors of final demand, supply, prices, wages and financial flows. The next to be mentioned are models relying on microeconomic foundations, whose equations were specified based on the neoclassical behaviour optimizing assumptions. These models distinguished the following sectors: households maximizing their preferences, producers maximizing profits or minimizing costs under imperfect competition, public institutions and a foreign sector. The next categories of models referred to general equilibrium theory, which became the basis for constructing the computable general equilibrium models (CGE), being static and partly deterministic. This class of models was more recently generalised by taking advantage of modern business cycle theory (Kydland and Prescott 1982), which resulted in the development of the Dynamic Stochastic General Equilibrium (DSGE) models. The DSGE models distinguish within the production sector the producers of intermediate commodities (including domestic producers) and the producers of final commodities that include with the suppliers of domestic and imported goods.

The above modelling frameworks are characterised more in detail in the next chapter, where the classifications of economic activities as used in macroeconometric models are additionally described.

In the next chapters equation specifications and the empirical results of parameter estimation arranged by the type of economic agent are discussed. Chapter 16 presents equations explaining households' activities, such as consumer demand and investment demand functions, and labour force supply equations, following the utility maximizing behaviour.

In Chap. 17 the specification of the equations explaining the behaviour of the profit maximizing (or cost minimizing) enterprises is discussed. Except for the equations explaining inventory changes, the most important role is played by the production functions. The functions are used in generating potential output and production supplies and (after conversion) in constructing functions explaining the demand for production factors. The investment demand functions are determined, likewise the employment demand functions. The chapter closes with a description of equations explaining producer prices and average wages.

Chapter 18 deals with issues concerning the modelling of equilibria and disequilibria in the commodity and services markets, and in the labour market. The quantitative and price adjustment mechanisms of market clearing are discussed in this chapter. The quantitative adjustments include inventory changes and changes in the capacity utilization rate, exports and imports. The descriptions presented in the previous chapter are made complete by introducing equations explaining public institutions' demand for consumer goods, the demand for inventory changes, and foreign trade. Labour market changes determining unemployment rates are also presented, together with the NAIRU and NAWRU characteristics. The chapter presents how final goods prices, production factor prices and average wages are determined in the public sector, thus closing the description of the price and wage system. Finally, the major feedbacks (multipliers) in the household and enterprise sectors are demonstrated.

Financial flows modelling is the subject of Chap. 19. The chapter presents equations explaining revenues (incomes), expenditures and balances (savings), which were built for institutional agents, e.g. households, enterprises, public institutions (mainly the state budget) and foreign agents (using the balance of payments). The specifications of equations explaining direct and indirect taxes are discussed in the chapter. Lastly, the interlinks between institutional agents and the links with real sectors are demonstrated.

Equations explaining the money and credit markets are distinguished and discussed separately in Chap. 20. Credits for particular groups of agents and their bank deposits are explained. The demand for money function is specified. The equations explaining the interest rates are shown. Special attention is addressed to central banks' decisions adjusting interest rates to enable the achievement of the inflationary target. The exchange rate specification is broadly discussed. The chapter ends with examples showing interrelationships concerning inflationary processes and specific interest and exchange rates, analysed using the cointegration method.

Final comments are presented in Chap. 21 ending the monograph.

References

Fair, R. C. (2004). *Estimating how the macroeconomy works*. Cambridge: Harvard University Press.

Hendry, D. H. (1995). *Dynamic econometrics*. Oxford: Oxford University Press.

Juselius, K. (2006). *The cointegrated VAR model*. Oxford: Oxford University Press.

Klein, L. R., Welfe, A., & Welfe, W. (1999). *Principles of macroeconometric modeling*. Amsterdam: North-Holland.

Kydland, F. E., & Prescott, E. C. (1982). Time to build and aggregate fluctuations. *Econometrica*, *50*, 1345–1370.

Majsterek, M. (2008). *Wielowymiarowa analiza kointegracyjna w ekonomii (Multidimensional cointegration analysis in economics)*. Łódź: Wydawnictwo UŁ.

Welfe, A. (2009). *Ekonometria. Metody i ich zastosowanie (Econometrics. Methods and their use)* (4th ed.). Warszawa: PWE.

Chapter 15
Macroeconometric Models—The Classification

15.1 Introduction

When a national economy is being modelled, complex economic and social processes as well as relationships between economic agents that enter into economic systems must be taken into account. The systems are composed of individual economic agents or of their interlinked bodies, such as their organizations. Based on the major functions played in economic processes, the following groups of economic agents can be distinguished: households, enterprises, public institutions (including financial ones) and foreign agents. Different types of markets are also distinguished—commodity and services markets, labour and money markets. Sellers and buyers in the markets participate in barter transactions, which are followed by flows of commodities and services, price adjustments and financial flows. The criterion that is used the most widely stresses the predominant kind of economic activity, leading to the distinction of sections and industries, such as agriculture or manufacturing industry (Bodkin et al. 1991).

By investigating the activities of the bodies of economic agents, the knowledge of the mechanisms underlying their functioning and growth can be extended. This is an important area of empirical analyses based on macroeconometric models.

Macroeconometric models were initially built using mainly classifications that distinguished the kind of economic activity (SNA). A more recent tendency is to apply the type of economic agent as the primary criterion, other criteria being given a secondary role.

15.2 Classifications of the National Economy in Macromodels

Economic agents are usually classified in macroeconometric models based on international statistics criteria and rules. They derive from the system of national accounts (SNA), which was developed at the UN Statistical Office as a result of R. Stone's pioneering efforts and has been used all over the world since the 1990s.

W. Welfe, *Macroeconometric Models*,
Advanced Studies in Theoretical and Applied Econometrics 47,
DOI 10.1007/978-3-642-34468-8_15,

Let us remind that until the end of the 1980s the former centrally planned economies (CPI) operated a special system of national economy balances, called Material Product System (MPS).

In the above systems particular economic agents are classified by their predominant economic activity. In the enterprise sector, the agents are usually enterprises (frequently conducting various activities) rather than plants representing technologically homogeneous units of a uniform profile. The SNA has been recently extended to accounts concerning the socio-demographic processes. As the processes were commonly presented as a flow matrix, the enlarged system is called a social accounts matrix (SAM).[1]

The early macroeconometric models used activity classifications which were close to SNA and accentuated the macroeconomic orientation. As a result, the mainstream models usually distinguished real and financial processes (flows), linked through (broadly understood) price adjustments.

The real processes include:

- generation of gross domestic output (GDP),
- production of commodities and services,
- employment and labour force,
- fixed capital and its replacement,
- GDP distribution and use, followed by the generation of final demand,
- export and import of commodities and services,
- household consumption,
- consumption of public institutions,
- investments in fixed capital and inventory changes.

Production and production factors may be further classified by the predominant technology, raw materials, etc., as well as decomposed into sheltered and open sectors. A more detailed decomposition of consumption, investment and foreign trade is also possible, using the common commodity classifications.

The financial processes include:

- the current transactions of economic agents:
  - revenues (incomes),
  - expenditures (user costs),
  - surplus or deficit,
- the capital transactions of economic agents:
  - assets and their changes,
  - liabilities and their changes,
  - capital transaction balances,
- wages and prices, including exchange rates and interest rates.

Particular kinds of financial flows are generally specific to different types of economic agents, so they are modelled separately for different institutional sectors.[2]

---

[1] A detailed description of national accounting systems and the SAM matrix is provided in Tomaszewicz (1994) and Zienkowski (2001). Cf. also Klein et al. (1999).

[2] Cf. Klein and Welfe (1982) and Klein et al. (1999).

Because of information scarcities, the modelling of capital flows has not been sufficiently developed.

Models using the above classification may be either one-sectoral and then they do not distinguish any kind of activity, or multi-sectoral, if such distinctions have been introduced. The multisectoral models may include many hundreds or thousands of subdivisions, hence large number of equations.

In the classifications with prevailing microeconomic criteria, the major sectors are households, enterprises, public institutions and foreign agents. Economic agents are assigned to the sectors based on the theoretical underpinning accepted by the authors of the models, who postulate the optimizing behaviour of economic agents rather than institutional criteria.

This type of classification can be illustrated using a grouping of economic activities assumed in the Fair model (Fair 2004).

Households

- consumption
- residential investment
- labour force supply
- financial assets (partly distinguished in the financial sector)

Enterprises

- producer prices
- production
- fixed capital and fixed capital investment
- employment, the number of employees and hours worked
- wages in the enterprise sector.

Further disaggregation is possible and then criteria based on the kind of economic activity are mostly used. For example, consumption may be decomposed into the consumption of durables, non-durables and services, production may be broken down into agriculture, manufacturing industry etc.

Financial Flows (Disaggregated as in the Previous Classification)
Markets are classified by

- commodities and services
- labour
- money.

A deeper decomposition of the markets is usually performed following special classifications, such as SITC in foreign trade. The models characterizing the markets' behaviour are typically composed of equation systems explaining demand from potential buyers, supply from potential sellers and prices clearing the markets. If interlinks in a market subsystem overlap the entire national economy, the computable equilibrium models (CGE) can be constructed. Developed at the World Bank since 1980s, the CGE models have spread all over the world.[3]

The above classifications were extended in models that introduced dynamic and stochastic relationships, i.e. in the dynamic, stochastic general equilibrium (DSGE) models. In particular, the domestic and imported commodities were clearly distinguished among the models' commodity flows. The domestic production was carried out within the "intermediate goods" production sector. Domestic and imported goods were combined into the "final goods" production sector, being in fact a representation of the domestic trade sector.

15.3 Macroeconometric Models. Their Major Types

The econometric models of national economies belong to the class of econometric models that have reached the highest degree of excellence.

Until quite lately, particular countries functioned in a state of relative 'splendid isolation', implementing their national economic policies and using uniform socio-economic statistics. The macroeconometric models in the countries were used not only to perform macroeconomic analyses, but also to conduct sectoral studies, analyses of particular markets, analyses of financial processes, etc. The limited availability of statistical data prevented the regional models from being broadly used in particular countries. On the other hand, though, the multicountry models have been gaining importance over the last 20 years. This particularly applies to the world economy models containing separate models for the main industrialized countries, as well as the models of the distinguished regions of the rest of the world.[4]

A special class of macromodels is input-output models that mostly represent interindustry and/or intersectoral commodity flows and the related price systems. They may be linked with econometric submodels that explain, on the one hand, income generation and the resulting final demand and, on the other hand, output generation and production factors. This provides grounds for constructing the integrated models of the national economy (cf. Tomaszewicz 1994).

In constructing macroeconometric models alternative assumptions about what major economic mechanisms define specific economic regimes can be used. The first group of the models consists of a system where the demand and supply of commodities and production factors effectively adjust to respective price and wage changes. These are the models of national economy in (full) equilibrium. The CGE models belong to this class of models. The class can be coherently represented by the following system of equations:

- a demand equation y_t^d:

$$y_t^d = d(p_t, \ldots) \tag{15.1a}$$

[3]Cf. Ginsburgh and Keyzer (1997) and for Poland Okólski and Timofiejuk (1978), Orłowski (1992), Żółkiewski (1995).

[4]Cf. Whitley (1994) and Klein et al. (1999), where the description of the Project LINK world economy model can be found.

- a supply equation y_t^s:

$$y_t^s = s(p_t, \ldots) \tag{15.1b}$$

- An identity assuming that demand equals supply:

$$y_t^d \equiv y_t^s \equiv y_t \tag{15.1c}$$

where y is the realization.

The solution of this three-equation system yields the market clearing price p_t^*:

$$p_t = p_t^* \tag{15.1d}$$

The above models are infrequently built in their pure form, because the assumption about prices being the only market adjustment instruments is rarely met and the adjustments are not necessarily effective. There are rigidities in price and wage behaviour, as well as cost adjustments. This neo-Keynesian point of view was accepted for the dynamic, stochastic general equilibrium (DSGE) models.

Most macromodels assume, however, that the prevailing economic mechanisms involve quantitative adjustments, such as inventory changes, changes in the capacity utilization levels and employees' working time, as well as in exports and imports. Price adjustments, if they occur, are either lagged or of small importance (this does not mean that prices are constant, but that they change mainly because of changing unit costs).

In the real world, the above adjustments are seldom effective. There are several markets where the equilibrium condition (15.1c) is not met. Macromodels assuming that this type of markets exists are called disequilibrium models in a broad sense. Disequilibria can take different forms and appear in different markets.

The first to be distinguished are the macromodels assuming excess supply of production factors, i.e. with non-fully utilized production capacities and unemployment, and with a likely excess supply of commodities and services. If these conditions are met, it is justified to assume that supply follows (constrained) demand. For this reason, the models are called demand determined or demand oriented. We can write:

$$\text{for } y_t^s > y_t^d \Rightarrow y_t^s \to y_t^d \text{ and } y_t^d = y_t \tag{15.2a}$$

where $y_t^s - y_t^d = y_t^s - y_t$ is excess supply.

If the supply of production factors or of commodities and services (or foreign money reserves) is restricted, then the market transactions will represent the realizations of supply. These models most frequently contain not only the supply equations, but also the estimates of excess demand and respective disequilibria. They are called the supply-constrained or supply-determined models. They can be presented as:

$$\text{for } y_t^s < y_t^d \Rightarrow y_t^s = y_t \tag{15.2b}$$

where $y_t^d - y_t^s = y_t^d - y_t$ is excess demand.

The disequilibria conditions may in fact differ between sections and markets. The simplest approach is to assume that the min-condition holds:

$$y_t = \min(y_t^d, y_t^s) \tag{15.3}$$

In the real world quantitative and price adjustments occur concurrently, but their intensities are different. The efficiency of these adjustments, especially in the short-run, is market specific. It is frequently assumed that the adjustments are effective in the commodity and services markets, i.e. where demand equals supply. This follows from the assumption that because of the non-fully utilized production capacities, unemployment and sufficient foreign reserves, demand changes will in the short-run cause adjustments in supply. The adjustments will not involve significant price changes (excluding commodities characterised by rigid supply, such as agricultural crops). The above conditions, marked by non-fully utilized production capacities and long-term unemployment, are described by the following class of models representing the regime of Keynesian unemployment:

$$X_t^s \geq X_t^d = X_t \quad N_t^s > N_t^d = N_t \quad \text{and} \quad K_t^s > K_t^d = K_t \tag{15.4}$$

where:

X_t is output (GDP),
N_t is employment,
K_t is fixed capital.

Under conditions characterized by excess demand for commodities and services, two economic regimes are generally distinguished. In the classical unemployment regime it is assumed that entrepreneurs will not decide to produce enough products to meet demand, if they do not find it profitable. Then restrictions in meeting demand are followed by employment restrictions. Hence:

$$X_t^s < X_t^d \Rightarrow X_t^s = X_t \text{ and } N_t^s > N_t^d = N_t \text{ and most frequently } K_t^s > K_t^d = K_t \tag{15.5}$$

According to the more frequently distinguished conditions, excess demand is caused by non-sufficient supply resulting from the shortage of one of the production factors. This economic regime is called a regime of suppressed inflation to underline the non-satisfactory efficiency of price adjustments. If labour supply is restricted, then excess demand affects both commodity and labour markets:

$$X_t^s < X_t^d \Rightarrow X_t^s = X_t \text{ and } N_t^s < N_t^d \Rightarrow N_t^s = N_t \tag{15.6}$$

Within the above regime, other sources of deficit (of insufficient supply) can be distinguished, such as restrictions in the delivery of energy or raw materials that at the macro-scale can be attributed to limited foreign reserves (necessary to finance the import of intermediate commodities (MZ)). These conditions were typical of the former centrally-planned countries. The first models of the developing countries assumed that foreign demand was met, while deliveries to the domestic markets were residual, because of deficiencies affecting the domestic production potential.

In practice, the pure forms of the above regimes are found rather rarely, so model builders construct mixed models assuming the coexistence of different regimes (Malinvaud 1977).

15.4 A Stylized Structure of the National Economy Models

Over the last 50 years the structure of the macroeconometric models has changed significantly. Its simplified, stylized version will be demonstrated below as a skeleton system.[5] The first to be presented is the stylized structure of macroeconometric models prevailing in the 1980s, where the kind of economic activity was used as the criterion for ordering equations.

The key role was played by the national accounts identity defining the demand for GDP (X_t):

$$X_t = C_t + G_t + J_t + (E_t - M_t) \tag{15.7}$$

The basic behavioural and technological relationships were as follows:

- consumption function

$$C_t = c(Y_t, r_t, C_{t-1}) \tag{15.8}$$

- investment function

$$J_t = j(X_t, r_t, K_{t-1}) \tag{15.9}$$

- exports function

$$E_t = e\left(WT_t, p_t^w / p_t, E_{t-1}\right) \tag{15.10}$$

- imports function

$$M_t = m\left(X_t, p_t / p_t^m, M_{t-1}\right) \tag{15.11}$$

- employment function

$$N_t = n(X_t, N_{t-1}) \tag{15.12}$$

- producer price equation

$$p_t = p\left(w_t, N_t / X_t, p_t^m\right) \tag{15.13}$$

- average wage equation

$$w_t = w(u_t, p_t) \tag{15.14}$$

- population activity equation

$$N_t^s / L_t = n(u_t, w_t / p_t) \tag{15.15}$$

- money demand equation

$$M_t^d = m(Y_t, p_t, r_t) \tag{15.16}$$

where the endogenous variables are:

C_t is household consumption (constant prices),
E_t is exports (constant prices),

[5] Cf. Klein et al. (1999) and Whitley (1994).

J_t is gross investment (constant prices),
K_t is fixed capital (period end, constant prices),
M_t is imports (constant prices),
N_t is employment,
N_t^s is labour force supply,
p_t is producer prices,
p_t^m is imports prices, $p_t^m = p(p_t^w)$,
u_t is unemployment rate, $u_t = (N_t^s - N_t)/N_t^s$,
w_t is nominal average wages,
Y_t is real disposable household incomes,

the exogenous variables are:

G_t is real expenditures of public institutions,
L_t is population size,
WT_t is global world exports,
p_t^w is world prices.

The specification of the above equations will be briefly commented on. Its major modifications which were introduced in the next years will be discussed in the paragraphs below. A full-length presentation will take place in the next chapters of this monograph.

The consumption function (15.8) represents a Keynesian orientation. Consumption is mainly determined by the real disposable household income. Lagged consumption introduced because of inertia (the Brown effect) can be also interpreted as a summary characteristic of the lag distribution of real incomes. The interest rate explains consumption changes caused by the changes in savings.

The investment function (15.9) is a version of a flexible accelerator function. It is assumed that the desired level of fixed capital K_t^* is determined by output level (X_t) and interest rate (r_t):

$$K_t^* = k_t(X_t, r_t) \tag{15.17}$$

and that fixed capital expands proportionally to the difference between the desired and effective stocks of fixed capital:

$$\Delta K_t = \lambda\left(K_t^* - K_{t-1}\right) \tag{15.18}$$

Hence investment will be equal to the sum of the increase in fixed capital and its depreciation $D_t = d_t K_{t-1}$:

$$J_t = \Delta K_t + D_t = i(X_t, r_t, K_{t-1})$$

In the stylized model the public investment demand is assumed to be exogenous.

The foreign trade equations are more or less standard. Exports depend on the world demand and imports on the total domestic demand. In both cases demand is appropriately adjusted by relative prices.

The employment function is most frequently generated from inverting the production function. An important role is played by the lags in employment adjustments to output changes. The labour force supply is determined by multiplying the number of population by the active population ratio. The ratio is affected by the labour market situation (unemployment level) and the attractiveness of the available jobs (real wages). The rate of unemployment was residual.

Producer prices depend on the unit costs represented by labour costs and import prices, and average wages are determined by the rate of inflation and unemployment rate.

Financial flows are described by the equation explaining money demand depending on real incomes, prices and interest rate. The remaining components of financial flows, including state budget revenues and expenditures, are explained by the relevant identities.

The above system of equations contains all major feedbacks characteristic of macroeconometric models. Expanding real disposable income is followed by a respective increase in GDP and, allowing for employment lags and predetermined wages, in real incomes. Therefore, the consumer multiplier can be presented as the following relations:

$$\Delta Y_t \rightarrow \Delta C_t \rightarrow \Delta X_t \rightarrow \Delta N_t \rightarrow \Delta Y_t$$

In the above models the accelerator appears, which is demonstrated through the following relationships:

$$\Delta J_t \rightarrow \Delta X_t \rightarrow \Delta J_t$$

The additional imports multiplier must be taken into account:

$$\Delta X_t \xrightarrow{+} \Delta M_t \xrightarrow{-} \Delta X_t$$

The models also contain an inflationary loop accentuating the role of wage inflation:

$$\Delta p_t \rightarrow \Delta w_t \rightarrow \Delta p_t$$

The above feedbacks are presented in Figs. 15.1 and 15.2. that outline the relations in the stylized model. Their empirical counterparts were obtained using multiplier analyses that had preceded numerous model simulations of the likely impacts of economic policy.

15.5 Model Dynamics. Rational Expectations. Long- and Short-Term Relationships

New tendencies that led to significant changes in the models' structure will be briefly described below. They were characterised in detail in part I of this monograph.

At the turn of the 1970s attention was paid to the necessary standardisation of techniques for introducing lags into particular equations. Following the LSE methodology, D. Hendry (1995) proposed to systematically test lags for significance, starting from the longest lags available (a top-down approach).

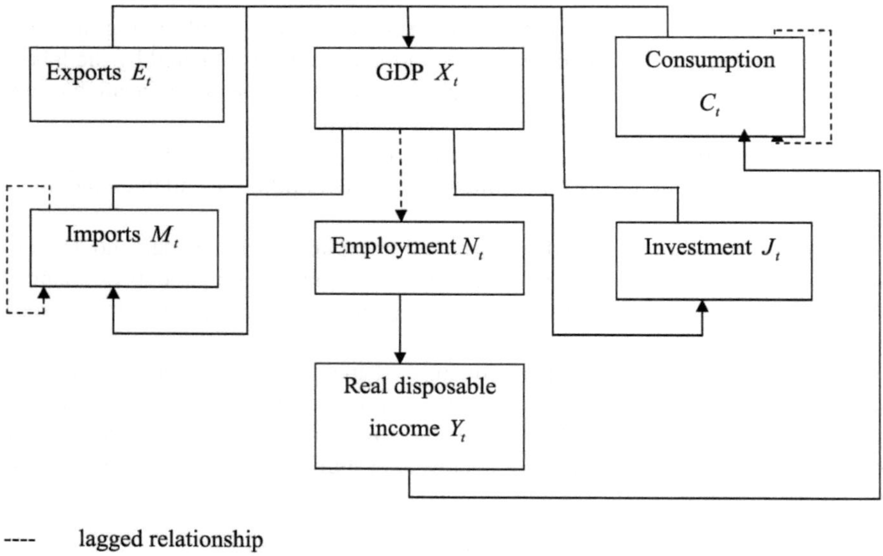

Fig. 15.1 Consumer multiplier and accelerator

The use of economic agents' expectations in macromodels has a long tradition. Expectations based on data regularly obtained from household and firm surveys were already introduced into the appropriate equations of the early models of the US economy.

However, it was not until the turn of the 1980s that following the Lucas critique the tendencies became strong enough to cause the introduction of economic agents' expectations, including rational expectations, into the appropriate equations. Estimation techniques for rational expectations were developed (Fair and Taylor 1983) and applied to many macromodels, mainly to the equations explaining prices, interest rates, wages and exchange rates. Yet, the rational expectation assumption was not generally accepted. It was argued that the economic underpinning was exaggerated and that in fact expectations might have been "model-consistent". The forward-looking specifications were introduced into the IFM MULTIMOD (Laxton et al. 1998). Nevertheless, many model builders believed that the (small) economic agents did not have knowledge necessary to make forecasts based on the theoretical models, so they used adaptive expectations or, more recently, expectations determined by the learning process.

In the 1980s attention was paid to the fact that many economic processes (starting with consumption) included relations that tended to be stable in the long-run. This led to a proposal to construct the equations in such a manner as to distinguish the long-term (equilibrium) relationships from the short-run adjustments that should lead to a steady state (equilibrium). It was generally accepted that the solution could be an appropriately transformed ADL equation with lags. The approach resulted in

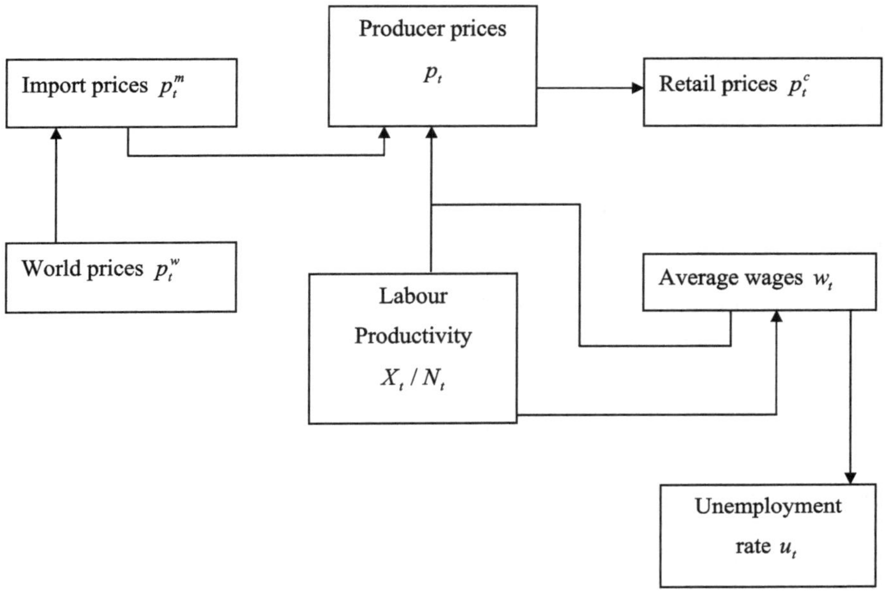

Fig. 15.2 Inflationary loop

the emergence of the error correction model (ECM). Its standard form for a one-period lag reads as follows:

$$y_t^* = \beta_0 + \beta_1 X_t \tag{15.19}$$

$$\Delta y_t = (1 - \alpha_1)\left(y_{t-1} - y_{t-1}^*\right) + \alpha_2 \Delta X_t + \varepsilon_t \tag{15.20}$$

where * is a long-term relationship and the variables represent either levels or logs.

The model builders applied the above specification techniques with different frequency in the 1990s, as they became aware that this approach had a solid theoretical and statistical underpinning in the cointegration theory, given that the statistical time series of economic variables are most frequently non-stationary (cf. Engle and Granger 1987).

15.6 The Microeconomic Foundations of Models' Specification

At the turn of the 1980s, attention was paid to the fact that the macroeconomic relationships presented in the models should be specified respecting the microeconomic foundations. Following the neoclassical theory households and enterprises were assumed to optimize their activities—households maximized utility and enterprises maximized profits (or minimized costs) under imperfect competition. Consumption, investment (residential) functions and labour force supply functions of households were derived from utility maximization. By solving the profit maximization problem of enterprises, the equations explaining the demand for production factors, producer price equations and wage equations can be built, assuming predetermined production functions.

The specification of the equations differs from that previously shown (the addition of expectations aside), because stocks being at economic agents' disposal were introduced besides flows. In particular, according to the Friedman permanent income hypothesis, the consumer demand function (15.8) was extended by introducing a variable that stood for the stock of household personal wealth V_t (initially containing financial assets, but subsequently also physical ones, mainly apartments and residential buildings). At a later time, the expected future real labour incomes called human wealth (H_t) were introduced, following the long-life Modigliani hypothesis. More details can be found in Chap. 16.

The demand for production factors was specified taking into account the impact of labour being substituted for machines and equipment. In the investment function (15.9) the investment user costs (their major component being the interest rate) were introduced as an additional variable. This allowed determining the expected profits from investments. In the next years attention was given to the necessity to allow for the installation costs of new equipment which involved appropriate lags in the investment process. Attempts were made to utilise the Tobin's "Q" concept.

In the specification of the employment function (15.12), real wages or wage-to-profits ratios were introduced. These specifications are broadly discussed in Chap. 17.

15.7 The Modelling of Supply

The initial specification of the structural equations generally assumed that as result of market transactions the demand for commodities and services and the demand for production factors were met. It was therefore assumed that the supply of commodities and services and the supply of production factors followed demand. For that reason, in macromodels the supply functions were not explicitly specified. An exception was the labour market where labour force supply was generated from a separate equation, which made it possible to estimate the unemployment rate characterising disequilibrium in this market.

The possibility of frictional disequilibria coming into existence in the commodity and labour markets attracted attention quite early. The disequilibria could be eliminated by inventory adjustments, so some models were provided with separate equations explaining changes in the inventories of final commodities, or by adjustments in exports or imports. In the latter case, the characteristics of the potential demand gaps had to be calculated, which were also used for determining producer prices.

These characteristics can be built in different ways. The most frequently used are the coefficients of capacity utilization WX_t, which can be obtained from appropriate surveys or computed from the deviations of output trends. The coefficients obtained as a ratio of effective output X_t to potential output X_t^p have the strongest theoretical underpinning:

$$WX_t = X_t / X_t^p \tag{15.21}$$

The potential output is typically generated from a production function. Let us assume that it will be the most frequently used Cobb-Douglas production function with constant returns to scale:

$$X_t^p = BA_t K_t^{\alpha} N_t^{(1-\alpha)} e^{\varepsilon_t} \tag{15.22}$$

where A_t is total factor productivity explaining the impacts of technological progress, α is output elasticity with respect to fixed capital, and ε_t is the disturbance term.

The production functions were explicitly specified in only several annual models and the impacts of technological progress were exogenous (represented mostly by an exponential trend). The precondition for introducing new specifications with endogenized total factor productivity (TFP) was the development of endogenous growth theory. It was assumed that TFP growth was dependent on an increase in knowledge capital represented by human capital and cumulative R&D expenditures, both domestic and foreign (W. Welfe 2009). The above functions were most commonly used in the disequilibrium models and in the long-term models.

The specification of the equations explaining the supply of particular groups of commodities and services could be found in computational general equilibrium models where the equilibrium prices had to be determined, and in the supply-determined models assuming realisation of supply and presence of excess demand in the markets.

The supply-determined models were usually constructed in countries with centrally planned economies and in developing countries going through the early phase of their development. The key role in these models was played by the production functions of industry and agriculture, generating commodity supplies. The allocation of supplies, including imports, was described by supply functions defined for particular groups of intermediate and final users. In most cases the supply of exports followed foreign demand, whereas domestic consumer demand and particularly investment demand were rationed.

Chronic unemployment present in industrialized countries was an object of numerous studies implying the use of the disequilibrium-type models (Barro and Grossman 1971). In the models, both demand and supply functions were specified (supply depended on fixed capital and employment). An economy was assigned to one of the competing economic regimes empirically, using the min-condition (Dreze et al. 1990). The enlarged disequilibrium models were used then to describe the functioning of the centrally planned economies, especially in the period when chronic disequilibria prevailed in the commodity and services markets (Davis and Charemza 1989; Welfe 1992). In the 1990s and afterwards, when economic regimes with constrained demand prevailed, the disequilibrium models' supply sectors retained only the production functions generating potential output.

In the supply sectors of the early demand-determined models the decisive role was played by the wage and price equations. Regarding the macroeconometric models functioning in the United Kingdom, the most important contributions were made by Layard and Nickell (1985), who stressed the role of negotiations between employers and employees' representatives in forming wages. As a result, the specification of wage equations was changed to explain real wage levels as the functions of the unemployment rate in the short-run and of labour productivity in the long-run. Taking into account that wages play a special role in forming prices in the imperfect markets, the concept of non-accelerating rate of unemployment (NAIRU) was formulated as an alternative to the natural unemployment rate concept.

In the 1980s, attention was focused on the finding that market pressures were as important in price determination as cost pressures. Market pressures representing the demand or supply gaps can be measured using several characteristics of the capacity utilization rate WX_t. Hence, this variable was introduced in different forms into the producer price equations in the majority of the models. It is also used as an additional variable in the equations explaining exports and imports, modifying their volumes in response to market pressures that are mainly generated by unexpected demand shocks.

With the relaxation of international financial flows and the associated abandonment of control over exchange rates, research attempts were made to endogenize their development. The most significant became the theory assuming that the exchange rate based on the purchasing power parity (PPP) exchange rate varies according to changes in the ratios between the interest rates of the compared countries (uncovered interest parity UIP) followed by respective capital flows and to changes in risk premium.

More recently, attention was paid to the wage, price and cost formation rigidities related to the institutional environment. This neo-Keynesian perspective is generally adopted in the DSGE models.

15.8 The Stylized Version of the Macroeconometric Model Structure in the Last Twenty Years

The above changes can by summarized by presenting an alternative schema of the stylized structure of the demand-determined macromodels. The presentation will concentrate on the specification of the long-term equations.[6]

The national accounts identity:

$$X_t = C_t + J_t + G_t + \Delta R_t + E_t - M_t \tag{15.23}$$

The equations of the demand sector:

- Consumer demand

$$C_t = c(H_t, V_t, r_t) \tag{15.24}$$

[6] An analogous schema for the early 1990s can be found in Whitley (1994, p. 51).

- Investment demand

$$J_t = j\big(X_t, r_t, w_t/p_t^s\big) \tag{15.25}$$

- Inventory increase

$$\Delta R_t = r(X_t, WX_t) \tag{15.26}$$

- Imports

$$M_t = m\big(X_t, WX_t, p_t/e_t p_t^w\big) \tag{15.27}$$

- Exports

$$E_t = e\big(WT_t, WX_t, p_t/e_t p_t^w\big) \tag{15.28}$$

Consumption depends on expected income H_t, personal wealth V_t and interest rate r_t. Investment depends on GDP, user costs represented by interest rate (r) and the ratio between prices of production factors (wages and fixed capital deflators). An inventory increase is dependent on GDP and the rate of capacity utilization (WX_t). In the foreign trade equations, the rate of capacity utilization adjusts the volumes of exports and imports in addition to relative prices.

The equations of the supply sector are as follows:

The equations explaining the demand for production factors are not significantly different from those discussed earlier (see (15.12) for employment). The other equations are:

- Real average wages

$$w_t/p_t = w(u_t, \pi_t, tx_t) \tag{15.29}$$

- Producer prices

$$p = p\big(WX_t, w_t/\pi_t, ep^w\big) \tag{15.30}$$

- Exchange rate

$$e = e\big(e^e, r_t/r_t^w, \gamma\big) \tag{15.31}$$

where

e^e is expected exchange rate,
$\pi_t = X_t/N_t$ is labour productivity,
u_t is unemployment rate,
tx_t is tax rate,
r_t^w is foreign interest rate,
γ_t is risk premium.

Real average wages depend on the rate of unemployment, labour productivity and tax rates. Producer prices derive from labour costs, import prices and the rate of capacity utilization. The exchange rate is determined by its expected value, the ratio of interest rates and the risk premium. The money-demand equation has a similar specification as in (15.16).

Let us note that in the short-run dynamic adjustments follow with certain lags, which justifies the introduction of lagged endogenous variables into the short-term equations. Expectations can be introduced in a similar manner.

The above stylized system of equations can be easily transformed into a structure where the point of departure is the criterion of the kind of economic agent. This structure will distinguish households, enterprises, public institutions and foreign agents. This is important, because the structure will be used in the following chapters of this book.

15.9 An Outline of Estimation Methods and Computational Techniques

The changes in the economic orientation and the structure of macromodels were associated with the changes in the model parameter estimation methods, precipitated by the computer revolution which abolished barriers limiting the numbers of equations and, more importantly, enabled the use of advanced numerical techniques. The development of iterative procedures used to solve large nonlinear systems (the Gauss-Seidel, Newton techniques) removed the necessity to linearize the equation systems and to transform them into quasi-recursive. These and the next comments are very introductory, as the descriptions of the estimation methods can be found in comprehensive monographs on the theory of econometrics (Hendry 1995; W. Welfe 2009).

The concepts and methods known as the Cowles Commission methods dominated in the mainstream models for many years. In constructing the macroeconometric simultaneous models particular structural equations had to be specified following the economic theory postulates. Variables whose introduction was theoretically unjustified were eliminated by imposing zero restrictions on the appropriate parameters. A broad spectrum of estimation methods was developed to ensure the consistency of parameter estimators in this class of simultaneous equations models. The methods were the two-stage (TSLS), three-stage (3SLS) least squares, maximum likelihood (ML), limited information likelihood (LIML), and instrumental variables method (IIV), as well as other special mutations of the procedures. They were initially used to estimate the parameters of the small models, but with the high-power computers becoming available they were also applied to estimate the parameters of the large-scale macromodels. The results of many empirical investigations showed, however, that parameter estimators obtained from the ordinary least squares (OLS) usually showed only a negligible bias. Because of that, the OLS remained the major parameter estimation technique for the equations of the macroeconometric simultaneous equations models.

In the last years estimation methods have been developed to address new needs in economic modelling. The relevant examples are the rational expectations models (Fair and Taylor 1983), on the one hand, and the disequilibrium models with unobservable variables (representing demand and supply), on the other. They were characterised in a monograph by Quandt (1988).

The procedures following the Cowles Commission principles were challenged by Sims (1980). He opposed the constructors of the models, claiming that the structural equations were specified in a way that could be precarious, because the zero restrictions were chosen arbitrarily. This critique led to the development of the vector autoregression (VAR) technique. In its initial version all model's variables were made dependent on all variables, but with lags. This eliminated the simultaneity of the variables, because the jointly interdependent model was transformed to its reduced form.

The equations of the above models did not have the economic interpretation, so the models were mainly used in forecasting. They were generalized, though, by imposing appropriate parameter restrictions, which ensured the possibility of returning to structural equations (SVAR). This approach was mainly adopted in sectoral analyses, for instance those dealing with inflationary processes (Staszewska-Bystrova 2009).

The last decade witnessed the development of investigations attempting to link the models with the cointegration analysis. Because of technical problems the investigations had to be limited to small systems containing several equations only. Hence they dealt with small segments of national economies, such as inflation processes in Norway and Poland or the UK money markets, in hope of returning to the idea of recursive segmentation of the macromodels.

An alternative to the traditional procedures was estimation methods built on the results of the time series analysis. The empirical investigations show that most economic time series are nonstationary (mainly I (1)) and that the use of the levels of variables in the regression analysis may lead to spurious regressions. Engle and Granger (1987) proposed a solution to this problem. It became the theoretical underpinning of the broadly use of the error correction models (ECM). The two-stage procedure they put forward first uses the OLS to estimate the parameters of the long-term static equations (the parameter values are frequently calibrated) and then the ECM is applied to the first differences of the variables (or their logs) in the equations representing the short-run adjustments. The ECM is applied together with lags and leads representing the respective expectations. The above techniques found a broad application in estimating the parameters of particularly the large macroeconometric models.

It must be finally mentioned that the Bayesian estimation methods associated with the development of the DSGE models started to be used.

References

Barro, R. J., & Grossman, H. I. (1971). A general disequilibrium model of income and employment. *American Economic Review*, *61*, 82–93.

Bodkin, R. G., Klein, L. R., & Marwah, K. (Eds.) (1991). *A history of macroeconometric model-building*. Aldershot: E. Elgar.

Davis, C. M. & Charemza, W. (Eds.) (1989). *Models of disequilibrium and shortage in centrally planned economies*. London: Chapman and Hall.

Dreze, J. H., Bean, C., Lambert, J. P., Metha, F., & Snessens, H. R. (Eds.) (1990). *Europe's unemployment problem*. Cambridge: MIT Press.

Engle, R. F., & Granger, W. J. (1987). Cointegration and error correction: representation, estimation and testing. *Econometrica, 55*, 251–276.

Fair, R. C. (2004). *Estimating how the macroeconomy works*. Cambridge: Harvard University Press.

Fair, R. C., & Taylor, J. B. (1983). Solution and maximum likelihood estimation of dynamic nonlinear expectations model. *Econometrica, 51*, 1169–1186.

Ginsburgh, V., & Keyzer, M. (1997). *The structure of applied general equilibrium models*. Cambridge: MIT Press.

Hendry, D. H. (1995). *Dynamic econometrics*. Oxford: Oxford University Press.

Klein, L. R., & Welfe, W. (1982). *Wykłady z ekonometrii (Lectures in econometrics)*. Warszawa: PWE.

Klein, L. R., Welfe, A., & Welfe, W. (1999). *Principles of macroeconometric modeling*. Amsterdam: North-Holland.

Laxton, D., Isard, P., Faruqee, H., Prasad, E., & Turlelboom, B. (1998). *MULTIMOD Mark III: the core dynamic and steady-state models* (IMF Occasional Paper No. 164).

Layard, P. R. G., & Nickell, S. J. (1985). The causes of British unemployment. *National Institute Economic Review, 111*, 62–85.

Malinvaud, E. (1977). *The theory of unemployment reconsidered*. Oxford: Basil Blackwell.

Okólski, M., & Timofiejuk, I. (1978). *Statystyka ekonomiczna. Elementy teorii (Economic statistics. Elements of theory)*. Warszawa: PWE.

Orłowski, W. (1992). *Opcje polityki monetarnej i fiskalnej okresu przejścia do gospodarki rynkowej. Analizy z wykorzystaniem Modelu Równowagi Ogólnej gospodarki polskiej (Options of monetary and fiscal policy in the period of transition. Analyses using the CGE model of the Polish economy)*. Ph.D. Dissertation, Łódź.

Quandt, R. E. (1988). *The econometrics of disequilibrium*. New York: Basil Blackwell.

Sims, C. A. (1980). Macroeconomics and reality. *Econometrica, 48*, 1–48.

Staszewska-Bystrova, A. (2009). *Wektorowe modele autoregresyjne w analizie makroekonometrycznych szeregów czasowych (Autoregression vector models in analyses of macroeconometric time series)*. Toruń: Dom Organizatora.

Tomaszewicz, Ł. (1994). *Metody analizy input-output (Methods of input-output analysis)*. Warszawa: PWE.

Welfe, A. (2009) *Ekonometria. Metody i ich zastosowania (Econometrics. Methods and their applications)* (4th ed.). Warszawa: PWE.

Welfe, W. (1992). *Ekonometryczne modele gospodarki narodowej Polski (Econometric models of the Polish economy)*. Warszawa: PWE.

Welfe, W. (Ed.) (2009). *Knowledge-based economies. Models and methods*. Frankfurt/Main: P. Lang.

Whitley, J. D. (1994). *A course in macroeconomic modelling and forecasting*. New York: Harvester/Wheatsheaf.

Zienkowski, L. (2001). *Co to jest PKB? Jego rola w analizach ekonomicznych i prognozowaniu (What is the GDP? Its role in economic analyses and forecasting)*. Warszawa: DW Elipsa.

Żółkiewski, Z. (1995). *Model równowagi ogólnej z uwzględnieniem zachowań monopolistycznych (The CGE model allowing for monopolistic competition)* (Studia i Prace, Z prac ZBSE, GUS i PAN No. 226).

Chapter 16
Modelling the Behaviour of Households

This chapter describes submodels explaining household behaviour that can be found in macroeconometric models. The submodels contain not only the consumption functions, but also investment demand functions concerning apartments and residential buildings (sometimes also consumer durables) and labour force supply functions. The modelling of households' financial assets will be postponed until the chapter on the modelling of financial flows.

16.1 The Consumer Demand Function

Consumption functions in the early macromodels explained the global household consumption C_t (frequently called individual or private consumption), sometimes consumption per capita (C_t/L_t), where L_t stood for the number of inhabitants. The global consumption was mostly identified with total real consumer expenditures, notwithstanding the fact that the value of the flows of services of consumer goods would be equal to the expenditures only in case of non-durables and services. The use of durables, especially of dwellings, was estimated in a few models only. Several other models omitted durables (Fernandez-Corugedo et al. 2007). Total consumption (expenditures) was frequently decomposed into durables (cars), non-durables and services. Numerous commodity groups were distinguished only in the multisectoral models, especially in those making use of the input-output blocks of equations—one of them was the Cambridge model of the UK (Stone 1954; Suchecki 2006).

The above functions had a Keynesian origin. The major explanatory variable was the real disposable income Y_t after tax or its value per capita (Y_t/L_t). To take account of the competitive role of savings, the real interest rate R_t (either short- or long-term) was introduced as an additional explanatory variable. This can be represented with the following equation:

$$\ln C_t^* = \alpha_0 + \alpha_1 \ln Y_t + \alpha_2(1 + R_t) + \xi_t \tag{16.1}$$

W. Welfe, *Macroeconometric Models*,
Advanced Studies in Theoretical and Applied Econometrics 47,
DOI 10.1007/978-3-642-34468-8_16,

Different categories of household income were initially distinguished, for instance labour income, farmers' incomes, entrepreneurs' profits, and capital gains, which were assigned specific propensities to consume. The later models abandoned income decomposition, because the income distribution differences became less pronounced and the distribution of income stabilized. In a larger number of cases real net credits were being added to real incomes to allow for the increasing availability of this additional source of purchasing funds. The above specification represents the long-run consumer demand function. Hence, it was frequently assumed that consumption elasticity with respect to the real disposable income could be calibrated at a level equal one ($\alpha_1 = 1$), which expressed the conviction that the savings ratio would stabilize in the long-run.

In the multisectoral models the consumer goods prices (indices, deflators for commodity groups) were introduced as explanatory variables. If the number of commodity groups was small, then the relative prices were used, i.e. the ratios between particular price indices and the general price index. If the number of commodity groups was large, then special "complete" models were used, beginning with R. Stone's linear expenditure system (LES), which was applied to the Cambridge model of the UK. A more recent example of a large multisectoral model is the DRI-WEFA model of the US economy containing 11 categories of durables, 9 groups of non-durables and 17 groups of services (DRI-WEFA 2002).

The inertia in households' behavioural responses to the impacts of changes in real incomes were observed relatively early. As proposed by the Canadian econometrician T.M. Brown, this habit persistence could be represented using lagged consumption C_{t-1} (Brown 1952). The modification is expressed by the following equation:

$$\ln C_t^* = \alpha_0 + \gamma \ln C_{t-1} + \alpha_1 \ln Y_t + \alpha_2 (1 + R_t) + \xi_t \tag{16.2}$$

Assuming that $C_t = C_{t-1}$, the long-term elasticities can be calculated. The elasticity with respect to Y_t is given by $\frac{\alpha_1}{1-\gamma}$.

For the sake of illustration, let us refer to the results obtained for Poland (based on the 1960–1998 sample): the estimate of the autoregression coefficient was $\gamma = 0.53$, whereas the short-run elasticity with respect to real disposable income was $\alpha_1 = 0.40$ and the long-term elasticity equalled 0.85 (Welfe 2001, p. 61).

The estimates for the industrialized countries were obtained, e.g. in the multicountry model MIMOSA, using a slightly shorter sample. The estimates of the parameter γ were within the range of 0.3–0.6, whereas the long-run elasticities with respect to real personal incomes per capita ranged from 0.85 to 1.02 (MIMOSA Modelling Group 1990, p. 10).

The habit formation coefficients which were calculated in the later studies in the 21st c. produced higher results, ranging from 0.6 to 0.8 for the UK, US and the EURO Area (Harrison et al. 2005, pp. 109–110, in BEQM model).

It is worth mentioning that the hypothesis postulating that demand depends on the real interest rate turned out to be non-satisfactory. The parameter estimates took non-acceptable positive values. As a result, many model builders assumed that consumer decisions were affected by money illusion and nominal interest rates were used instead (cf. Fair 2004). Sometimes the α_2 parameter values were calibrated or the interest rate was omitted.

Table 16.1 Household consumption elasticities

Country	W.r.t. real disposable income		W.r.t. real interest rates	
	Short-term	Long-term	Impulse multipliers	Sustained multiplier
Canada	0.58	1.0	−0.13	−0.57
France	0.73	1.0	−0.23	
Germany	0.78	1.0	−0.23	−0.76
Italy	0.10	1.0	−0.05	−2.4
Japan	0.62	1.0	−0.23	−1.5
United Kingdom	0.28	1.0	−0.19	−0.65

Source: FRB Global (Levin et al. 1997, p. 7)

In the 1990s the long-term and short-term demand functions started to be distinguished by the model builders. The short-term equations were dynamic and the ECM were used. The point of departure was the static long-term equation (16.1). The standard form of the short-term equation is:

$$\Delta \ln C_t = \alpha_0 + \beta_1\left(\ln C_{t-1} - \ln C^*_{t-1}\right) + \beta_2 \Delta \ln Y_t + \beta_3 \Delta(1 + R_t) + \xi_t \qquad (16.3)$$

As an illustration, the results obtained with the multicountry model FRB Global (Levin et al. 1997) will be shown. In the model the long-term elasticity with respect to real income was calibrated at a level equal to one. The real long-term expected interest rate was included, as well as a variable representing labour activity.

A special property of the above equations that macromodels used for many years was that consumer demand depended on the current real disposable income only, which means that households were liquidity constrained. In the 1980s, after the failures in forecasting UK consumption, attention was directed to the previously ignored impact of the real personal wealth of households W_c, modifying household behaviour. Hendry and von Ungern-Sternberg (1981) introduced financial wealth WL_t, and in the mid-80s the appropriateness of adding physical wealth WN_t (housing wealth) as an additional variable was recognized. In the case of countries (or periods) where the appropriate information was not available, the inflation rate was employed as a symptomatic variable. This happened in the early UK models, in the INTERLINK model (OECD 1993), the MIMOSA models and, more recently, also in the MESANGE model for France (Allard-Prigent et al. 2001). As well as being a major source of real changes in the value of personal wealth, this variable also ensured the transmission of financial sector shocks to the real sector.

It was argued that with the introduction of personal wealth as an additional explanatory variable competitive to real personal income, the homogeneity restriction should be imposed. The sum of consumption elasticities with respect to real disposable income and real personal wealth should be one. The following long-term equation meets the homogeneity restriction:

$$\ln C^*_t = \alpha_0 + \alpha_1 \ln Y_t + (1 - \alpha_1) \ln W_t + \alpha_2(1 + R_t) + \xi_t \qquad (16.4)$$

This function is still used in many macroeconometric models. For the sake of illustration, let us present the estimates of the long-run elasticities obtained for

Table 16.2 Long-term elasticities of personal consumption w.r.t. the Y, W, and R variables

Country	EUROMON			NIGEM		
	Y	W	R	Y	W	R
France	0.87	0.13	−0.004[b]	0.86	0.14	−2.89[b]
Germany	0.88	0.12	−0.004[b]	0.80	0.20	
Italy	0.88	0.12	−0.002[a]	0.80	0.20	−0.002[a]
Spain	0.93	0.07	−0.002[a]	0.91	0.09	
United Kingdom	0.85	0.15	−0.003[a]	0.84	0.16	−0.007[a]

Source: de Bondt et al. (1997)

[a]Short-term

[b]Long-term

the models EUROMON (de Bondt et al. 1997) and NIGEM (NIGEM 2002) (Table 16.2).

It must be noted that the above estimates of elasticities with respect to interest rates are not comparable with the estimates shown for the GLOBAL model.

Since Davidson et al. (1978), the parameters of the dynamic, short-run equations have been typically estimated using the error correction model (ECM). It has been so, because the variables in (16.4) are usually non-stationary, but their first differences are mostly stationary. This supposition is subject to testing.

This yields a conventional short-term equation, where a one-period lag is assumed (one quarter in the quarterly models):

$$\begin{aligned}\Delta \ln C_t = \beta_0 + \beta_1\left(C_{t-1} - C^*_{t-1}\right) + \beta_2 \Delta \ln Y_t + (1-\beta_2)\Delta \ln W_t \\ + \beta_3 \Delta(1+R_t) + \xi_t\end{aligned} \tag{16.5}$$

where C^* is the long-term consumer demand function as in (16.4).

Table 16.3 presents parameter estimates of the consumer demand function for developing countries. The estimates were obtained using the Bank-Gem model of the World Bank. It is noteworthy that the homogeneity assumption was not introduced.

Further dynamization of the consumer demand function implied the introduction of expectations of explanatory variables, including rational expectations. This particularly applied to the UK models, but also to the US economy models built at the FRB. Let us present a simple specification of the short-term equation used in the NIGEM model as an example (NIGEM 2002). To the equation being a mutation of (16.5) the expected consumption with one period lead was introduced as an additional explanatory variable:

$$\Delta \ln C_t = \lambda\left[\ln C_{t-1} - \alpha \ln Y_{t-1} - (1-\alpha)\ln W_{t-1}\right] + \delta \Delta \ln C_{t+1} + \varepsilon_t \tag{16.6}$$

where δ is the time preference rate.

In this forward looking equation, the parameter was calibrated at a level $\delta = 0.97$. The following estimates of the elasticities α were obtained: 0.83 for France, Germany and Italy, and 0.86 for the United Kingdom (Barrell et al. 2004).

Table 16.3 Parameter estimates of the consumption function for developing countries

Country groups	Long-term with respect to		Short-term w.r.t. real disposable income	Error correction coefficient
	Personal wealth	Real disposable income		
Group 1	0.16	0.88	0.49	−0.24
	(4.9)	(27.9)	(8.9)	(4.0)
Group 2	0.16	0.70	0.32	−0.19
	(8.1)	(28.8)	(7.4)	(6.0)
Group 3	0.13	0.66	0.35	−0.15
	(3.7)	(16.6)	(8.2)	(3.5)
Group 4	0.11	0.60	0.36	−0.31
	(3.4)	(9.3)	(7.8)	(3.6)
Group 5	0.04	0.85	0.51	−0.41
	(1.7)	(70.1)	(12.6)	(8.7)
Group 6	0.07	0.76	0.38	−0.30
	(7.6)	(46.9)	(18.0)	(8.9)

Note: t-Student's statistics are given in the brackets

Source: Petersen et al. (1991, p. 37)

Group 1: countries of middle incomes, demand constrained, no major debt overhang, strong institutions, less open

Group 2: countries of middle income, demand constrained, no major debt overhang, strong institutions, more open

Group 3: countries of middle income, supply constrained, indebted, strong institutions, more open

Group 4: countries of middle income, supply constrained, indebted, strong institutions, less open

Group 5: countries of middle income, supply constrained, indebted, weak institutions, less open

Group 6: countries of low income, supply constrained, indebted, weak institutions, less open

In the models covering several countries additional explanatory variables were added, i.e. an unemployment rate whose increase would postpone current consumption (the models of Belgium and Lithuania), the level of professional activity or even the output gap (GLOBAL-FRB for USA).

The consumer demand functions that have been characterized so far were important components of macroeconometric models describing market economies where the consumer demand was met. Market clearing was ensured by market mechanisms. However, in the consumer goods markets of many countries, including centrally planned economies (CPE), temporary or chronic deficits were observed. In other words, an unobservable excess demand CE_t appeared:

$$CE_t = CD_t - CS_t > 0 \tag{16.7}$$

where CD_t is consumer demand, CS_t is supply of consumer goods.

In these circumstances the specification of the consumer demand function had to be modified.[1] The most important modifications included the extension of the notion of consumer demand by adding postponed demand (transferred from the previous periods), its modification related to forced substitution and the extension of disposable income to take account of forced savings (the inflationary gap). The observable disequilibrium indicators were proposed, their functions representing the unobservable excess demand. Their introduction as an additional explanatory variable allowed estimating the parameters of the equation explaining observed consumption and thus consumer demand. We have:

$$CD_t = CS_t + CE_t = C_t + f(IN_t) \tag{16.8}$$

where IN_t is the disequilibrium indicator, and $C_t = CD_t - f(IN_t)$.

In the early period of CPEs' transition towards market economies, deficits in consumer goods were still found in several countries. This caused that the model builders started to look for special disequilibria indicators. For instance, in the model of the Russian Federation the role was given to the output gap characteristics (the wage payment lags were additionally introduced into this equation) (Basdevant 2000).

In the 1980s the specification of the consumer demand functions was substantially changed in many market economy models, following the principles of the neoclassical theory of consumer behaviour. Friedman (1957) formulated the permanent income hypothesis on the assumption that unobservable permanent income was an adequate determinant of consumption (Yaari 1965). Modigliani (1957, 1966) proposed the life-cycle hypothesis of savings, referring to the results of the previously conducted utility analysis (Modigliani and Brumberg 1954; cf. also Ando and Modigliani 1963 and Modigliani 1975).

The permanent income hypothesis assumes that consumption can be split into permanent consumption dependent on permanent income and transitory consumption related to transitory income composed of irregular and incidental elements. The permanent income which is an unobservable variable can be approximated—as suggested by Muellbauer and Luttimore (1995)—using a weighted average of the real disposable income and personal wealth. As a matter of fact, this would mean a return to the initial specification (16.4).

The life-cycle hypothesis based on the concept formulated by Blanchard (1985) found a more general application. It utilizes a model of overlapping generations where consumers are assumed to optimize consumption over their lifetimes, given a fixed time horizon (consistent with the predetermined probability of death). This model refers to rational behaviour of households. Let us mention that Hall (1978) earlier formulated the hypothesis of intertemporal optimization, yielding the Euler equation.

However, following the Campbell and Mankiw study (1991), it had to be recognized that a certain proportion of households is liquidity (credit) constrained, so

[1] Most comprehensive description of the modifications can be found in Welfe (1991) and Welfe and Welfe (2004).

they take their decisions based on their current real disposable incomes. The average shares of such households were estimated at 44 % in the MULTIMOD model and at 30 % in the QUEST model, with particular countries being considerably different in this respect. The FRB/US model estimated their share in the USA at only 10 %.

In the theory, a rational household (consumer) is assumed to maximize its discounted utility over the life cycle, given its predetermined life expectancy. Aggregation leads to the following long-run consumption function, where consumption is related to real wealth W_t:

$$C_t^W = \alpha W_t \tag{16.9}$$

where $W_t = V_{t-1} + H_t$, and H_t is the present expected human real wealth and V_t is the real financial wealth.[2]

The expected human real wealth can be calculated from the equation:

$$H_t = \int_{t=0}^{\alpha} (Y_t) e^{-(r+\lambda+n)t} \, dt \tag{16.10}$$

where

r is real interest rate,
λ is the probability of death,
n is the rate of population growth,
Y_y is the real net labour income.

This equation can be approximated as follows:

$$H_t = \sum_{i-0}^{\alpha} \left(\frac{1-\lambda}{1-r} \right) Y_t \tag{16.11}$$

The assumed discount rate was rather high. In the FRB/US model it was 25 % annually.

The real financial wealth gained a broad meaning. It should include the market value of enterprises, government debt and net foreign assets. Following Masson et al. (1990), its use is justified by the assumption that the household sector has full control over domestic financial assets. Several authors treat these components separately, looking for different marginal propensities to consume.

The parameter α which represents the long-run marginal propensity to consume can be treated as a function of relative risk aversion, of the rate of time preference δ depending on the intertemporal elasticity of substitution, of the real interest rate and of the probability of death:

$$\alpha = 1 - (1-\lambda)(1-\delta)(1-r) + \cdots \tag{16.12}$$

The equation that should be estimated is obtained by adding the component determined by expected wealth and the component related to the liquidity-constrained households, dependent on the current real disposable income:

$$C_t = C_t^W + f(Y_t) \tag{16.13}$$

[2]Several authors look at the impact of the consumption-total wealth ratio, i.e. of α, on the expected stock returns (cf. Lettau and Ludvigson 2001, and Zachłod-Jelec 2010).

In practice, the double-log approximation is most frequently used and a distinction is drawn between real human and financial wealth:

$$\ln C_t = \alpha_0 + \alpha_1 \ln H_t + (1 - \alpha_1) \ln V_t + \alpha_2 \ln Y_t + \xi_t \tag{16.13a}$$

In the applications it was generally assumed that $\lambda = 0.02$, which is equivalent to assuming life expectancy of 50 years. The time preference rate was assumed to be 0.009 on average (from 0.005 in Japan to 0.01 in the US, following the QUEST model). The elasticity range of intertemporal substitution was wide, from low 0.2 for the UK to 0.35 for the US as reported in the description of the BEQM model (Harrison et al. 2005, p. 109) and 0.6 for the US according to Smets and Wouters (2003).

Although the calculation of financial wealth is quite complicated, the above concepts were used to construct the consumption functions for many multicountry models, such as MULTIMOD or QUEST, the country models in Belgium, Finland (BOF 5), Spain and also for the recently built DSGE models.

16.2 Equations Explaining Household Investment Demand

Investment expenditures of households are commonly understood as amounts spent to purchase or construct apartments and dwelling houses. Only rarely does the notion include expenditures on durables, mainly because of data scarcity. It is usually assumed that real investment expenditures on dwelling construction (JBD_t) depend on real disposable income and the interest rates charged on long-term credits. Hence, the standard form of the long-term equation is:

$$\ln JBD_t = \alpha_0 + \alpha_1 \ln Y_t + \alpha_2(1 + R_t) + \xi_t \tag{16.14}$$

where R_r is the interest rate.

It is frequently assumed that the share of investment expenditures stabilizes in the long-run, hence the elasticity α_1 is calibrated at a level of 1 (for instance in the model MESANGE).

The above specification can be regarded as insufficient, because it fails to include the impacts of the relative prices of investment goods (PJ_t/PC_t) and ignores replacement demand driven by the aging of dwellings. Fair (2004) suggested introducing the difference between the depreciation of houses and lagged investment expenditures as an additional explanatory variable. We obtain then:

$$\begin{aligned}\ln JBD_t = {} & \alpha_0 + \alpha_1 \ln Y_t + \alpha_2(1 + R_t) + \alpha_3 \ln(PJ_t/PC_t) \\ & + \alpha_4(\delta KH_t - JBD_{-1}) + \xi_t \end{aligned} \tag{16.15}$$

where

KH_t is the stock of dwellings at period end,
δ is the depreciation rate.

Particular models, for instance in MIMOSA, explain the ratio of investment to dwelling stock (JBD_t/KH_t), allowing for changes in real disposable income and relative prices. To address cyclical fluctuations, the unemployment rate changes were introduced. MIMOSA produced the following estimates of the long-term elasticities: 0.6–0.9 w.r.t. real disposable income, −0.3, −0.7 w.r.t. relative prices and between −0.3 and −1.8 w.r.t. interest rate.

Investment expenditures in the multisectoral models are disaggregated and the above subsector is extended to include equations explaining the investment process, such as the number of buildings in process of construction, the duration of construction activities, etc.

16.3 Equations Explaining Labour Force Supply

The labour force supply functions were specified already in the early macroeconometric models. A direct explanation of the global labour force supply was only rarely provided—most frequently the labour activity coefficients were used as the explained variables. The labour activity (AK_t) coefficient is defined as a ratio between labour force supply (NS_t) and population size (L_t):

$$AK_t = NS_t/L_t \tag{16.16}$$

Therefore, the estimates of labour force supply were obtained in the macromodels from multiplying the estimated coefficients of labour activity by the population size estimate, using mainly the results of demographic forecasts.

In the multisectoral models, the coefficients of labour activity were disaggregated allowing for gender and age.

Let us mention that the underlying data on labour force supply were not very accurate. The supply was estimated by adding up the working-age population and the number of the registered unemployed. Doubts have been raised more recently, whether the total number of unemployed workers should be added or only their number representing the natural rate of unemployment.

Beginning with the early macroeconometric models, for example the Wharton quarterly model, labour activity started to be viewed as a variable strongly influenced by cyclical fluctuations. As a result, lagged unemployment rates were introduced as explanatory variables. Their impact accounted for job-seekers' inclination to give up job searches when the unemployment rate was growing (the discouraged worker hypothesis). Their significance was highly diversified, inducing weak reactions in France and Italy and strong responses in Japan (MIMOSA 1990).

More recently, economic conditions causing changes in labour activity and labour force supply were deduced from the maximization of the household utility function, the time offered to potential entrepreneurs being treated as an alternative to consumption. Solving this optimization problem yields the labour force supply function, where the supply (after converting the hours worked into the number of workers) depends on real wages. This specification prevails in the recently built models, including the RSGE models. It is commonly believed that a real wage increase has a positive impact—the prospects of having higher incomes are likely to induce the non-employed persons to seek jobs. However, an alternative interpretation is also possible—an increase in employed persons' real wages may encourage their family members to withdraw from the labour force.

Taking into account both these factors, the equation explaining labour activity will have the following form:

$$AK_t = f(AK_{t-1}, Z_t, U_t, U_{t-1} \ldots) \tag{16.17}$$

where Z_t is the real wage, U_t is additional variable.

An approximately similar specification can be found in the models by Fair (2004) and Coen and Hickman (2006). They differentiate the labour activity coefficients by gender and age.

In several macromodels the social, institutional determinants of labour force supply were addressed. The determinants include requirements regulating workers' rights to retirement, maternal leaves, etc. They change stepwise, so their impacts are most frequently represented by introducing respective dummies.

On the other hand, the role of the non-wage income sources is emphasized. An increase in personal wealth, equivalent to being better-off, may make a family member resign from a job (W_t); this option was introduced into the Fair model (Fair 2004). An increase in social contributions can also be observed. Unemployment benefits growing vis-à-vis the minimum wage may discourage job searches. To allow for these impacts, an additional variable, YBG_t/YP_t, was introduced into the W-models for Poland, where *YBG* represents social contributions, including unemployment benefits, and *YP* stands for personal disposable income. The elasticities were close to -0.1 (Welfe 2009).

References

Allard-Prigent, C., Auderis, C., Berger, K., Carrot, N., Duchene, S., & Presin, F. (2001). *Presén-tation du modéle MÉSANGE, modéle economiétrique de simulation et d'analyse générale de l'economic* (Document de travail). Paris: Ministére de l'Économic, Direction de la Prevision.

Ando, A., & Modigliani, F. (1963). The life cycle hypothesis of saving: aggregate implications and tests. *American Economic Review*, *53*, 55.

Barrell, R., Becker, B., Byrne, J., Gottschalk, S., Hurst, I., & van Welsan, D. (2004). Macroeconomic policy in Europe, experiments with monetary response and fiscal analyses. *Economic Modelling*, *21*, 877–931.

Basdevant, O. (2000). An econometric model of the Russian Federation. *Economic Modelling*, *17*, 305–336.

Blanchard, O. (1985). Debt, deficits and finite horizons. *Journal of Political Economy*, *93*, 223–247.

Brown, T. M. (1952). Habit persistence and lags in consumer behaviour. *Econometrica*, *20*, 355–371.

Campbell, J. Y., & Mankiw, N. G. (1991). The response of consumption to income, a cross country investigation. *European Economic Review*, *35*, 723–767.

Coen, R. M., & Hickman, B. G. (2006). An econometric model of potential output, productivity growth, and resource utilization. *Journal of Macroeconomics*, *28*, 645–664.

Davidson, J. E. H., Hendry, D. H., Srba, E., & Yeo, S. (1978). Econometric modelling of the aggregate time-series relationship between consumers. Expenditures and income in the United Kingdom. *Economic Journal*, *89*, 661–692.

de Bondt, G. J., van Els, P. J. A., & Studeman, A. C. J. (1997). *EUROMON: a macroeconometric multicountry model for the EU* (DNB Staff Reports No. 17). Amsterdam: De Nederlandsche Bank.

Fair, R. C. (2004). *Estimating how the macroeconomy works*. Cambridge: Harvard University Press.

Fernandez-Corugedo, E., Price, S., & Blake, A. P. (2007). The dynamics of aggregate UK consumers' non-durable expenditure. *Economic Modelling*, *24*, 453–469.

Friedman, M. (1957). *A theory of the consumption function*. Princeton: Princeton University Press.

Global Insight Co. (2002). DRI-WEFA's macroeconometric models. http://www.iccfglobal.org/pdf.

Hall, R. E. (1978). Stochastic implications of the cycle permanent income hypothesis: theory and evidence. *Journal of Political Economy*, *86*, 971–987.

Harrison, R., Nikolov, K., Quinn, M., Ramsay, G., Scott, A., & Thomas, R. (2005). *The Bank of England quarterly model*. London: Bank of England.

Hendry, D. H., & von Ungern-Sternberg, T. (1981). Liquidity and inflations effects in consumers' expenditures. In A. S. Deaton (Ed.), *Essays in the theory and measurement of consumer behaviour*. Cambridge: Cambridge University Press.

Lettau, M., & Ludvigson, S. (2001). Consumption, aggregate wealth and expected stock returns. *The Journal of Finance*, *16*, 815–849.

Levin, A. T., Rogers, J. H., & Tryon, R. W. (1997). *A guide to FRB/Global* (Mimeo, International Finance Discussion Papers No. 588). Washington: FRS.

Masson, P., Symansky, S. A., & Meredith, G. (1990). *MULTIMOD MARK II: a revised and extended model* (IMF Occasional Paper No. 71). Washington: IMF.

MIMOSA Modelling Group (1990). *MIMOSA, a model of the world economy* (OFCE Working Paper No. 90-02). Paris.

Modigliani, F. (1966). The life cycle hypothesis of saving, the demand for wealth and the supply of capital. *Social Research*, *33*, 160–217.

Modigliani, F. (1975). The life cycle hypothesis of saving, twenty years later. In M. Parkin & A. R. Nobay (Eds.), *Contemporary issues in economics*. Manchester: Manchester University Press.

Modigliani, F., & Ando, A. (1957). Test of the life cycle hypothesis of saving. *Bulletin Oxford University Institute of Statistics*, *19*, 99–124.

Modigliani, F., & Brumberg, R. H. (1954). Utility analysis and the consumption function: an interpretation of cross-section data. In K. K. Kurihara (Ed.), *Post Keynesian economics* (pp. 388–436). New Brunswick: Rutgers University Press.

Muellbauer, J., & Luttimore, R. (1995). The consumption function: a theoretical and empirical overview. In M. H. Pesaran & M. Wickens (Eds.), *Blackwell Handbook of Economics* (pp. 221–311). Oxford: Blackwell.

NIGEM (2002). *The world model manual*. London: NIESR.

OECD (1993). *The OECD INTERLINK system* (Mimeo).

Petersen, C. E., Pedersen, K. N., Riordan, E. J., Lynn, R. A., & Bradley, T. (1991). *BANK-GEM: a World Bank global economic model* (Mimeo). Washington: The World Bank. SECAP.

Smets, F., & Wouters, R. (2003). An estimated dynamic stochastic general equilibrium model of the Euro area. *Journal of the European Economic Association*, *1*, 1123–1175.

Stone, J. R. N. (1954). Linear expenditure system and demand analysis: an application to the pattern of British demand. *Economic Journal*, *64*, 511–527.

Suchecki, B. (2006). *Kompletne modele popytu (Demand systems)*. Warszawa: PWE.

Welfe, A. (1991). Modelowanie nierównowagi (Disequilibria modelling). *Przegląd Statystyczny*, *38*, 117–133.

Welfe, W. (Ed.) (2001). *Ekonometryczny model wzrostu gospodarczego (Econometric model of economic growth)*. Łódź: Wydawnictwo Uniwersytetu Łódzkiego.

Welfe, W. (2009). Makroekonometryczny model gospodarki opartej na wiedzy (Macroeconometric model of a knowledge-based economy). *Acta UŁ, Folia oeconomica, 229*. Łódź: Wydawnictwo Uniwersytetu Łódzkiego.

Welfe, W., & Welfe, A. (2004). *Ekonometria stosowana (Applied econometrics)*. Warszawa: PWE.

Yaari, M. E. (1965). Uncertain lifetime, life insurance, and the theory of the consumer. *Review of Economic Studies*, *32*, 137–150.

Zachłod-Jelec, M. (2010). Interrelations between consumption and wealth in Poland. *Central European Journal of Economic Modelling and Econometrics*, *2*, 37–57.

Chapter 17
The Models of the Enterprise Sector

17.1 Introduction

The modelling of the enterprise sector has a long tradition and a rich literature. Let us begin by describing the modelling process as applied in the demand-determined models. It will be assumed that enterprises function in the commodity and labour markets under imperfect competition. They assume that the expected demand for their products is predetermined and take into account the external conditions of their activities, including the availability of labour force, credits, the competitiveness of the surrounding firms, etc. (Welfe 2000).

In the one-sectoral models the macro-scale demand for firms' commodities and services will be obtained from a well-known accounting identity (all variables in constant prices):

$$X_t = C_t + G_t + J_t + \Delta R_t + (E_t - M_t) \tag{17.1}$$

where:

X_t is GDP,
C_t is personal household consumption,
G_t is real expenditures of public institutions,
J_t is real gross investment expenditures,
ΔR_t is inventory increase,
E_t is exports,
M_t is imports.

In the multisectoral models either bridge equations will be constructed by linking the final demand components with sectoral output through the input-output submodels or the submodels' approximations will be applied. This procedure usually consists of two steps. In the first step the demand for gross output by industries Q_{ti} is obtained and in the second step the demand for value added (net output) X_{ti} is determined. Hence, we have (in a matrix notation):

$$Q_t = \Gamma_t^C C_t + \Gamma_t^G G_t + \Gamma_t^J J_t + \Gamma_t^R \Delta R_t + \Gamma_t^E E_t - \Gamma_t^M M_t \tag{17.2}$$

W. Welfe, *Macroeconometric Models*,
Advanced Studies in Theoretical and Applied Econometrics 47,
DOI 10.1007/978-3-642-34468-8_17,

where $Q_t, C_t \ldots$ are the vectors of gross output and final demand components, Γ_t^i is the matrix linking the i-th component of final demand with the gross output of particular industries, and

$$X_t = A_t Q_t \tag{17.3}$$

where A_t is the technical coefficient matrix (the unit use of intermediate commodities).

The estimates of the elements of the above matrices and of the Γ^i vectors can be obtained every couple of years or, rather infrequently, once a year. Hence, they are often assumed constant, especially with respect to the quarterly models. Otherwise they are updated, being either treated as the functions of time or the functions of relative prices (cf. Welfe and Welfe 2004).

The systems of equations constructed for the enterprise sector will be shown below in some detail, assuming predetermined demand for the sector's output. The early approach will be demonstrated first, where the point of departure is the relations in the production process represented by the production functions (Welfe 2001). The appropriate inversion of the production functions opened the way for constructing functions explaining the demand for production factors: fixed capital and employment. Then the impact of the neoclassical concepts will be shown. It will be assumed that the equation systems for the enterprise sector are obtained by solving the optimization problem under imperfect competition. The equations will explain output, the demand for the production factors, i.e. fixed capital and investment, employees and their working time, as well as producer prices and average wages (Klein et al. 1999; Fair 2004).

17.2 Determination of Output

As it has been emphasized above, macroeconometric models most frequently assume that, according to Eq. (17.1) or (17.2), the demand for domestic production is met, allowing for the current output of enterprises. An intermediate step in realizing this demand has been introduced in a few models by analysing the adjustments in the finished goods inventories. It is assumed that the firms planning their output take into account, in addition to the expected demand for their products, also the necessary increase (decrease) in their existing inventories. This is represented by the following identity:

$$X_t = S_t + \Delta V_t$$

where S_t is sales (constant prices) determined from (17.1) or (17.2), V_t is inventories of finished commodities (constant prices).

The level of inventories is most frequently assumed to adjust to the level of sales. Hence we have:

$$\ln V_t = \beta \ln S_t + \zeta_t \tag{17.4}$$

The demand for output can be approximated by the following non-linear equation:

$$\begin{aligned}\ln X_t &= \alpha_0 + \alpha_1 \ln S_t + \alpha_2 \Delta \ln V_t \\ &= \alpha_0 + (\alpha_1 + \alpha_2 \beta) \ln S_t - \alpha_2 \ln V_{t-1} + \zeta_t \end{aligned} \tag{17.5}$$

The expected sales are adjusted here according to the planned inventory changes.

The above relationships are sometimes extended by an explicit introduction of output and inventory expectations. Fair (2004) suggested introducing the adaptive expectations. The expected output will be obtained from the following equation:

$$\ln X_t - \ln X_{t-1} = \lambda \left(\ln X_t^* - \ln X_{t-1} \right) + \mu_t \tag{17.6}$$

where X_t^* is the expected output determined from (17.5).

The expected volume of inventories V_t^* is determined from (17.4):

$$\ln V_t^* = \beta \ln S_t$$

Hence, after substituting, the following equation being a dynamic extension of (17.5) will be obtained:

$$\begin{aligned}\ln X_t = {} & \lambda \alpha_0 + (1 - \lambda) \ln X_{t-1} + \lambda (\alpha_1 + \alpha_2 \beta) \ln S_t \\ & - \lambda \alpha_2 \ln V_{t-1} + (\mu_t + \zeta_t) \end{aligned} \tag{17.6$'$}$$

The estimate of $\beta = 1.2$ that R. Fair obtained for the model (17.6′) indicated that inventories grew more than proportionally compared with the growth rate of sales; $\lambda = 0.68$ was showing that the differences between the expected and actual output in the previous quarter significantly affected the current level of output. An alternative procedure would involve further dynamization of the above equation and the use of ECM to estimate the parameters of the short-term equation.

17.3 Equations Explaining the Demand for Production Factors and the Production Function

In the early macroeconometric models the assumption about output being demand determined was followed by the supposition that the next step in the modelling of the production process should involve the derivation of the demand for production factors, i.e. of the demand for fixed capital, employment and raw materials, and for other intermediate inputs, including imported commodities (Welfe 2005).

The point of departure was the production functions where the potential output is related to the inputs of production factors, given the particular technology. If the functions are inverted, i.e. solved for the particular production factors, the respective demand functions can be derived. The functions explain the demand for fixed capital and, indirectly, for its increase (investments), the demand for employment and working time, and also demand for raw materials, materials and energy (including imports of intermediate commodities). In the multisectoral models, demand will be appropriately disaggregated (Welfe 2005).

In the world literature many forms of the production functions have been proposed and discussed (cf. Welfe and Welfe 2004). Our discussion will be confined to those most frequently used within the macroeconometric models.

The first to be presented is the Cobb-Douglas production function assuming constant elasticities with respect to the production factors:

$$X_t = BA_t K_t^{\alpha} N_t^{\beta} e^{\varepsilon_t} \tag{17.7}$$

where:

A_t	is total factor productivity (TFP),
K_t	is fixed capital (constant prices),
N_t	is employment,
ε_t	is disturbance term,
$\alpha > 0$	is elasticity with respect to fixed capital,
$\beta > 0$	is elasticity with respect to employment,
$\alpha + \beta = \nu$	the homogeneity level; if $\nu = 1$, then no returns to scale.

The second is the constant elasticity of substitution (CES) function:

$$X_t = \gamma\left[\delta K_t^{-P} + (1-\delta)N_t^{-\rho}\right]^{-v/\rho}$$

where:

γ	is efficiency of the production process parameter,
δ	is the parameter measuring the intensity of fixed capital impact,
$\rho > 0$	is the parameter related to the elasticity of substitution, $\delta_{K,N} = 1/(1+\rho)$; if $\rho \to 1$, the function is reduced to (17.7),
ν	is homogeneity level indicating returns to scale.

As mentioned, the demand functions for a particular production factor will be derived from an appropriate inversion of the production function. For the sake of simplicity, we shall use the Cobb-Douglas production function transformed by taking logs of both its sides (the small letters indicate the logarithms of the variables):

$$x_t = b + a_t + \alpha k_t + \beta n_t + \varepsilon_t \tag{17.7$'$}$$

(a) The demand function for fixed capital has the following form:

$$k_t = c - \gamma a_t + \gamma x_t - \beta\gamma n_t - \gamma\varepsilon_t \tag{17.8}$$

where $\gamma = 1/\alpha$, $c = -b/\alpha$.

A direct use of this form in the estimation process has a disadvantage, because a high collinearity between output and employment can be expected. If no returns to scale are assumed, i.e. $\alpha + \beta = 1$, then collinearity can be avoided by making appropriate transformations of the above function. They lead to the determination of either the capital/output function (K_t/X_t) or the capital/labour function (K_t/N_t). Taking logs, there will be determined:

(aa) the capital/output function:

$$k_t - x_t = c - \gamma a_t - \beta\gamma(n_t - x_t) - \gamma\varepsilon_t \tag{17.8$'$}$$

where the capital-output ratio depends on the labour-output ratio; the reciprocal of this function represents fixed capital productivity that depends on labour productivity;

(ab) the capital/labour function:

$$k_t - n_t = c - \gamma a_t - \gamma(n_t - x_t) - \gamma\varepsilon_t \tag{17.8''}$$

where the capital-labour ratio increases, if labour productivity shows an increase.

(b) The demand function for employment has the following form:

$$n_t = d - \mu a_t + \mu x_t - \alpha\mu k_t - \mu\varepsilon_t \tag{17.9}$$

where $d = -b/\beta$, $\mu = 1/\beta$.

The collinearity between output X_t and fixed capital K_t can be avoided by assuming that there are no returns to scale. With this assumption, it is possible to determine the equations explaining either the employment-output ratio (N_t/X_t) or the employment-fixed capital ratio (N_t/K_t) that will be used in estimating the number of jobs:

(ba) The employment/output function has the following form:

$$n_t - x_t = d - \mu a_t - \alpha\mu(k_t - x_t) - \mu\varepsilon_t \tag{17.9'}$$

The employment-output ratio declines following the increase in the capital-output ratio and in TFP. The reciprocal of this function defines the labour productivity function.

(bb) The employment/fixed capital function reads as follows:

$$n_t - k_t = d - \mu a_t - \mu(k_t - x_t) - \mu\varepsilon_t \tag{17.9''}$$

The number of jobs grows smaller, if the capital-output ratio and TFP increase. Inverting the function allows an alternative derivation of the ratio of fixed capital to employment, relating this ratio to the capital-output ratio.

In the above discussion only the technological properties of the production process have been taken into account. However, it was acknowledged already in the early models that entrepreneurs making decisions about a production process, including the generation of the demand for production factors, follow the results of optimization, i.e. of profit maximization (or cost minimization), under imperfect competition.

The profit maximization process can be simply shown as:

$$\max \sum_{t=0}^{\infty} \beta^t (P_t X_t - WP_t N_t - R_t K_t) \tag{17.10}$$

assuming that X_t is determined from the production function (17.7) and P_t is producer prices, R_t is the price of fixed capital, and WP_t is the average nominal wage.

The solution of this optimization problem yields the specifications of the long-term equations explaining the demand for production factors that additionally accounts for the impacts of prices. The demand function for fixed capital has now the following form:

$$k_t^* = c - \gamma a_t + \gamma x_t - \beta\gamma n_t - \nu(r_t - p_t) + \varepsilon_t \tag{17.11}$$

where the additional explanatory variable is the real price of fixed capital R_t/P_t—its higher level is associated with lower demand.

The demand function for employment reads as follows:

$$n_t^* = d - \mu a_t + \mu x_t - \alpha\mu k_t - v(wp_t - p_t) + \xi_t \qquad (17.12)$$

where the additional explanatory variable is the real wage—its increase reduces the demand for employment.

The above equations constitute the point of departure for specifying the extensions of the demand functions for fixed capital and its increase (investments) and of the employment demand functions discussed in the next sections.

17.4 The Demand Functions for Fixed Capital and Business Investment

The enterprises' demand for fixed capital mostly concerns its global volume, which is treated as homogeneous. In the multisectoral models, though, it is disaggregated into the demand for machinery, equipment, buildings and structures (Eisner and Strotz 1963).

The demand for fixed capital is firstly confronted with its existing stock. The realization of the demand may initially increase the utilization rate of fixed capital. However, in most cases this involves the necessity to increase its stock due to new investments. This process can be approximated by assuming that there are adaptive adjustments, which lead to the generation of the fixed capital stock at the end of the period K_t.

Let us define:

$$K_t - K_{t-1} = \gamma\left(K_t^* - K_{t-1}\right) \qquad (17.13)$$

Following (17.11), we can write:

$$K_t^* = CA_t^{-\gamma} X_t^{\gamma} N_t^{\beta\gamma} (R_t/P_t)^{-v} e^{\varepsilon_t} \qquad (17.14)$$

Hence, the increase in fixed capital will equal:

$$\Delta K_t = \gamma K_t^* - \gamma K_{t-1} \qquad (17.13')$$

On the other hand, the increase in fixed capital can be obtained by taking into account the supply of investment goods understood as a difference between gross business investment (installed equipment) I_t and liquidation (scrapping) of fixed capital D_t:

$$\Delta K_t = I_t - D_t \qquad (17.15)$$

The scrapping rate being generally assumed to be constant is frequently substituted by the rate of depreciation. We have:

$$D_t = \delta K_{t-1} \qquad (17.16)$$

Hence

$$\Delta K_t = I_t - \delta K_{t-1} \qquad (17.15')$$

Comparing Eqs. (17.13′) and (17.15′) we obtain:

$$\gamma K_t^* - \gamma K_{t-1} = I - \delta K_{t-1}$$

Solving this equation for gross investment produces gross investment demand as a function of the demand for fixed capital:

$$I_t = \gamma K_t^* + (\delta - \gamma) K_{t-1} \tag{17.17}$$

where K_t^* is determined from Eq. (17.14).

Most important in this chain of relationships is the impact of output being the essence of the accelerator rule (Chirinko 1992).

Macromodels generally assume that the gross investment is equal to real investment spending J_t ($J_t \equiv I_t$). However, several country models show differences between the above investment indicators (this happens if the investment expenditures are used to finance production in progress). Then additional bridge equations are introduced:

$$J_t = \sum_{i-0}^{m} \omega_i I_{it} \tag{17.18}$$

where $0 < \omega_i < 1$ are the weights.

The specification of the business investment functions in the macroeconometric models often differs from that described above. This will be presented below in some more detail.

Many model builders linearize the functions determining the demand for fixed capital. Then the demand has the following form (cf. i.e. Klein et al. 1999):

$$K_t^* = \beta_0 + \beta_1 X_t + \beta_2 (R_t / P_t) + \varepsilon_t \tag{17.19}$$

The impact of employment has been omitted, because it is partially captured by the changes in output X_t, and also the effects of technical progress have been ignored.

Assuming an adaptive adjustment mechanism:

$$K_t = \lambda K_t^* + (1 - \lambda) K_{t-1}$$

we have

$$\Delta K_t = \lambda \beta_0 + \lambda \beta_1 X_t + \lambda \beta_2 (R_t / P_t) - \lambda K_{t-1} \tag{17.20}$$

Using then the identity (17.15)

$$\Delta K_t = I - \delta K_{t-1}$$

and Eqs. (17.20) and (17.15), we derive the equation explaining investment I_t:

$$I_t = \lambda \beta_0 + \lambda \beta_1 X_t + \lambda \beta_2 (R_t / P_t) + (\delta - \lambda) K_{t-1} \tag{17.21}$$

The above equation or its logarithmic representation was used in many, mainly early macroeconometric models. It was usually extended by introducing relevant lag distributions The lag distributions of output, i.e. of the flexible accelerator, were introduced to allow for lagged deliveries, installation work, etc., and also for investors' expectations derived from past experiences (Koyck 1954).

In the 1960s D. Jorgenson generalized the above model by introducing, as an explanatory variable, broadly understood investment user costs KU_t to the investment function, which replaced the interest rate in the function. He also accentuated that because of the length of the investment process the relevant lag distributions must be introduced (Jorgenson 1965).

The investment user cost variable was constructed in such a manner as to include the fiscal components allowing the analysis of the likely fiscal policy impacts. We have:

$$KU_t = \frac{P_{jt}}{P_t}(r_t - \delta)(1 - m_t - z_t)/(1 - t_t) \tag{17.22}$$

where:

P_{jt} is investment goods prices,
P_t is GDP deflator,
r_t is interest rate,
δ is the rate of depreciation,
m_t is the tax rate on investment credit,
z_t is the tax rate on depreciation,
t_t is corporate income tax rate.

Taking into account the lags in the investment process, we obtain:

$$I_t = \sum_{J=0}^{J} \alpha\beta_j \Delta\left(X_{t-j} KU_{t-j}^{-\sigma}\right) + u_t \tag{17.23}$$

where σ is the elasticity of substitution between fixed capital and the other production factors and u_t is a disturbance term. In applications, it is frequently assumed that $\sigma = 1$.

The ratios of the investment and fixed capital $(i_t - k_{t-1})$ are sometimes explained. They depend on the factors listed above (cf. Dreger and Marcellino 2007).

D. Jorgenson's neoclassical concept was applied in many macroeconometric models. Some of them used its simplified versions. In the models (a) the explanatory variables were treated as separable; (b) the sums of the first differences ΔX_t were substituted by a weighted sum of the lagged investment I_{t-1} (using the Koyck's transformation) and an increase in the current output ΔX_t, and (c) to allow for the irregularity of output increases they were substituted by the output levels X_t. The user costs were frequently represented by their major components—the interest rates.

It has been noticed, that this specification does not sufficiently explain the observed large differences between the rates of growth of investment and rates of growth of the GDP, especially in the peaks and bottoms of the business cycle. They can be attributed to changes in business expectations of investment risks, represented by changes in the risk premium (RP_t). Hence, in the majority of the recent applications the risk premium characteristics were added. Sometimes it is regarded as a component of the investment user cost. When the above simplifications and adjustments are taken into account, the investment function becomes:

$$I_t = \alpha_0 + \alpha_1 I_{t-1} + \alpha_2 X_t + \alpha_3 KU_t + \alpha_4 RP_t + \varepsilon_t \tag{17.24}$$

Table 17.1 The parameter estimates of the investment demand functions

Countries	Elasticity of investment with respect to			A relative increase caused by unit change in the nominal long-term interest rate	
	Lagged investment	GDP			
		Short-term	Long-term	Short-term	Long-term
France	0.955	0.021	0.47	−0.0025[a]	−0.056
Germany	0.893	0.088	0.82	−0.0027	−0.021
Italy	0.914	0.051	0.59	−0.0017[a]	−0.020
Japan	0.923	0.045	0.58	−0.0026	−0.034
Germany	0.893	0.088	0.82	−0.0027	−0.021
UK	0.840	0.153	0.96	−0.0042[a]	−0.026

Source: Fair (2004, p. 269)

[a]Variable lagged by one quarter

In many macroeconometric models the exponential representations frequently prevail. Then we have (small letters stand for the logarithms):

$$i_t = \alpha_0 + \alpha_1 i_{t-1} + \alpha_2 x_t + \alpha_3 ku_t + \alpha_4 rp_t + \zeta_t \tag{17.24'}$$

For the sake of illustration, the parameter estimates obtained by R. Fair for the functions explaining investment demand will be shown (see Table 17.1). The user costs were represented by the long-term interest rate. The parameters were estimated using TSLS.

It must be stressed that the estimates of the autoregression coefficients are extremely high. This fact explains the considerable differences between the estimates of the short-run and long-run elasticities with respect to GDP; the long-run elasticities are close to 1 only for the United Kingdom. For the other countries they range from 0.5 to 0.6. This result does not justify the proposals to calibrate this elasticity at the level equal to one.

The next example refers to the parameter estimates of the investment demand functions obtained from the multicountry model MEMMOD (see Table 17.2). The estimates were achieved by means of a common two-stage estimation procedure, where the second stage was the ECM. The investment user costs were represented by interest rates. The long-term equation was as follows:

$$\ln I_t^* = \alpha_0 + \alpha_1 \ln X_t + \alpha_2 0.01(R_t/P_t)$$

and the short-run equation:

$$\Delta \ln I_t = \beta_1 \Delta \ln I_{t-1} + \beta_2 \Delta \ln X_t + \beta_3 0.01 \Delta(R_t/P_t) + \beta_4 \left(\ln I_t - \ln I_{t-1}^*\right)$$

The estimated long-term elasticities with respect to GDP are close to one or are calibrated at this level (except for Italy); the short-run elasticities range from 0.37 to 0.7 (excluding Germany). The main reason for the differences is the large (0.5–0.7) autoregression coefficients (defined differently than in the previous example). The adaptation to the equilibrium level is very slow.

Table 17.2 The elasticity estimates of the investment demand functions. The 1974–1997 sample

Countries	Long-term elasticities w.r.t.		Short-term elasticities w.r.t.			Estimates of the error correction term β_4
	GDP α_1	Real interest rates α_2	Lagged investment β_1	GDP β_2	Real interest rates β_3	
Canada	1.17	−1.00	0.73	0.45	−0.83	−0.15
	(30.2)		(10.2)	(3.5)	(2.5)	(3.3)
France	1.00	−0.02	0.67	0.49	−0.20	−0.09
		(4.4)	(13.3)	(5.6)	(3.6)	(4.0)
Germany	1.30		0.49	1.21	−0.41	−0.10
	(26.6)		(7.5)	(7.0)	(3.1)	(2.7)
Italy	0.44	−0.55	0.68	0.34	−0.07	−0.20
	(14.0)	(3.1)	(14.6)	(4.9)	(0.7)	(4.4)
Japan	1.0	−1.97	0.67	0.56	−0.10	−0.11
		(4.7)	(13.1)	(5.9)	(7.1)	(3.5)
UK	1.04	−0.30	0.51	0.73	−0.15	−0.29
	(31.5)		(7.6)	(5.4)	(0.6)	(4.4)
USA	0.98		0.66	0.66	−0.28	−0.20
	(30.0)		(13.9)	(6.5)	(1.8)	(4.9)

The values of t-Student statistics are given in the brackets. Source: Macro-Econometric Multi-Country Model (Deutsche Bundesbank 2000)

In several macroeconometric models the attempts were made to extend the above specifications by introducing explicitly the capacity utilization rates WX_t; their high values encourage enterprises to undertake new investment projects. The attempts date back to the early, quarterly Wharton models using the special Wharton capacity utilization indices (Evans and Klein 1968; cf. also Harrison et al. 2005; Welfe 2009a). In this case we have:

$$i = \alpha_0 + \alpha_1 i_{t-1} + \alpha_2 x_t + \alpha_3 ku_t + \alpha_4 rp_t + \alpha_5 wx_t + \zeta_t \qquad (17.24'')$$

In the late 1960s it was pointed out that the investment process implies additional installation costs (changes to equipment, personnel education, etc.). A proposal followed that the specification of the investment function be extended by adding the variable KA_t for the installation costs.

According to the proposals put forward by Lucas (1967) and Treadway (1969), it was assumed that installation costs are a quadratic function of the difference between the investment-fixed capital ratio and its long-run level:

$$KA_t = \frac{\chi}{2}\left[\frac{I_t}{K_{t-1}} - (\delta + g)\right]^2 K_{t-1} \qquad (17.25)$$

where:

δ is the rate of depreciation or liquidation,
g is the long-run rate of GDP growth.

In the simplified version the installation costs only depend on the investment-fixed capital ratio, a case in point being the QUEST II model (Dramais et al. 1997; Roeger and in't Veld 1997).

In several macromodels, especially those distinguishing investment demand for machinery and equipment, the substitution of labour for fixed capital is taken into account. To account for the impact of substitution, an additional explanatory variable representing the ratio between average wages (WP_t) and the investment expenditures deflator (PJ_t) is introduced. Then we have the following equation, where the installation costs are ignored:

$$i_t = \alpha_0 + \alpha_1 i_{t-1} + \alpha_2 x_t + \alpha_3 ku_t + \alpha_4 rp_t + \alpha_5 wx_t + \alpha_6 (wp_t - jp_t) + \xi_t \qquad (17.24''')$$

The parameters of the above investment demand function for machinery and equipment were estimated for Poland using the W8 D-2002 model (Welfe 2004). The following estimates were obtained:

$$\hat{\alpha}_2 = 0.61, \quad \hat{\alpha}_3 = -0.26, \quad \hat{\alpha}_5 = 0.90, \quad \hat{\alpha}_6 = 0.19$$

all being statistically significant, including the positive impact of substitution.

The second major tendency in modelling investment demand stemmed from the Q theory developed by J. Tobin (Brainard and Tobin 1968; Tobin 1969). This theory assumes that enterprises take investment decisions only if the market value of the enterprise exceeds the replacement costs of its fixed capital. This means that the decisions depend on the value of the coefficient Q^* defined as:

$$Q_t^* = V_t / PJ_t K_t \qquad (17.26)$$

where V_t is the market value of the enterprise.

The above relations are derived from the solution of a dynamic programming exercise that postulates the maximization of future profits (Chirinko and Fazzari 1988). Assuming the Cobb-Douglas production technology, the presence of installation costs and the condition (17.15′) defining the dynamics of fixed capital, the following long-term investment demand function can be obtained:

$$I_t = \left(\delta + g + \frac{Q_t - 1}{\chi} \right) K_{t-1} \qquad (17.27)$$

where:

δ is the scrapping or depreciation rate,
g is the rate of GDP growth.

In the above formula, the coefficient Q_t stands for the relation between the market value of a marginal increase in fixed capital and its replacement costs. In applications, because the value of this indicator is not observable, it is replaced by the average value of Q_t^* calculated from formula (17.26). Because an investment process takes time, the respective lags are introduced. This will be illustrated by the equation specified in the model MULTIMOD MARK III (1998):

$$\frac{I_t}{K_{t-1}} - \delta - g = k_1 Q_t^* + k_2 Q_{t-1}^* + \varepsilon_t \qquad (17.28)$$

that provided the following estimates: $k_1 = 0.033$, $k_2 = 0.048$.

The above concept found a restricted use despite its theoretical attractiveness (e.g. it defines market expectations).

In the early stages of modelling the investment demand, the importance of access to investment financing sources for taking investment decisions was stressed. Consequently, financial constraints were introduced into the investment demand equations. The investment financing sources included firms' own funds (depreciation and partly profits), borrowed amounts (bank credits), and subsidies. In the models of developed market economies financial constraints were abandoned, following the assumption that effective investment demand can be fully met there and that the financial constraints are well represented by the interest rates on credits.

Notwithstanding, the above explanatory variables were long used in the models of centrally planned economies and developing countries, where bank credits were usually rationed. These concepts have somewhat revived in the recent years, when the asymmetry of information available to banks and the borrowing firms has been stressed, following the neo-Keynesian theory. In taking investment decisions banks usually have less information, which makes them suspicious and may lead to credit rationing. To protect themselves against likely losses, banks tend to seek better securities. This strengthened the argument to introduce into the investment demand functions the risk premium. Variables directly representing the availability of credits are still introduced into the developing—country models.

17.5 The Demand Functions for Employment and Working Time

The demand for employees was specified in macromodels in an alternative way. The inverted production function was employed, so that the category 'hours worked' (H_t) could be used as an explained variable. This approach was chosen because hours worked adjusted to output changes faster than employment. It prevailed in the early macroeconometric quarterly models, for instance in the Wharton models. However, because of the data availability constraints, the approach was frequently used only in the manufacturing industry (cf. Evans and Klein 1968). Hence, the explained variable 'employment' (the number of employees N_t) found broader use. However, employment changes followed output changes with some delay. They took place if the entrepreneurs recognized output changes as permanent. The transitory output changes were mainly followed by changes in the working hours. When output was declining, the entrepreneurs preferred to shorten the working time and to keep their personnel (labour hoarding), because of the high costs of transitory dismissals and of the recruitment of new employees. When output was expanding, their first decision was to add more overtime hours. To allow for these adjustments, either lagged employment was introduced as an additional explanatory variable or an adaptive expectation equation was used (Welfe 2004). Then, separate equations explaining changes in the average number of hours worked by an employee (h_t) had to be built.

As mentioned, the point of departure for specifying the equations explaining employment demand is the inverted production functions. Using the Cobb-Douglas function in the same way as it was used in Sect. 17.2, we obtained the formulas (17.9) and (17.12); after allowing for the inertia, they take the following form.

The long-run equation explaining the employment demand reads:

$$\ln N_t^* = \alpha_0 + \alpha_1 \ln N_{t-1} + \alpha_2 \ln X_t + \alpha_3 \ln A_t + \varepsilon_t \tag{17.29}$$

Frequently $\alpha_2 = 1$; fixed capital was omitted, because it was collinear with output; it was also assumed that the changes in the fixed capital-output ratio were absorbed by A_t, a variable representing total factor productivity.

The short-term equation explaining the employment demand with ECM is as follows:

$$\begin{aligned}\Delta \ln N_t = {} & \beta_1 \Delta \ln N_{t-1} + \beta_2 \big(\ln N_{t-1} - \ln N_{t-1}^*\big) \\ & + \beta_3 \Delta \ln X_t + \beta_4 \Delta \ln A_t + \beta_5 \Delta \ln XW_{t-1} + \varepsilon_t \end{aligned} \tag{17.30}$$

In the short run, the changes in the capacity utilization rate are important—its increase makes labour demand grow. For this reason, the variable (XW_t) was introduced into the above equation.

For the sake of illustration, let us present the estimation results that the Bank of England (Harrison et al. 2005) obtained using the quarterly model of the UK. The estimates confirm the hypothesis about the adjustments being lagged ($\hat{\beta}_1 = 0.65$) and very slow ($\hat{\beta}_2 = -0.03$); the impact of an increase in the capacity utilization rate was positive.

More recently, the neoclassical concepts that allow for changes in the employee retention costs have been applied in the majority of macroeconometric models in specifying the employment demand functions. Initially, the point of departure was the maximization of profits, but in the recent years the cost minimization rule is used in parallel (mainly in the RSGE models), leading to slightly different results.

In the case of profit maximization (see formula (17.10)), the long-term demand function can be obtained by making use of marginal labour productivity. Assuming the technology represented by the Cobb-Douglas production function, we have:

$$\frac{\partial X_t}{\partial N_t} = \beta \frac{X_t}{N_t} = \frac{WP_t}{P_t} \tag{17.31}$$

where WP_t is the before-tax nominal wage, β is output elasticity with respect to employment.

Solving this identity for employment we arrive at the following long-term function of employment demand:

$$N_t^* = \beta X_t \frac{P_t}{WP_t}$$

Table 17.3 Estimates of the parameters of the employment demand functions. The 1974–1997 sample

Countries	Long-term elasticities w.r.t.		Short-term elasticities w.r.t.			Error correction coefficient
	GDP α_1	Real wages α_2	GDP β_1	Real wages β_2	Lagged investment β_3	β_4
Canada	0.61	0.23	0.26	0.09	0.59	−0.27
	(25.9)	(3.6)	(8.7)	(3.0)	(11.7)	(2.0)
France	0.19	0.13	0.19	0.13	0.39	−0.95
	(8.7)	(4.6)	(12.6)	(6.8)	(7.3)	(5.9)
Germany	0.52	0.72	0.17	0.24	0.55	−0.29
	(10.3)	(13.4)				(3.1)
Italy	0.21	0.18	0.20	0.13	0.44	−0.24
	(3.2)	(2.0)	(4.7)	(2.1)	(4.6)	(4.1)
Japan	0.30		0.09		0.64	−0.22
	(98.0)		(5.7)		(11.1)	(3.9)
Netherlands	0.47	0.12	0.28		0.12	−0.83
	(26.4)	(5.0)	(7.6)		(2.7)	(7.4)
United Kingdom	0.40	0.40	0.11	0.08	0.8	−0.10
	(13.9)		(7.4)	(5.7)	(27.7)	(3.8)
USA	0.82	0.72	0.39	0.24	0.45	−0.21
	(47.6)	(13.9)	(15.2)	(6.3)	(11.8)	(3.8)

The absolute values of t-Student statistics are shown in the brackets. Source: Macro-Econometric Multi-Country Model (Deutsche Bundesbank 2000)

This result, implying that employment changes are proportional to those in the output-real wages ratio, is generally regarded as too strong. Hence, it is approximated by the long-run equation:

$$\ln N_t^* + \alpha_0 + \alpha_1 \ln X_t - \alpha_2 \ln W_t + \varepsilon_t \tag{17.32}$$

where $W_t = WP_t / P_t$ is real wages.

In the short-run, using ECM and allowing for inertia and changes in the capacity utilization rate, we have:

$$\begin{aligned} \Delta \ln N_t = {} & \beta_1 \Delta \ln X_t - \beta_2 \Delta \ln W_t + \beta_3 \Delta \ln A_t + \beta_4 \Delta XW_{t-1} \\ & + \beta_5 \Delta \ln N_{t-1} + \beta_6 \left(\ln N_{t-1} - \ln N_{t-1}^*\right) + \varepsilon_t \end{aligned} \tag{17.33}$$

Several macroeconometric models do not utilize all of the explanatory variables listed above. Let us illustrate this with the estimation results obtained with the MEMMOD model (Deutsche Bundesbank 2000), where the impacts of the capacity utilization rate and of technical progress were ignored (Table 17.3).

The long-term elasticities with respect to GDP are greatly diversified and far from one, which seems to result from the omitted impact of technical progress. The short-term elasticities are generally much lower. Employment sensitivity to the changes in real wages is significant and highly diversified. The adjustment process is on the whole slow, the exceptions being France and Netherlands. Similar specifications can be found in many other macromodels of particular countries, such as the BEQM (Harrison et al. 2005), or of regions, e.g. the AWM (Fagan et al. 2005), or in the world multicountry models MZE (Beffy et al. 2003), QUEST II (where the real labour costs include the costs of employee vacations).

In several macromodels attempts were undertaken to make the specification of the employment demand functions even richer. Profits were included as additional explanatory variable in the long-term equations:

$$\ln N_t^* = \alpha_0 + \alpha_1 \ln X_t - \alpha_2 \ln W_t$$

the short-term equations:

$$\Delta \ln N_t = \beta_1 \Delta \ln X_t - \beta_2 \Delta \ln W_t + \beta_3 \Delta \ln N_{t-1} + \beta_4 \left(\ln N_{t-1} - \ln N_{t-1}^*\right)$$

variables, the lagged and especially expected output was introduced (for instance, in the NIESR model of the UK). It was stressed, though, that with the introduction of output expectations the accuracy of estimates and forecasts did not greatly improve (Wallis et al. 1987). The impact of the changes in the NAIRU level was also analysed (Dreger and Marcellino 2007).

The approach assuming production cost minimization instead of profit maximization gained many supporters, especially among the constructors of the DSGE models. The approach seems closer to the concepts where the levels of wages, one of the major cost components, depend on the outcomes of the negotiations conducted by the representatives of entrepreneurs and employees (Layard and Nickell 1985).

The minimum of the cost function K_t will be defined:

$$\min K_t = \min(WP_t N_t + C_t K_t) \tag{17.34}$$

where C_t is the unit cost of fixed capital services and WP_t is a nominal wage.

The solution of this optimization problem leads to the following long-term employment demand function:

$$\ln N_t^* = \alpha_0 + \alpha_2 \ln X_t + \alpha_3 \ln(WP_t/C_t) + \alpha_4 \ln A_t + \alpha_5 \ln N_{t-1} + \varepsilon_t \tag{17.35}$$

Fixed capital was omitted from this equation for the reasons mentioned above; TFP (variable A_t) is frequently approximated by an exponential trend. Sometimes the ratio between wages and the unit costs of fixed capital services is substituted by the two variables, which are introduced separately and represent real terms. Indeed, the demand for employees increases if real wages decline (the elasticity may equal 1) and if the real costs of hiring capital services increase (Smets and Wouters 2003). For the MIMOSA model (Delessy et al. 1996) a joint variable being a weighted sum of the logarithms of the ratios between wages and the unit costs of the services of hired fixed capital was constructed; the weights were estimated.

In the short-term equations, other explanatory variables are also introduced, in addition to the variables listed above, similar to those specified for the equations obtained for profit maximisation.

In the more recent years, the equations explaining the total number of hours worked by employees had the form of identities:

$$H_t = N_t h_t \tag{17.36}$$

where h_t is the working time (hours) per employee.

As a result, the equations explaining working time per employee are generally specified in the macromodels. The long-term equations explain changes in the length of the working day. Because they follow legislative changes, they often are the functions of time:

$$\ln h_t^* = \gamma_0 + \gamma_t t + \varepsilon_t \tag{17.37}$$

In the short-run, they are assumed to depend on the cyclical output variations and to adjust slowly, with some delay. This is reflected in the following equation:

$$\Delta \ln h_t = \beta_1 \Delta \ln h_{t-1} + \beta_2 \Delta \ln X_t + \beta_3 \left(\ln h_{t-1} - \ln h_{t-1}^*\right) + \zeta_t \tag{17.38}$$

The Bank of England's quarterly model of the UK (Harrison et al. 2005) provided the following estimates: $\hat{\beta}_1 = 0.56$ and $\hat{\beta}_3 = -0.048$. These results confirm the above hypothesis. The impacts of output changes were characterized by the respective elasticities, being 0.037 for the current GDP increase and 0.055 for the lagged GDP.

17.6 The Production Supply and Production Function

In the previous sections, enterprises' behaviour was characterized assuming that the demand for their production was predetermined. This means that their decisions to adjust supply to the given demand for output influenced their decisions to generate a demand for production factors such that their supply ensured the realization of output at the expected level. The production functions—after conversion—were the instruments for generating demand for production factors.

However, the question that can be raised for the market economies is whether producers' expectations are accurate enough and thereby whether the production offered corresponds to the effective demand. But the most important question concerns the size and mobility of the not fully used capacities, i.e. fixed capital and labour, when the existing unemployment allows hiring additional personnel. Therefore, in the model construction process the level of potential output must be defined and estimated, assuming the full utilization of production factors. This enables the estimation of the utilization rate of potential output that exerts a significant impact on the short-term adjustments.

In the supply determined economies with chronic scarcity of products, particularly in the economies with persistent disequilibria in the commodity and services markets (in some periods in the centrally planned economies), the domestic output usually falls short of meeting final demand. It is supply and not demand that is realized in transactions. In order to describe these situations properly the specification of the extended production functions including the likely sources of scarcities of the production factors, not only of fixed capital and employment, but also of energy and raw materials, especially from imports, is necessary.

The issues in the specification of the production functions for demand-determined market economies will be discussed first. They concern the domestic enterprises using the services of fixed capital and employees to deliver commodities and services, given the technology described by the relevant production function.[1] This function, specified for value added (GDP of the national economy), has the following general form:

$$X_t^* = x_t(K_t, N_t, A_t, \varepsilon_t) \tag{17.39}$$

As mentioned, the most frequently used are the double-log function, i.e. the Cobb-Douglas function (17.7) and the constant elasticity of substitution function (CES) (17.8). The time series based estimation of the functions' parameters series involved many problems that did not find full satisfactory solutions. Firstly, the explanatory variables—fixed capital and employment—were often collinear, so it was frequently assumed that there were no returns to scale. Using the Cobb-Douglas production function we have that the elasticities $\alpha + \beta = 1$. This allows us to use transformations defining either fixed capital productivity:

$$x_t - k_t = b + a_t + (1 - \alpha)(k_t - n_t) + \varepsilon_t \tag{17.39$'$}$$

or labour productivity:

$$x_t - n_t = b + a_t + \alpha(k_t - n_t) + \varepsilon_t \tag{17.39$''$}$$

The next empirical question is that the observed data on the effective output (X_t) represent the realizations of the demand for production, whereas the production function (17.39) defines the potential output, i.e. production capacities that are usually not fully utilized. Based on the definition of the capacity utilization rate $WX_t = X_t / X_t^*$, the modified production function (frequently called short-term) explaining the effective output X_t will take the following form:

$$X_t = WX_t X_t^* = w(K_t, N_t, WX_t, A_t, \varepsilon_t) \tag{17.40}$$

Macroeconometric models used various procedures to estimate the rates of capacity utilization WX_t. The major approaches will be listed below. In the early models of the USA economy the Wharton Index of Capacity Utilization was broadly used. It was generated by comparing the current, effective output with the output in the peak quarters of the business cycle (Klein 1966). In many European countries the summary indicators based on the data on the capacity utilization rates reported in the business surveys or firms statistics are used (Grzęda Latocha 2005). Partial information on shift utilization or employees' time worked was used less frequently (Welfe 1992). To model the inflationary processes special indicators were applied.

[1] Following the DSGE models typology they are the components of the sector of domestic producers of intermediate goods.

They were based on output deviations from its trend; treating them as the measures of the capacity utilization rate is rather unjustified.

Further, there are the problems of measuring fixed capital and employment. Several model builders rely on the statistical information about the stock of fixed capital as on period ends (typically year). The average stock is obtained as a simple arithmetic average interpolated for quarterly periods. In the models, the period-end stock is calculated using the balance equation:

$$K_t = K_{t-1} + I_t - \delta K_{t-1} \tag{17.41}$$

where δ is the scrapping rate and where necessary

$$I_t = \sum_{j=0}^{J} \omega_j J_{t-j} \tag{17.42}$$

ω_j being the parameters of the lag distribution of investment expenditures.

Many model builders frequently criticized the use of the statistical data on fixed capital because of their low reliability. They preferred instead to generate time series independently by adding up the "more reliable" data on the level of investment, allowing for fixed capital depreciation. This means that Eq. (17.41) as applied many times onwards, starting with an arbitrarily selected early period. While offering higher accuracy, this technique is far from being sufficiently satisfactory. Both the techniques demand additional adjustments to account for changes in the services rendered by fixed capital.

In most cases fixed capital is treated as homogeneous. However, this assumption is abandoned in many models where machines and equipment as well as buildings and structures are distinguished at the minimum, which perceive different functions in the production process. In many macromodels (for instance the NIESR model of the UK economy) attempts were undertaken to distinguish fixed capital representing different technological levels, i.e. the different generations of machinery and equipment (assuming higher productivity of the new equipment). A simple approximation of this approach was the attempts to separate the impacts of the most recent generation, for instance of equipment used less than 5 years. Specifications applied particularly in the developing country models to distinguish the imported machines and equipment likely to have higher productivity than the domestic ones had a similar meaning (Welfe 1992). More recently, the tendencies to separate computers, software and telecommunication equipment from the total fixed capital so that the impacts of computerized production processes could be quantified became stronger (Jorgenson 2000; Jorgenson et al. 2003). The above extension of the specification of the production functions was intended to help isolate the effects of computer-aided production and management processes (Collechia and Schreyer 2002; Van Leeuwen and Van der Wiel 2003). It must be noted, however, that the extensions reduced the spectrum of the likely effects of TFP increase.

Many macroeconometric models use the working time of labour (H_t) as an explanatory variable. However, in several countries these statistical data are not available, especially for quarterly periods, so the employment data are mainly applied to this end.

17.7 Total Factor Productivity

The effects of technological progress represented by total factor productivity (TFP), i.e. by variable A_t in Eq. (17.40), were initially considered exogenous and mostly presented as the exponential functions of time:

$$A_t = \lambda_0 e^{\lambda_1 t + \zeta_t} \tag{17.43}$$

The assumption about the effects being constant in time and having a constant rate of growth, usually not exceeding 1 %, seemed too strong in the last years, especially in the long term analyses and forecasts. The increasing role of knowledge capital in economic growth, outperforming the impacts of investment, contributed to the development of studies measuring the effects of technological progress and its sources in the last fifteen years (Smith 2002; Welfe 2006).

Total factor productivity is an unobservable variable. Its dynamics, following the broadly accepted views, can be represented by the Solow's residual (Solow 1957, 1962). TFP will be obtained as a difference between the growth rate of output generated from the production function and the growth rate of output generated from the same production function but ignoring technological progress. Given the Cobb-Douglas production function (17.7) with constant returns to scale extended for the capacity utilization rate *WX*, we have in logs:

$$\Delta x_t = \Delta wx_t + \left[\alpha \Delta k_t + (1-\alpha)\Delta n_t + \Delta a_t\right] \tag{17.44}$$

Having neutralized the effects of technological progress, i.e. assuming $\Delta a_t = 0$, we have:

$$\Delta x^0 = \Delta wx_t + \left[\alpha \Delta k_t + (1-\alpha) n_t\right] \tag{17.44'}$$

Subtracting both sides yields:

$$\Delta a_t = \left(\Delta x_t - \Delta x_t^0\right) = \Delta x_t - \Delta wx_t - \left[\alpha \Delta k_t + (1-\alpha)\Delta n_t\right] \tag{17.45}$$

To estimate the rate of TFP growth one must know the rate of productive capacity utilization and the estimates of output elasticity with respect to fixed capital (Florczak and Welfe 2000).

The issues involved in the determination of the rate of capacity utilization were discussed in the previous section. It must be stressed, though, that many macromodels ignore this variable, which leads to biased estimates of TFP dynamics. Specifically, the rates of output growth are falling during recessions: if a decrease in capacity utilization is then ignored, the TFP growth will be underestimated; on the other hand, ignoring the rising rates of capacity utilization during recovery with increasing rates of growth causes the overestimation of the TFP growth (cf. Welfe 2007, 2009b).

The output elasticity α with respect to fixed capital can be directly estimated, if the function explaining TFP is specified, i.e. if the variable A_t depends on factors determining technological progress. This procedure is rather complex and infrequently used in the macromodelling practice (cf. for example the early versions of the W8D model of the Polish economy, in Welfe 2004). More frequently, the values of the elasticities are calibrated. As far as the Cobb-Douglas function is concerned, the conclusions derived from the neoclassical theory of production can be used, stating that the parameter estimates of the production function are equal to the respective shares of profits and labour costs in GDP. As the shares are not uniquely defined, the parameter estimates differ between macromodels. In the MEMMOD model the estimates for the major industrialized countries ranged from 0.38 for France to 0.51 for Germany (Deutsche Bundesbank 2000). It was stressed, however, that the errors in the determination of the estimates may seriously impact the evaluation of TFP dynamics (cf. Welfe 2002).

Despite the above problems, the TFP dynamics has been recently introduced, especially into the long-term models, as an explanatory variable associated with the intensifying attempts to explain this dynamics (Richards 2000). It is convenient to decompose TFP dynamics representing the dynamics of knowledge capital into three factors, which are respectively linked to the dynamics of fixed capital (rather investment) A_t^K, the dynamics of employment, i.e. broadly understood human capital A_t^N, and the dynamics of disembodied knowledge capital A_t^W.

Based on the Cobb-Douglas production function, this decomposition reads as follows:

$$\Delta a_t = \Delta a_t^W + \alpha \Delta a_t^K + (1-\alpha)\Delta a_t^N \tag{17.46}$$

It is generally assumed that the dynamics of the disembodied, freely available knowledge capital is stable in time and that it can be represented by an exponential trend:

$$\Delta a_t^W = \mu_0 + \mu_1 t \tag{17.47}$$

Sometimes it is linked with employment dynamics or patent data (Mac Garvie 2005).

The dynamics of the effects of technological progress embodied in fixed capital is related to the dynamics of cumulative real R&D expenditures, both domestic and foreign, transferred to the country in question (Saggi 2000). This is represented by the following equation:

$$\ln A_t^K = \beta_1 \ln S_t^k + \beta_2 \gamma \ln S_t^m \tag{17.48}$$

where:

S_t^k is the cumulative real domestic R&D expenditures,
S_t^m is the cumulative real foreign R&D expenditures,
γ is a weight representing the role of imports, i.e. the openness of the economy.

For the first time the impact of the real domestic R&D expenditures was directly introduced into the production functions in the DRI models of the US economy. The above concepts for using the cumulative R&D expenditures were broadly applied in research projects based on the international cross-section time series data, the first of which was reported by Coe and Helpman (1995) and raised a discussion between Keller (2004) and Coe and Hoffmaister (1999). It was followed by projects reported in Engelbrecht (1997, 2002), Bayoumi et al. (1999), cf. also Welfe (2003, 2009b).

Domestic cumulative R&D expenditures are obtained from adding up the deflated current R&D expenditures SB_t, after allowing for the knowledge capital depreciation:

$$S_t^k = S_{t-1}^k + SB_t - \delta S_{t-1}^k \tag{17.49}$$

where δ is the rate of knowledge capital depreciation, frequently assumed at the 5 % level.

The specification of the knowledge capital transfer from abroad is more complicated. It is represented by the foreign cumulative real R&D expenditures. It is commonly confined to the expenditures of major industrialized countries. The direct and indirect transfers of knowledge capital are distinguished (Jaffe 1986; Eaton and Kortum 1996). The direct transfer involves telecommunication lines, country's technological proximity, the availability of knowledge contained in patents and licenses, etc. (Lee 2005). The indirect transfer takes place through imports. The alternative variants use either imports of intermediate commodities (new technologies) or imports of investment goods (new machines); the latter turned out to be more efficient (Xu and Wang 1999, 2000). Most recently, attempts were developed to use the weighted imports of commodities classified by their technological level; the weights were increasing along with rising technological maturity (Welfe 2009a).

The indirect transfer of foreign knowledge capital will be represented by the weighted sum of real current R&D expenditures:

$$SB_t^m = \sum_j w_j SB_j \tag{17.50}$$

where $0 < w_j < 1$ the weight as is defined above, staying with knowledge capital transferred from the country j.

The cumulative real expenditures transferred from abroad are obtained from the balance equation:

$$S_t^m = S_{t-1}^m + SB_t^m - \delta S_{t-1}^m \tag{17.51}$$

In the last years it was stressed that the economy to which foreign knowledge capital is transferred must be adequately prepared to absorb it (Lichtenberg and van Pottelsberghe de la Potterie 1998). This justifies extending the above equations by introducing variables representing the maturity of the economies in question, such as the minimum level of real domestic R&D expenditures or the level of human capital (Cincera and van Pottelsberghe de la Potterie 2001). The impact of FDI, an important channel of knowledge transfer in the emerging markets, is also analysed (Borensztein et al. 1998; Van Pottelsberghe de la Potterie and Lichtenberg 2001).

The empirical results of the above investigations are commonly described as TFP elasticities with respect to domestic and foreign real cumulative R&D expenditures (Crispolti and Marconi 2005). The estimation results based on the international cross-section time series data which were obtained by the authors mentioned above do not show considerable differences. The highly industrialized countries had the highest elasticities with respect to domestic knowledge capital. For the G7 countries β_1 was in the 0.14–0.16 range, for the other industrialized countries $\beta_1 = 0.06$–0.10, whereas for developing countries the elasticities were close zero. The elasticities showing the impact of knowledge capital transfers β_2 ranged from 0.08 to 0.10, but for developing countries they were ca. 0.5 (Welfe 2009b).

Knowledge capital embodied in labour force is commonly represented by human capital per employee. The notion of human capital is understood broadly—as a

Table 17.4 Output elasticities with respect to the average number of schooling years

Author	Levels		First differences	
Nehru et al. (1995)	0.078	(2.02)	0.079	(0.70)
Barro and Lee (1996)	0.165	(4.82)	0.083	(1.47)
Cohen and Soto (2001)	0.397	(7.98)	0.525	(2.57)
de la Fuente and Domenech (2000)	0.407	(7.76)	0.520	(2.17)

The values of t-Student statistics are given in the brackets. Source: de la Fuente (2004, Table 4.103)

summary characterization of particular employees' properties determining their efficiency. Because of that, a variety of human capital indicators were proposed (Benhabib and Spiegel 1994). However, they all have a common property—the level of education is regarded as crucial. Taking this characteristic as a point of departure, the global level of human capital H_{it} can be defined as follows:

$$H_{it} = \sum_i \mu_i N_{it} \tag{17.52}$$

where N_{it} is the number of employees having the i-th education level.

Human capital per employee will be obtained using the following formula:

$$h_t = H/N_t = \sum_i \mu_{it} N_{it} / \sum_i N_{it} \tag{17.53}$$

The above formula can be extended by introducing other groups of employees, distinguished according to their gender, age, position and industry (Jorgenson 2000; Jorgenson and Stiroh 2000). This extension has been enabled in most developed countries by the enlargement of the databases containing the demographic structures of population.

The major problem that still remains to be solved is how to define the weights μ_i. It is most frequently postulated that they should represent the length of education, i.e. the number of schooling years. The length of education was initially estimated indirectly, using the enrolment ratios. Recently this information has become directly available, which has significantly improved its accuracy (de la Fuente 2004).

It has been recently stressed that the weights defined above do not necessarily reflect the market efficiency of employees representing different levels of educational attainment. More adequate would be weights reproducing the relations between the average wages of employees with different levels of education. According to the Mincer equation, these relations are also dependent on the differences in the education levels. In this case, we have:

$$\mu_i = WP_i / WP_0 \tag{17.54}$$

where WP_i is average wages of employees with different educational levels.

The investigations aimed to identify human capital impacts on TFP growth were yielding uncertain and contradictory results for many years, including negative results. It was realised, however, that the main reason why such results were obtained were databases containing low-accuracy data on the length of schooling at different levels (de la Fuente 2004; Soto 2001).

The empirical investigations based on the international 1960–1990 samples covering the OECD countries used the information on the number of schooling years. The production functions treating human capital as a separate explanatory variable were applied. The results of these investigations are shown for the levels and first differences of the logs of the variables in Table 17.4. They characterize output elasticities with respect to the average number of schooling years.

All estimates based on the levels were statistically significant, but those based on first differences were significant only in the more recent investigations. These results may be challenged, because the regressions did not allow for the impacts of R&D expenditures. See also results for D.C. (Engelbrecht 1997).

Empirical investigations using as the weights the ratios of average wages paid to employees with different educational levels have been rather scarce (cf. Krueger and Lindahl 2001). The results obtained for Poland indicate that the choice of the weights is by no means an academic problem. For the period 1991–1998 the average annual rates of human capital growth were respectively 0.54 % or 0.78 %, depending on whether the wage relations or the number of schooling years was used as the weights (Welfe et al. 2002).

In the empirical research projects covering particular countries the impacts of the human capital changes can be analysed using a large spectrum of employees' characteristics, for instance their gender, age, position etc. (cf. for the USA—Jorgenson et al. 2003).

Another postulate has been to extend the notion of human capital by including additional characteristics, such as work experience, level of health, and migration (Benabou 2002). These extended human capital characteristics can be found in the submodel of knowledge capital for Poland, where work experience was approximated by employee age and health status by average life expectancy of women and men in employment (Florczak 2009).

17.8 Equations Explaining Producer Prices

The equations explaining prices were important components of all early macroeconometric models (Courbis 1977). In the Klein-Goldberger model price changes depend on the changes in the unit costs of production reduced to labour costs (Klein and Goldberger 1955). Inflation induced by these changes is called cost-push inflation. In the next models the price equations were extended by allowing for other cost items and the characteristics of market tensions. The main reason for changes in the latter component was the demand shocks, so inflation they caused is called demand-pull inflation.

In the price determination process producer prices play a major role. Following the neoclassical concepts, their equations are derived from the solution of the profit maximizing problem. Having determined their production programs, the entrepreneurs are in a position to set prices allowing them to cover production costs and to ensure profits.

The unit production costs (ku_t) are composed of:

(a) unit labour costs (kw_t) that can be calculated by dividing the average wages (WP_t) by the labour productivity (WY_t) and by allowing for the social contributions (v_t), i.e.:

$$kw_t = WP_t / WY_t (1 + v_t) \tag{17.55}$$

(b) the costs of using the intermediate commodities (raw materials, materials and energy) that at the macro-scale can be reduced to the costs of using imported materials (km_t) and energy (ke_t); they can be computed by multiplying the unit use coefficients, respectively (a_t^m) and (a_t^e), by the import prices of materials (P_t^m) and energy (P^e). The import prices in domestic currency are determined by multiplying the world prices (P_t^w) by the exchange rate (ER_t). Hence, we have:

$$km_t = a_t^m P_t^m = a_t^m ER_t P_t^{wm} \tag{17.56}$$

and

$$ke_t = a_t^e P_t^e = a_t^e ER_t P_t^{we} \tag{17.56'}$$

(c) the remaining costs, including those resulting from the purchase of services, debt service (dependent on the interest rates) and depreciation.

The unit costs are fully represented in the input-output models (Welfe 1992). In the macroeconometric models this happens only infrequently, mainly because of scarce information concerning the changes in the unit use of materials and other costs. Hence, the unit costs are usually reduced to their major components, namely labour costs and import prices (assuming constant import-output ratios).

The labour cost components, i.e. average wages and labour productivity, are treated separately in most cases. Average wages depend, *inter alia*, on retail prices and these on producer prices. This leads to a simultaneous feedback, i.e. an inflationary loop (see sections below). This may result in the construction of a reduced equation explaining producer prices, where the variables explaining wages, especially the unemployment rate, are introduced instead of average wages.

The second major factor determining the dynamics of producer prices is tensions in the commodity markets that affect output or the demand gap. The tensions are mainly represented by the characteristics of the capacity utilization rate (WX_t). Many different indicators have been used to this end. For instance, in addition to the indicator constructed as a ratio between the effective and potential output the Fair models (Fair 2004) also used the deviations from the GDP trend, as well as the rate of unemployment (the latter indicator was also applied in the FRB model of the US economy).

In the specification of the producer price equations the dynamic adjustments play an important role. The neo-Keynesian models stress rigidities in price adjustments, which implies the introduction of appropriate lags. Expectations are also introduced, particularly into the quarterly models, and occasionally the price inflation targets.

Table 17.5 Parameter estimates of the producer prices equations

Countries	Elasticities of producer prices with respect to				
	Lagged prices α_2	Labour costs α_3	Import prices α_4	Demand gap α_5	Time α_6
Canada	0.726	0.214	0.28	−0.164[b]	0.00025
	(7.0)	(2.6)	(1.2)	(2.4)	(1.0)
France	0.806	0.057	0.023	−0.0444	0.00002
	(3.0)	(2.1)	(1.9)	(4.5)	(0.2)
Germany	0.984		0.008	−0.150[a]	0.00008
	(57)		(1.3)	(2.3)	(0.7)
Italy	0.942		0.033	−0.210	0.00050
	(140)		(7.3)	(5.8)	(3.9)
Japan	0.937		0.016	0.080	0.00035
	(45)		(2.3)	(3.4)	(2.0)
Netherlands	0.816		0.050	−0.0563	0.00086
	(16)		(4.2)	(1.9)	(3.4)
UK	0.829	0.136	0.063	−0.302	−0.00034
	(18)	(2.9)	(6.2)	(4.6)	(1.8)

The absolute values of t-statistics are given in the brackets. Note: the variable measuring the demand gap was defined as [a]an unemployment rate, [b]a capacity utilization rate, [c]deviations from the GDP trend. Source: Fair (2004, Table B5, p. 295)

The long-term equations of producer prices may have the following form:

$$\ln P_t^* = \alpha_0 + \alpha_1 \big(\ln WP_t(1+v_t) - \ln WY_t\big) + \alpha_2 \ln P_t^m + \varepsilon_t \tag{17.57}$$

whereas the short-term equations:

$$\begin{aligned}\Delta \ln P_t = {} & \beta_1 \Delta \ln P_{t-1} + \beta_2\big(\ln P_t - \ln P_{t-1}^*\big) + \beta_3\big(\Delta \ln WP_t(1+v_t) - \Delta \ln WY_t\big) \\ & + \beta_3 \Delta \ln P_t^m + \beta_4 \Delta \ln WX_t + \Delta\varepsilon_t \end{aligned} \tag{17.58}$$

The results of the empirical investigations generally confirm the opinion about the leading role of the labour costs. The estimates of the long-term elasticities of producer prices with respect to labour costs are mostly close to one (for instance in the models FRB/US and MIMOSA). This is why in some models (e.g. the MZE model) they are calibrated at a level equal to one. The short-term elasticities are much lower, which can be attributed to the impacts of lags and leads (cf. Fair 2004; MEMMOD 2000). The short-term elasticities with respect to import prices are, on the whole, statistically significant, being rather below 0.1. Sometimes homogeneity is imposed in the long-term equation such that $\alpha_2 = 1 - \alpha_1$. The impact of the market tensions is usually significant. The above will be illustrated using the parameter estimates of the traditional producer price equations that Fair (2004) obtained with TSLS (Table 17.5).

It must be stressed that the inertia were high in nearly all countries. Hence the estimates of the short-term elasticities were comparatively low, likewise the autonomous price increase.

17.9 Equations Explaining Average Wages

Wages are one of the most significant components of production costs. The before-tax wages are an important subject of negotiations conducted by entrepreneurs and employees or their representatives—the labour unions. The latter are mainly interested in maximizing after-tax wages. This gives primary importance to the wage bargaining concept that the English econometricians Nickell (1984) and Layard developed at the end of the 1980s (Layard et al. 1991).

However, in the early macroeconometric models the most important observation was that the level of wages changes following unemployment rate changes. It was formulated for the first time in the specification of the wage rate equations in the Klein-Goldberger model (1955). Its international reputation came with the studies on wage development in the United Kingdom carried out by Phillips (1958). In the concept he formulated, thereafter called a Phillips curve, the rate of wage growth is related to the rate of unemployment. The algebraic formula is as follows:

$$\mathring{W}P_t = BU_t^{\gamma} e^{\varepsilon_t} \tag{17.59}$$

or in logs:

$$\Delta wp_t = \beta + \gamma u_t + \varepsilon_t \tag{17.59$'$}$$

This relation has been modified many times; its generalized form is shown below:

$$\Delta wp_t = \beta + f(U_t) + \varepsilon_t \tag{17.59$''$}$$

The modifications consisted, *inter alia*, in predetermining the value of the parameter γ ($\gamma = 1, 0, -1/2, -1$) in formula (17.59′) or in substituting the unemployment rate by its inverse:

$$\Delta wp_t = \beta + \gamma u^{-1}, \quad \gamma > 0 \tag{17.59$'''$}$$

The use of the "pure" form of the above concept was broadly criticized, especially from the neoclassical position. Firstly, it was emphasized that entrepreneurs and potential employees have different expectations, so there must be some minimum period between leaving a job and taking a new one. This justifies the existence of friction unemployment, whose impact on the wage level is practically meaningless. Therefore, the unemployment rate in the above equations must be appropriately modified by excluding either friction unemployment or natural unemployment.

Secondly, because wage negotiations are central to wage determination, it should be assumed that the expectations presented by entrepreneurs and employees are the most important. Because the expectations primarily concern price changes, the negotiations must protect the levels of real wages from falling. In the long-run, the negotiations would involve the acceptable distribution of the effects of a labour productivity increase.

The introduction of price expectations leads to the expectations-augmented Phillips curve:

$$\Delta wp_t = \beta_0 + \beta_1 f(U_t) + \beta_2 \Delta pc_t^e + \varepsilon_t \tag{17.60}$$

where pc_t^e is the expected retail price.

The expected prices are either retail prices or employees' living costs. Expectations are constructed as rational expectations or as adaptive expectations formed based on the observed tendencies:

$$\Delta pc_t^e \approx \Delta pc_{t-1} \tag{17.61}$$

In several countries suffering from high rates of inflation (such as Italy and Poland) special regulations were introduced, tying the rates of wage growth (mainly in the state-owned firms) to the rates of price growth. The wage indexation procedures thus prescribed required the use of constant coefficients (elasticities) whose values were lower than one (for instance 0.9). This was intended to weaken wage pressures. The introduction of the coefficients into the wage equations did not make wage forecasts more accurate, though (Welfe and Welfe 2004).

A wage equation constructed to expresses the outcomes of negotiations over the impacts of living cost changes and employees' shares in the results of productivity increase (WY_t) conducted between entrepreneurs and employees must take into account that the negotiations deal with the nominal wage levels and not with their rates of growth.

Hence the generalized specification of the equations explaining nominal wages (cf. Tobin 1972) takes the following form in the long run:

$$wp_t^* = \alpha_0 + \alpha_1 f(u_t) + \alpha_2 \ln pc_{t-1} + \alpha_3 \ln wy_t + \alpha_4 t + \xi_t \tag{17.62}$$

In the short-run, the rates of wage growth will be explained:

$$\begin{aligned} \Delta wp_t = \beta_0 + \beta_1 \left(wp_{t-1} - wp_{t-1}^*\right) + \beta_2 \Delta f(u_t) \\ + \beta_3 \Delta pc_{t-1} + \beta_4 \Delta wy + \xi_t \end{aligned} \tag{17.63}$$

The introduction of a constant rate of growth (trend) is justified, because of the impacts of autonomous wage growth factors (due to longer stay in the job, etc.). The long-term effects of price changes are represented by the constant elasticity α_2. If these elasticities (frequently measured with respect to current prices) are close to one, then many models calibrate their values, assuming $\alpha_2 = 1$. Then the subject of analyses and estimations will be the real wages $W_t = WP_t / PC_t$ and the long-term equation will be reduced to:

$$w_t^* = \alpha_0 + \alpha_1 f(u_t) + \alpha_3 wy_t + \alpha_4 t + \zeta_t \tag{17.62'}$$

Likewise, the reduced short-term equation will have the form:

$$\Delta w_t = \beta_0 + \beta_1 \left(w_{t-1} - w_{t-1}^*\right) + \beta_2 \Delta f(u_t) + \beta_4 \Delta wy + \zeta_t \tag{17.63'}$$

Table 17.6 The elasticities of average wages. The 1965–1992 sample

Countries	Elasticities of average wages with respect to					
	Consumer goods prices CPI		Labour productivity		Unemployment rate	
	Long-run	Lagged	Long-run[a]	Lagged	Long-run	Lagged
France	0.82	4 months			−0.58	−0.58
West Germany	0.79	3 months	0.37	6 months	−1.24	−0.30
Italy	0.92	3 months	0.42	2.5 months	−0.33	−0.33
Japan	1.00	4 months	0.65	6 months	−1.45	−1.45
UK	1.00	3 months	0.38	12 months	−0.41	0.0
USA	0.93	9 months			−0.55	−0.55

Source: MIMOSA (Delessy et al. 1996)
[a]Minimum wage level added

It has been demonstrated that constant and positive elasticity of wages with respect to labour productivity implies systematically increasing productivity. It is hardly believable, though, that should productivity be declining the employees would accept the related impacts, i.e. wage cuts. Hence, if $\Delta WY_t / W_t < 0$, it seems reasonable to assume that $\beta_4 = 0$.

In the macroeconometric models of particular countries and in the multicountry models the wage equations are generally specified following the above rules, however some differences and special properties can be spotted. As mentioned, in the equations explaining average wages the estimates of the long-term elasticities of average wages with respect to consumer prices were frequently close to one or slightly smaller. This is illustrated in Table 17.6. Hence, in many other models the elasticities were calibrated at a level equal to one, i.e. the equations explaining real wages were used.

The short-term elasticities with respect to consumer prices show impacts lagged by 3–4 months. The elasticities with respect to labour productivity are diverse. Their long-run estimates are close to 0.4 for the West European countries, while the Japanese value of 0.65 is similar to that for Poland (0.5–0.7). The correlation with the rate of unemployment is negative. In Germany and Japan wage sensitivity was much higher than in the other European countries and the USA.

The above characterizes a demand-determined market economy, where enterprises negotiating wage levels maximize their profits under imperfect competition. In the supply-determined economies scarcities of production factors, including labour, may appear along with disequilibria. As a result, excess demand for employees showing differences in intensities is observed in the labour market. It can be written as the negative rate of unemployment (U_t^n). It follows from the presentation that the specification of the wage equation shown above must be modified, so that the unemployment rate is substituted by the rate of excess demand for employees. The long-run equation will have the following form:

$$\ln W^* = \alpha_0 + \alpha_1 \ln f\left(U_t^n\right) + \alpha_3 WY_t + \alpha_4 t + e^{\zeta_t} \tag{17.64}$$

In the empirical investigations the question about how to define a suitable measure of the excess demand U_t^n must be answered. The demand is directly non-observable. The disequilibrium indicators can be represented by the characteristics of employment changes, i.e. the ratios between the number of vacancies and the number of registered unemployed job-seekers or the ratios between recruitments and dismissals (Welfe 1993). The above equations were used in estimating the wage equation parameters for Poland in the period of centrally planned economy (Welfe and Welfe 2004).

References

Barro, R. J., & Lee, J. W. (1996). International measures of schooling years and schooling quality. *American Economic Review, Papers and Proceedings*, *86*(2), 218–223.

Bayoumi, T., Coe, D. T., & Helpman, E. (1999). R&D spillovers and global growth. *Journal of International Economics*, *47*, 399–428.

Beffy, P.-O., Bonnet, X., Paries, M. D., & Montfort, B. (2003). *MZE: a small macro-model for the Euro area* (Document de travail, G 2003/11). Paris: INSEE.

Benabou, R. (2002). *Human capital, technical change and the welfare state* (Temi di Discussione del Servizio Studi 465). Roma: Banca d'Italia.

Benhabib, J., & Spiegel, M. (1994). The role of human capital in economic development, evidence from aggregate cross-country data. *Journal of Monetary Economics*, *34*(2), 143–174.

Borensztein, E., De Gregorio, J., & Lee, J. W. (1998). How does foreign direct investment affect economic growth. *Journal of International Economics*, *45*, 115–135.

Brainard, W. C., & Tobin, J. (1968). Pitfalls in financial model building. *American Economic Review*, *58*, 99–122.

Chirinko, R. S. (1992). *Business fixed investment spending: a critical survey of modeling strategies, empirical results, and policy implications* (Working Paper No. 27). Munich: Center for Economic Studies, University of Munich.

Chirinko, R. S., & Fazzari, S. M. (1988). Tobin's Q, non-constant returns to scale, and imperfectly competitive product markets. *Recherches Economiques de Louvain*, *54*, 259–279.

Cincera, M., & van Pottelsberghe de la Potterie, B. (2001). International R&D spillovers: a survey. *Cahiers Economiques de Bruxelles*, *169*(1), 1–20.

Coe, D. T., & Hoffmaister, A. W. (1999). *Are there international R&D spillovers among randomly matched trade partners? A response to Keller* (Working Paper). International Monetary Fund.

Coe, D. T., & Helpman, E. (1995). International R&D spillovers. *European Economic Review*, *39*, 859–887.

Cohen, D., & Soto, M. (2001). *Growth and human capital: good data, good results* (CEPR Discussion Paper No. 3025).

Collechia, A., & Schreyer, P. (2002). ICT investment and economic growth in the 1990s': is the United States a unique case? A comparative study of nine OECD countries. *Review of Economic Dynamics*, *5*, 408–442.

Courbis, R. (1977). *Les modeles de prix. Pour la prevision et la planification etude comparative*. Paris: Dunod.

Crispolti, V., & Marconi, D. (2005). *Technology transfer and economic growth in developing countries: an econometric analysis* (Temi di Discussione del Servizio Studi 564). Roma: Bank of Italy.

de la Fuente, A., & Domenech, R. (2000). *Human capital in growth regressions: how much differences does data quality make* (Working Paper No. 262). OECD, Economics Department.

de la Fuente, A. (2004). Human capital and growth: some results for the OECD. In *Proceedings of OeNB workshops: Vol. 2. Current issues of economic growth*.

Delessy, H., Sterdyniak, H., et al. (1996). *The new version of MIMOSA, a macroeconomic model of the world economy (version 1995)* (Mimosa Team No. 96-01). CEPII/OFCE.
Deutsche Bundesbank (2000). *Macro-econometric multi-country model MEMMOD* (Mimeo). Frankfurt/Main: Deutsche Bundesbank.
Dramais, A., Roeger, W., & in't Veld, J. (1997). *QUEST II. A multi-country business cycle and growth model* (Mimeo). Paper presented at the Project LINK meeting, Malaysia. European Commission.
Dreger, C., & Marcellino, M. (2007). A macroeconometric model for the Euro economy. *Journal of the Policy Modelling*, *29*, 1–13.
Eaton, J., & Kortum, S. (1996). Trade in ideas, patenting and productivity in the OECD. *Journal of International Economics*, *40*, 251–278.
Eisner, R., & Strotz, R. H. (1963). Determination of business investment. In Commission on Money and Credit, *Impacts of monetary policy*. Englewood Cliffs: Prentice Hall.
Engelbrecht, H. J. (1997). International spillovers, human capital and productivity in OECD economies. An empirical investigation. *European Economic Review*, *41*, 1479–1488.
Engelbrecht, H. J. (2002). Human capital and international knowledge spillovers in TFP growth of a sample of developing countries: an exploration of alternative approaches. *Applied Economics*, *34*, 831–841.
Evans, M. K., & Klein, L. R. (1968). *The Wharton econometric forecasting model* (2nd enlarged ed.). Philadelphia: Economics Research Unit, University of Pennsylvania.
Fagan, G., Henry, J., & Mestre, R. (2005). An area-wide model for the Euro-area. *Economic Modelling*, *22*, 139–159.
Fair, R. C. (2004). *Estimating how the macroeconomy works*. Cambridge: Harvard University Press.
Florczak, W. (2009). Modelowanie kapitału ludzkiego i struktura ludności (Modelling human capital and population composition). In W. Welfe (Ed.), *Makroekonomiczny model gospodarki opartej na wiedzy* (pp. 241–279). *Acta UŁ, Folia oeconomica, 229*. Łódź: Wydawnictwo UŁ.
Florczak, W., & Welfe, W. (2000). Wyznaczanie potencjalnego PKB, łącznej produktywności czynników produkcji (Potential GDP and TFP determination). *Gospodarka Narodowa*, *11–12*, 10–55.
Grzęda Latocha, R. (2005). Ekonometryczna analiza koniunktury gospodarczej w krajach strefy Euro (Econometric analyses of business cycle in the Euro-area). *Ekonomista*, *5*, 621–643.
Harrison, R., Nikolov, K., Quinn, M., Ramsay, G., Scott, A., & Thomas, R. (2005). *The Bank of England quarterly model* (Mimeo). Bank of England.
Jaffe, A. B. (1986). Technological opportunity and spillovers of R&D: evidence from firms patent, profits and market values. *American Economic Review*, *76*, 984–1001.
Jorgenson, D. W. (1965). Anticipation and investment behaviour. In J. G. Duessenberry et al. (Eds.), *The Brookings quarterly econometric model of the United States* (pp. 35–92). Chicago and Amsterdam: Mc Nally & Co. and North-Holland.
Jorgenson, D. W. (2000). Information technology and the US economy. *American Economic Review*, *91*, 1–32.
Jorgenson, D. W., & Stiroh, K. J. (2000). Raising the speed limits: US economy growth in the information age. *Brookings Papers on Economic Activity*, *1*, 125–211.
Jorgenson, D. W., Ho, M. S., & Stiroh, K. J. (2003). Lessons from the US growth resurgence. *Journal of Policy Modeling*, *25*, 453–470.
Keller, W. (2004). International technology diffusion. *Journal of Economic Literature*, *42*, 752–782.
Klein, L. R. (1966). *Studies in quantitative economics: Vol. 1. The Wharton index of capacity utilization*. Philadelphia: University of Pennsylvania.
Klein, L. R., & Goldberger, A. S. (1955). *An econometric model of the United States, 1929–1952*. Amsterdam: North-Holland.
Klein, L. R., Welfe, A., & Welfe, W. (1999). *Principles of macroeconometric modeling*. Amsterdam: North-Holland.

Koyck, L. M. (1954). *Distributed lags and investment analysis*. Amsterdam: North-Holland.
Krueger, A., & Lindahl, M. (2001). Education for growth: why and for whom? *Journal of Economic Literature*, *39*, 1101–1136.
Laxton, D., Isard, P., Faruqee, H., Prasad, E., & Turlelboom, B. (1998). *MULTIMOD MARK III: the core dynamic and steady state models* (IMF Occasional Papers No. 164).
Layard, P. R. G., & Nickell, S. J. (1985). The causes of British unemployment. *National Institute Economic Review*, *111*, 62–85.
Layard, P. R. G., Nickell, S. J., & Jackman, R. (1991). *Unemployment: macroeconomic performance and the labour market*. Oxford: Oxford University Press.
Lee, G. (2005). Direct versus indirect international R&D spillovers. *Information Economics and Policy*, *17*, 334–348.
Lichtenberg, F. R., & van Pottelsberghe de la Potterie, B. (1998). International R&D spillovers, a comment. *The European Economic Review*, *42*, 1483–1491.
Lucas, R. E. Jr. (1967). Optimal investment policy and the flexible accelerator. *International Economic Review*, *8*, 78–85.
Mac Garvie, M. (2005). The determinants of international knowledge diffusion measured by patent citations. *Economic Letters*, *87*, 121–126.
Nehru, V., Swanson, E., & Dubey, A. (1995). New data base on human capital stocks in developing and industrial countries: sources, methodology and results. *Journal of Development Economics*, *46*, 379–402.
Nickell, S. J. (1984). The modelling of wages and employment. In D. H. Hendry & K. F. Wallis (Eds.), *Econometrics and quantitative economics*. Oxford: Basil Blackwell.
Phillips, A. W. (1958). The relation between unemployment and the rate of change of money wage rates in the United Kingdom, 1961–1957. *Economica*, *25*(4), 283–299.
Richards, R. G. (2000). Endogenous technological advance in an econometric model (Implications for productivity and potential output in the United States). *Economic Modelling*, *17*, 13–34.
Roeger, W., & in't Veld, J. (1997). *QUEST II. A multicountry business cycle and growth model* (Economic Papers No. 123). European Commission.
Saggi, K. (2000). *Trade, foreign direct investment and international technology transfer: a survey* (World Bank, Policy Research Papers, 2379).
Smets, F., & Wouters, R. (2003). An estimated dynamic stochastic general equilibrium model of the Euro area. *Journal of the European Economic Association*, *1*, 1123–1175.
Smith, K. (2002). *What is the knowledge economy: knowledge intensity and distributed knowledge bases* (Discussion Paper 2002-6). Maastricht: The UN University, INTECH.
Solow, R. (1957). Technical change and aggregate production function. *Review of Economics and Statistics*, *39*, 312–320.
Solow, R. (1962). Technical progress, capital formation and economic growth. *American Economic Review*, *52*, 76–86.
Soto, M. (2001). *Growth and human capital: good data, good results* (CEPR Discussion Paper No. 3025).
Tobin, J. (1969). A general equilibrium approach to monetary theory. *Journal of Money, Credit and Banking*, *1*, 15–29.
Tobin, J. (1972). The wage-price mechanism: overview of the conference. In O. Eckstein (Ed.), *The econometrics of price determination conference*. Washington: Federal Reserve System.
Treadway, A. B. (1969). On rational entrepreneurial behavior and the demand for investment. *The Review of Economic Studies*, *36*, 227–239.
Van Leeuwen, G., & Van der Wiel, H. (2003). Spillovers effects of ICT. *CBP Report*, *3*, 24–40.
Van Pottelsberghe de la Potterie, B., & Lichtenberg, F. R. (2001). Does foreign direct investment transfer technology across borders? *Review of Economics and Statistics*, *83*, 490–497.
Wallis, K. F., Fisher, P. G., Longbottom, J. A., Turner, D. S., & Whitley, J. D. (1987). *Models of the UK economy: a fourth review by the ESRC Macromodelling Bureau*. Oxford: Oxford University Press.
Welfe, A. (1993). *Inflacja i rynek (Inflation and markets)*. Warszawa: PWE.

Welfe, W. (1992). *Ekonometryczne modele gospodarki narodowej Polski (Econometric models of Poland's national economy)*. Warszawa: PWE.

Welfe, W. (2000). Empiryczne modele wzrostu gospodarczego (Empirical models of economic growth). *Ekonomista, 4*, 483–499.

Welfe, W. (Ed.) (2001). *Ekonometryczny model wzrostu gospodarczego (An econometric model of economic growth)*. Łódź: Wydawnictwo UŁ.

Welfe, W. (2002). Łączna produktywność czynników produkcji a postęp techniczny (TFP and technical progress). *Studia Ekonomiczne, 36–37*(1–2), 94–115.

Welfe, W. (2003). Długookresowy, makroekonometryczny model W8D gospodarki polskiej (Long-term macroeconometric model W8D of the Polish economy). *Prace IEiS UŁ, 141*. Łódź: Wydawnictwo UŁ.

Welfe, W. (Ed.) (2004). Długookresowy, makroekonometryczny model WD-2002 gospodarki polskiej (Long term macroeconometric model W8D-2002 of the Polish economy). *Acta UŁ, Folia oeconomica, 172*. Łódź: Wydawnictwo UŁ.

Welfe, W. (2005). Stylized empirical model of economic growth. In W. Welfe & A. Welfe (Eds.), *Macromodels 2004, Problems of building and estimation of econometric models* (pp. 109–125). *Acta UŁ, Folia oeconomica, 190*. Łódź: Wydawnictwo UŁ.

Welfe, W. (2006). Towards modelling knowledge-based economies. In *Second international symposium in management, engineering and informatics*, Orlando, FL, 16–19.07.2006. Proceedings.

Welfe, W. (Ed.) (2007). *Gospodarka oparta na wiedzy (Knowledge-based economy)*. Warszawa: PWE.

Welfe, W. (Ed.) (2009a). Makroekonometryczny model gospodarki polskiej opartej na wiedzy (Macroeconometric model of the knowledge economy for Poland). *Acta UŁ, Folia oeconomica, 229*. Łódź: Wydawnictwo UŁ.

Welfe, W. (Ed.) (2009b). *Knowledge-based economies. Models and methods*. Frankfurt/Main: P. Lang.

Welfe, W., & Welfe, A. (2004). *Ekonometria stosowana (Applied econometrics)* (2nd ed.). Warszawa: PWE.

Welfe, W., Florczak, W., & Sabanty, L. (2002). Kapitał ludzki i jego endogenizacja (Human capital and its endogenization). *Przegląd Statystyczny, 50*(2), 7–36.

Xu, B., & Wang, J. (1999). Capital goods trade and R&D spillovers in the OECD. *Canadian Journal of Economics, 32*(5), 1258–1274.

Xu, B., & Wang, J. (2000). Trade, FDI and international technology diffusion. *Journal of Economic Integration, 15*, 585–601.

Chapter 18
Modelling Disequilibria and Equilibria

18.1 Introduction

Previous chapters presented models describing the activities of households and enterprises, where the main stress was placed on real processes. The models allowed generating the demand and supply of goods and services in the commodity markets, and demand and supply of labour force in the labour markets. In this chapter the equilibration mechanisms in these markets will be discussed. Firstly, the mechanisms of the quantitative, mainly short-term adjustments will be described in terms of changes in: (a) inventories, (b) the rates of capacity utilization (production factors), (c) foreign trade. Then the mechanisms of price and wage adjustments will be discussed, linked to the unemployment rate analysis.

The point of departure will be the basic national account identity that for the past behaviour has the following well known form:

$$X_t = H_t + E_t - M_t \tag{18.1}$$

and

$$H_t = C_t + G_t + J_t + \Delta R_t \tag{18.2}$$

where (all variables in constant prices):

C_t is personal consumption from disposable income,
E_t is exports,
G_t is consumption of public institutions,
H_t is absorption by domestic final users,
J_t is investment expenditures,
M_t is imports,
ΔR_t is changes in inventories and reserves,
X_t is GDP.

To analyse the potential states of equilibria (or disequilibria) the description of economic agents' behaviour must be complemented first by specifying the public sector equations (to generate its consumption G_t, employment and wage funds) and

W. Welfe, *Macroeconometric Models*,
Advanced Studies in Theoretical and Applied Econometrics 47,
DOI 10.1007/978-3-642-34468-8_18, © Springer-Verlag Berlin Heidelberg 2013

the foreign sector equations (to generate exports E_t and imports M_t). Secondly, the equations explaining producer prices must be supplemented using the equations for other agents, i.e. a comprehensive price system needs to be constructed. But most of all, it seems indispensable that adequate economic regimes be distinguished, so that the particular components of identities (18.1) and (18.2) can be alternatively identified either as the demand of economic agents or supply from them.

18.2 Modelling the Global Disequilibria

In macroeconometric models, global demand for domestic production (GDP) at the macroscale (X_t^D) is usually determined as a sum of its components, as in (18.1) and (18.2). In the disequilibrium models that do not assume that demand and the supplies delivered to final users (X_t^S) could be balanced even temporarily, i.e. assume that $X_t^D \neq X_t^S$, the supply equations are specified for X_t^S.

In the disequilibrium models the central role is played by potential output. Its volume may vary depending on the assumptions concerning the availability of production factors. If the availability is unconstrained, the final demand X_t^D can be met. Otherwise, if there are constraints on the availability of fixed capital, the potential output X_t^K can be generated; if there are scarcities in labour force availability, the potential output X_t^N will be determined (Barro and Grossman 1971). There may also exist deficits in energy or raw materials supplies, their most frequent causes being constraints in the availability of imports (or in the balance of payments) that generate the potential output X_t^M (Florczak and Welfe 2000; Welfe and Welfe 2004).

The generation of potential output implies the use of production functions (Welfe 2000). As mentioned in the previous chapter, the Cobb-Douglas and CES functions were used the most frequently (Dreze et al. 1990). For the sake of simplicity, the Cobb-Douglas production function will be used below.

Potential output generated with a fixed capital constraint, i.e. assuming its full utilization, will be obtained by using the following formula:

$$X_t^K = BA_t K_t^{\alpha} N_t^{K(1-\alpha)} \tag{18.3}$$

where N_t^K is the number of jobs, $N_t^K < N_t^S$.

Potential output generated under constrained availability of labour force, i.e. assuming full employment, will be obtained with the following formula:

$$X_t^N = BA_t K_t^{\alpha} N_t^{U(1-\alpha)} \tag{18.4}$$

where $N_t^U = N_t^S - U_t^N$ and N_t^S is the labour force supply, U_t^N is natural unemployment.

Full employment is understood here as the potential number of working-age people reduced by the amount of natural unemployment resulting from the conditions associated with the job-seeking process (Coen and Hickman 1976).

The generation of potential output constrained by inadequately available energy or raw materials implies that the production function should be extended by introducing additional explanatory variables. A disequilibrium indicator was used in most cases, showing the degree to which the demand for energy or raw materials was met. At the macrolevel these constraints mostly translated into constraints affecting imports or the current account in the balance of payments.

The potential output for the above variants established, the global supply of domestic products to the final users X_t^S can be obtained from the following equation:

$$X_t^S = \left(X_t^{K-\rho} + X_t^{N-\rho}\right)^{-1/\rho} \tag{18.5}$$

where $1/\rho$ is a mismatch coefficient or if the constraints in the supply of energy or raw materials are additionally considered:

$$X_t^S = \left(X_t^{K-\rho} + X_t^{N-\rho} + X^{M-\rho}\right)^{-1/\rho} \tag{18.5'}$$

Using the general assumptions concerning the aggregation of economic agents' activities, the total output provided for final users, i.e. GDP, can be determined from the following formula:

$$X_t = \left(X_t^{D-\rho} + X_t^{K-\rho} + X_t^{N-\rho}\right)^{-1/\rho} \tag{18.6}$$

For the West-European countries and Poland the mismatch coefficient estimates were close to 0.02 (Dreze et al. 1990; Welfe et al. 1996).

The empirical results of the presented investigations show that constraints in the availability of machinery and equipment played an important role in the 1970s and 1980s, but in the subsequent years the final demand constraints were decisive. The results obtained for Poland were approximately similar, however in the late 1970s and in the early 1980s the restrictions on the availability of raw materials, including imported ones, were strongly accentuated.

In the dynamic stochastic general equilibrium (DSGE) models the equilibrium conditions are determined for the major economic agents participating in final demand. The agents are households (private consumption), public institutions (social consumption), investors and exporters. The demand of these groups of economic agents is compared with the supply of products directed to them. The supply is represented by the output of the "final goods producers". This rather strange category may contain trading and logistic firms and exporter firms. This "output" is composed of domestic products (manufactured by domestic "producers of intermediate products") and imported products. As a rule, the supply flows are aggregated using the CES function. For example, the supplies of consumer goods directed to households, C_t^S, are obtained from:

$$C_t^S = \gamma\left(\delta X_t^{C-\rho} + (1-\delta) M_t^{C-\rho}\right)^{-V/\rho} \tag{18.7}$$

where:

X_t^C is supply of domestic products directed to households,
M_t^C is imports of consumer goods directed to households,
$\rho > 0$ is a parameter linked to the elasticity of substitution $\sigma_{X,M} : \sigma_{X,M} 1/(1+\rho)$.

This approach is particularly attractive from the perspective of analysing commodities flows going to final users. However, its implementation encountered serious informational barriers, especially in the case of the flows of imported commodities. The constructors of the supply-determined macromodels of the former CPE countries faced similar difficulties (Welfe 1992).

18.3 Modelling Demand of Public Institutions (Social Consumption)

For the description of the domestic final demand to be complete, the functions representing public institutions' demand for consumer and investment goods must also be specified. The most important here is the demand financed from the government budget.

In general, a simple hypothesis is proposed, assuming that the public institutions' demand for consumer goods (G_t) is a non-linear function of the real current expenditures of the government budget. The respective inertia must be introduced, recognizing the dependence on the existing infrastructure. Hence, for the long-run we have:

$$\ln G_t^* = b_0 + b_1 \ln(BCP_t/PG_t) + \varepsilon_t \tag{18.8}$$

where

BCP_t is the current expenditures of the government budget,
PG_t is the deflator of public consumption;

in the short-run we have:

$$\Delta \ln G_t = a_0 + a_1 \big(G_{t-1} G_{t-1}^*\big) + a_2 \Delta \ln(BCP_t/PG_t) + \alpha_3 \Delta \ln G_{t-1} + \varepsilon_t \tag{18.9}$$

In several models the explanatory variable is equal to the current expenditures adjusted for debt service and/or military expenditures, if the latter are ranked higher by the government.

Investment demand is usually assumed to be exogenous. Regarded as an important instrument of economic policy, it expresses the more or less arbitrary decisions of the government.

18.4 Modelling Inventories

In the demand-determined macromodels where demand, including final demand, is met, its component, i.e. the demand for an increase in inventories and reserves ΔR_t, is distinguished and special equations are built. In the disequilibrium models where the supply of commodities was directly specified, inventory changes were residual and played an equilibrating role. They retained the role also in the equations of the aforementioned demand-determined models.

The macroeconometric models of less-developed countries and the former CPE countries used production functions to generate output and, consequently, to allocate the supplies of commodities and services among final users. At the macro-scale, the global supply X_t^S was determined as a sum of value added in particular sections. Its volume usually did not match the volume of global demand—market equilibrium was ensured by appropriate inventory changes ΔR_t. In these models, the following identity was defined:

$$\Delta R_t = C_t^D + G_t^D + J_t^D + \left(E_t^D - M_t^D\right) - X_t^S \qquad (18.10)$$

At present, it is used rather rarely. Other adjustments mechanisms predominate in macromodels. The accuracy of inventory changes calculated from an identity is non-satisfactory. This makes the attempts to explain the demand for an inventory increase more appealing.

In the specification of the equation explaining the demand for inventory increase the main stress must be placed on the major functions of inventories and reserves. They must ensure the continuity of production and trade. Hence, they should expand correspondingly to growing output and trade. The volume of the inventories of finished commodities depends i.a. on how precisely enterprises can predict the demand for output, as indicated in Chap. 17. Regarding raw materials and materials, their volume is determined by the frequency of their periodical supply. Moreover, they function as a buffer that protects production and trade processes against the possible demand shocks or commodity supply shocks (Lovell 1961, 1962).

The above premises were applied in macromodels to specify the inventory functions. The desirable level of inventories depends on the level of output (or sales). It is assumed that this adjustment process follows with a lag:

$$R_t^* = b_0 + b_1 X_t - b_2 R_{t-1} + \xi_t \qquad (18.11)$$

where: R_t^* is the desired level of inventories at period end, $b_1 > 0$ and $0 < b_2 < 1$.

The impacts of shocks, whether they follow from changes in the factors determining demand $D_t(\cdot)$ or from changes in the factors determining supply $S_t(\cdot)$, have short-term implications. They are introduced into the short-term equation:

$$\begin{aligned}\Delta R_t = {} & \alpha_0 + \alpha_1\left(R_{t-1} - R_{t-1}^*\right) + \alpha_2 \Delta X_t + \alpha_3 \Delta R_{t-1} \\ & + \alpha_4\left(D_t(\cdot) - S_t(\cdot)\right) + \xi_t \qquad (18.12)\end{aligned}$$

The above shocks are frequently characterised using the rate of capacity utilization WX_t. Its increase is associated with inventories decline and its decrease is followed by their expansion ($\alpha_4 < 0$).

In general, the prediction accuracy of an inventory increase is not high. Hence, only a few models treat the results of forecasts as important signals for enterprises, heralding the likely business cycle changes.

In the multisectoral models attempts were made to distinguish equations explaining particular kinds of inventories, such as finished commodities, work in progress, raw materials and materials (Juszczak 1982). The equations were specified for particular sections of the national economy, including the inventories of the trading organizations (cf. equations in the W-5 model of the Polish economy, Welfe 1992).

18.5 Modelling Adjustments in the Production Sector

In the demand-determined macroeconometric models output is determined by demand directed to enterprises (allowing for changes in the inventories of finished goods). This demand influences the demand for production factors that basically falls short of the full utilisation of their resources, so some or all of the factors remain in excess. The changes in this demand can be partly handled by adjusting appropriately capacity utilization and/or the utilization of equipment and employment.

Hence the obvious need that changes in enterprises' capacity utilization rates and/or in particular production factors be monitored, as mentioned in the previous chapter.

Solving this problem effectively poses several difficulties. Firstly, in many countries the information on the capacity utilization rate (WX_t) is limited. The Wharton Index used in the quarterly Wharton models of the US economy that was developed the earliest by comparing output with its values in the peak years was applied in the manufacturing industry (Klein and Summers 1966). In the European countries surveys making direct enquiries to enterprises about their capacity utilization rates were used (Grzęda-Latocha 2005). The application of these indicators is somewhat inconvenient, because to use them in forecasting equations explaining their dynamics must be built.

The concepts where the effective output (X_t) is compared with potential output derived from the production function seem to be the most appropriate theoretically and free of the above disadvantage. If the potential output were determined to ensure the full utilization of the fixed capital (X_t^K) as in (18.3), then the rate of capacity utilization will be obtained from:

$$WX_t^K = X_t / X_t^K \tag{18.13}$$

If the potential output were defined to ensure full employment (X_t^N), then the rate of capacity utilization will be calculated from:

$$WX_t^N = X_t / X_t^N \tag{18.14}$$

In this case, the precision of the estimates of the capacity utilization rates will depend on the computation accuracy of potential output. The above characteristics of the rate of potential output were broadly used in the macroeconometric models of the USA (Coen and Hickman 1976) and other countries, including Poland (Welfe 2009).

The rates of capacity utilization play a far-reaching role in macroeconometric models. They are introduced as explanatory variables representing market tensions not only into the equations explaining prices, but also into the foreign trade equations. Let us mention that in the equations explaining prices central banks often use a simplified procedure for capturing the effects of market tensions, which involves the construction of an indicator comparing effective GDP with GDP trajectory represented by an appropriate trend, which was mentioned in the previous chapter.

Macroeconometric models rather infrequently use the scarce information on the utilization rate of machinery and equipment as opposed to the use of information on employees' time worked. The latter allows analysing the utilization rate of employees' potential working time (Fair 2004), as well as the utilization rate of production potential by adjusting the number of shifts (Welfe 1992).

18.6 Modelling Unemployment

In the market economies labour market disequilibria are represented by unemployment, i.e. by the difference between labour force supply and employment. This understanding of unemployment can be found in the demand-determined macroeconometric models. In the supply-oriented models the excess demand for labour predominates, frequently related to hidden unemployment.

The level of unemployment is understood as registered unemployment or unemployment estimated through special surveys, such as LFS. Macromodels mostly use registered unemployment.

In macromodels, the level of unemployment UN_t and the rate of unemployment U_t are most frequently generated from respective identities:

$$UN_t = N_t^S(\cdot) - N_t^D(\cdot) \tag{18.15}$$

and

$$U_t = UN_t / N_t^S(\cdot) \tag{18.16}$$

where

$N_t^D(\cdot)$ is the demand function for employees,
$N_t^S(\cdot)$ is the labour supply function.

The above functions were specified for labour supply in Chap. 16 and for the demand of the enterprise sector in Chap. 17 (cf. Phillips 1958). The total demand for employees involves also public sector employees. Their number is either treated as exogenous or, preferably, generated from the equations that take into account the available infrastructure (frequently represented by lagged employment) and the likely changes in the level of total salaries assumed in the government budget. In this case, we have:

$$N_t^{PD} = \alpha_0 + \alpha_1 N_{t-1}^{PD} + \alpha_2 \Delta\left(F_t^{DP} / WP_t^P\right) + \varepsilon_t \tag{18.17}$$

where

N_t^{PD} is public sector' demand for employees,
FP_t^P is total salary fund for the public sector officers,
WP_t^P is average salary in the public sector.

In the long-run, average salaries paid in the public sector vary like average wages in the enterprise sector:

$$WP_t^P = AWP_t^{\alpha} e^{\xi_t} \tag{18.18}$$

Generating the rate of unemployment as a residual from identity (18.16) usually yields results that are not sufficiently accurate. The estimation (and prediction) errors of the demand and supply functions add up. Hence, attempts were made in several macromodels to estimate the changes in unemployment rates using the reduced specifications of the demand functions $N^D(\cdot)$ and of the labour supply functions $N^S(\cdot)$. It can be assumed in the short-run that labour supply changes slowly, so the impact of the changes can be either ignored or represented using a trend function. The most important are factors affecting employment demand, i.e. output and the effects of technological progress (Nickell 1984). This means that the rate of unemployment can be determined using the utilization rate of production capacities WX_t and the rate's fluctuations can be explained by the GDP fluctuations in the business cycle:

$$U_t = BWX_t^{\beta} e^{\mu t + \xi}$$

This approach is far from being excellent. As mentioned, the estimates of the capacity utilization rate can be obtained using competitive methods and are not highly accurate.

The observed unemployment rate represents the effects of changes in the labour market caused by employees changing employers (frictional unemployment) and the long-term tendencies arising from the permanent unavailability of jobs (structural unemployment). In the literature, Friedman (1968) and Phelps (1970) views were prevalent. They proposed distinguishing the category of natural unemployment, representing labour market equilibrium. They found that the economic policy instruments designed to reduce the rate of unemployment made it lower only temporarily and were followed by an increase in inflation (cf. Blanchard and Quah 1989).

The NAIRU (non-accelerating inflation rate of unemployment) concept derived from the analysis of the relations between entrepreneurs and employees (Layard et al. 1991) found more general applications. The NAIRU is an "equilibrium" unemployment rate such that exceeding it makes the rate of inflation go down and its decrease is followed by an increase in the rate of inflation. The values of the NAIRU computed for the USA oscillated around 6 % whereas for Western Europe were higher between 8–9 % (Elmeskov 1993).

To compute the NAIRU several simplifying assumptions concerning the equations explaining average wages and prices that serve as a point of departure must be introduced. For the sake of demonstration, let us derive the NAIRU using the long-run equations of average wages and prices. The equations will be shown by means of the logs of variables.[1]

The equation explaining the nominal average wages before tax reads as:

$$wp_t = \alpha_0 + \alpha_1 p_{t-1}^c + \alpha_2 wy_t - \alpha_3 u_t + \alpha_4 t_t^w \tag{18.19}$$

where p^c is a consumption deflator, t^w is social contributions, wy_t is labour productivity.

[1] This example follows the modified derivation shown by Whitley (1994, pp. 102–103).

The long-run real average wages will be obtained assuming that the elasticities with respect to prices will be calibrated at levels equal to one and that no lags are present. Hence:

$$wp_t - p_t^c = \alpha_2 wy_t - \alpha_3 u_t + \alpha_4 t_t^w \tag{18.20}$$

Considering that the logarithm of the consumption deflator can be presented as the difference between the logarithms of producer prices (p_t) and indirect taxes (t_t^i):

$$p_t^c = p_t - t_t^i \tag{18.21}$$

after substituting to (18.20) we shall have:

$$wp_t - p_t = \alpha_2 wy_t - \alpha_3 u_t + \alpha_4 t_t^w + t_t^i \tag{18.20$'$}$$

The initial producer price equation, following (17.55), reads as follows:

$$p_t = \beta_0 + \beta_1 (wp_t - wy_t + t_t^w) + \beta_2 p_t^m + t_t^z \tag{18.22}$$

where

p_t^m is import deflator,
t_t^z is corporate income tax.

The long-term producer price equation will be obtained based on a simplifying assumption that price elasticity with respect to labour costs is one, $\beta_1 = 1$, and the import deflator can be obtained by multiplying the world prices (p_t^w) and the exchange rate (er_t): $p^m = p^w + er_t$.

As a result, we have:

$$p_t - wp_t = wy_t + t_t^w + \beta_2 (p_t^w + er_t) + t_t^z \tag{18.22$'$}$$

hence

$$wp_t - p_t = -wy_t - t_t^w - \beta_2 (p_t^w + er_t) - t_t^z \tag{18.22$''$}$$

The NAIRU defining equation will be obtained by comparing (18.20′) and (18.22″):

$$-\alpha_3 u_t = (\alpha_2 - 1) wy_t - (1 - \alpha_4) t_t^w - t_t^i - t_t^z - \beta_2 (p_t^w + er_t)$$

hence

$$\alpha_3 u_t = (1 - \alpha_2) wy_t (1 - \alpha_4) t_t^w + t_t^i + t_t^z + \beta (p_t^w + er_t) \tag{18.23}$$

This derivation of NAIRU shows that its level depends on the tax rates and the exchange rate, and on level of productivity.

Alternative methods were applied. The first uses the rates of growth of variables, the second is based on the reduced Phillips formula, utilizing the Kalman filter (Elmeskov and MacFarland 1993).

A simplified approach was elaborated similar to that applied in calculating NAWRU as shown below.

The NAWRU (non-accelerating wage rate of unemployment) is an alternative concept that has found a broad use. This unemployment rate is not followed by an increase in average wages. It was originally developed for the OECD models (Elmeskov and MacFarland 1993). Its estimates can be found in many European macromodels, including those for the OECD member countries (Bårdsen et al. 2005). It can be derived from the following equation:

$$U_t^{NAWRU} = U_t - \left(\Delta U_t / \Delta^3 w_t\right) \Delta^2 w_t \tag{18.24}$$

where

$$w_t = \ln WB_t$$

To avoid irregular shocks the HP filter was commonly applied.

18.7 Modelling Foreign Trade

18.7.1 Introduction

The equations explaining exports and imports have been present in all macroeconometric models since their beginning. Many models initially treated foreign demand as exogenous (in the world economy models it was generated for particular country by the system). Later on, endogenized foreign demand served as a characteristic of the links with the world economy that could be shaped by the improving competitiveness of the exporting country.

In the supply-determined models and the general equilibrium models the export supply functions were also specified.

The macroeconometric models' equations explaining domestic demand for imported commodities and services are specified in most cases. This demand depends on the level of domestic economic activity, as well as being sensitive to the level of imports' competitiveness vis-à-vis domestic production. In a few models attempts were made to explain the supply of imported commodities. A relevant example is the DSGE models, where the sources of imported products were established.

The foreign trade equations cover not only its total volumes. The most frequent disaggregation distinguishes commodities and services. The commodity imports are decomposed by their use, i.e. into consumer goods, investment goods, and intermediate commodities, with fuels being frequently distinguished. The world economy models frequently use the UN SITC classification of exports and imports.

In many macroeconometric models the foreign trade flows were decomposed using also the geographical and political criteria (Bożyk et al. 1973; Czyżewski and Welfe 1990).

The discussion in this section will concentrate on the presentation of the exports and imports equations, bearing in mind their role in balancing the economy; the review of the structure of the import and export price equations will be postponed until the next section and the role of the above equations in modelling the balance of payments will be presented in one of the next chapters.

18.7.2 Equations Explaining the Demand for Exported Goods

Foreign demand for commodities exported by a country is composed of the demand of particular countries represented by the appropriate fractions of their global imports. Macromodels distinguish the most significant importers or their groups, such as EU members, or use total world imports (equal to total world exports). Exports volumes are treated as exogenous, except that they are explained by the pertinent variables like in the multicountry models, first of all by GDP. The models that belong to the above class usually assume that the exports E_{it} of the i-th country can be obtained as a sum of the products of the components of the international trade matrix a_{ij} by the volumes of foreign countries' imports M_{jt}, i.e. $E_{it} = \sum_j a_{ij} \cdot M_{jt}$. The matrix components were often assumed to have constant values, which implied the necessity of updating them (Hickman and Lau 1973; Klein et al. 1999).

The exporters' reactions to foreign demand changes were not immediate, which seems to be due to the traditional conclusion of contracts with an appropriate lead. Hence the suitable lags in the adjustments to changes in foreign demand.

The empirical long-term relationships show certain stability, as well as revealing that growth tendencies are stronger for exports than for global imports as implied by the elasticity of exports with respect to the summary imports of foreign countries which is higher than one. This result does not seem justified, being rather an indication that some important factor has been omitted. It was therefore suggested that the long-term elasticity of exports (calibrated) should be equal to one, meaning a stable share of the given country's exports in global imports. The introduction of the structural changes and qualitative improvements could be represented through an exponential trend, the use of which could bring down the aforementioned elasticity to a level close to one (Whitley 1994).

Foreign demand is modified along with changing competitiveness of exported products. This competitiveness is most frequently expressed through the exporting country's relative prices, or the relative labour costs or relative unit costs which are used more rarely. Price competition has been recently giving way to quality-based competition, so the absolute values of estimates of the price export elasticities have been declining in the last years.

In the short-run, unexpected shocks in the domestic markets disturb export activities in several ways. Shocks, such as floods, that impose constraints on the supply to the domestic markets or induce increases domestic demand are compensated for by lower supplies for exports. To represent the disturbances the common characteristics of the market tensions are used, such as the utilization rate of productive capacities.

The above examination leads to the following long-run equation explaining the foreign demand function for exported commodities:

$$\ln E_t^{D*} = \alpha_0 + \alpha_1 \ln M_t^W + \alpha_2 \ln\left(P_t^E / ER_t\right) / P_t^W + \varepsilon_t \tag{18.25}$$

The short-run foreign demand function is specified as follows:

$$\begin{aligned}\Delta \ln E_t^D = {} & \beta_0 + \beta_1\left(\ln E_{t-1}^D - \ln E_{t-1}^{D*}\right) + \beta_2 \Delta \ln M_t^W + \beta_3 \Delta \ln\left[\left(P_t^E / ER_t\right) / P_t^W\right] \\ & + \beta_4 \Delta \ln WX_t + \beta_5 \Delta \ln E_{t-1}^D + \varepsilon_t \end{aligned} \tag{18.26}$$

where

ER_t is exchange rate—domestic currency/world currency,
M_t^W is imports of the distinguished territory,
P_t^E is export deflator in domestic currency,
P^W is import deflator of the distinguished territory,
WX_t is the utilization rate of productive capacity.

As mentioned, the estimate of the long-run elasticity α_1 exceeded one in many models (for Poland it was 1.2). Hence it was frequently calibrated, assuming $\alpha_1 = 1$. Then the exponential trend was introduced into the long-term equation (18.25). Notice, that the equation could be reduced by introducing the ratio between exports and global imports as the explained variable.

In the major UK models the long-term export elasticities with respect to relative prices were between −0.28 and −0.40 (Whitley 1994); for Poland they were substantially lower (−0.11) (Welfe 2002). The short-term elasticities with respect to global imports were usually smaller than one. However, the short-term export elasticities with respect to relative prices were higher in absolute terms than the long-term elasticities, thus displaying stronger short-run sensitivity of the world markets.

18.7.3 The Exports Supply Functions

Commodities intended for exports were modelled mainly in the supply-determined macromodels. Regarding the demand-determined models, their builders most frequently gave up the construction of separate exports supply functions, assuming that supply followed effective demand. The models built in the former CPE countries usually contained the supply functions that represented the export potential of the economy, without assuming that foreign demand would be met. More recently the exports supply functions could be found in the CGE models, and particularly in the DSGE models. The equations are special in that they decompose the supply of export commodities by its origin, i.e. into the supply of domestic products and of imported commodities (re-exports).

The initial specification of the exports supply function started with the potential domestic output, the aim being to estimate the amount by which the domestic production potential was larger than domestic demand. The domestic production potential was represented by either global output (Q_t) or global domestic output GDP (X_t).

The offer of commodities for exports depends also on the profitability of exports. Its changes are related to the changes in the relative export prices, i.e. in the relations between the prices paid for exports (P_t^E) and those obtained in the domestic markets (P_t).

Hence the long-run exports supply function (E_t^S) will be specified as follows:

$$\ln E_t^{S*} = \alpha_0 + \alpha_1 \ln X_t + \alpha_2 \ln\left(P_t^E / P_t\right) + \xi_t \tag{18.27}$$

In the short-run, the effects of shocks produced by varying supply to the domestic users must be introduced. They are typically represented by the rate of capacity utilization (WX_t). Therefore, the short-run equation will have the following form:

$$\Delta \ln E_t^S = \beta_0 + \beta_1 \left(E_{t-1}^S - E_{t-1}^{S*}\right) + \beta_2 \Delta \ln X_t + \beta_3 \Delta \ln\left(P_t^E / P_t\right) + \beta_4 \Delta \ln WX_t + \xi_t \tag{18.28}$$

In the DSGE models the notion of exports potential is frequently extended by the explicit introduction of reexports into the supply equation. Let the global volume of the commodities for export be S_t^E; then, using the DSGE models terminology, the supply from "the producers of final goods" intended for export will equal:

$$S_t^E = \gamma \left(\delta X_t^{E^{-p}} + (1-\delta) M_t^{E^{-p}}\right)^{-v/p} \tag{18.29}$$

where

X_t^E is the output for exports from the "producers of intermediate goods",
M_t^E is imports for reexports.

The supply of commodities intended for exports redefined as above will substitute GDP in both the long-term equation (18.27)

$$\ln E_t^{S*} = \alpha_0 + \alpha_1 \ln S_t^E + \alpha_2 \ln\left(P_t^E / P_t\right) + \xi_t \tag{18.27'}$$

and the short-term equation (18.28) explaining the supply of exports:

$$\Delta \ln E_t^S = \beta_0 + \beta_1 \left(E_{t-1}^S - E_{t-1}^{S*}\right) + \beta_2 \Delta \ln S_t^E + \beta_3 \Delta \ln\left(P_t^E / P_t\right) + \beta_4 \Delta \ln WX_t + \xi_t \tag{18.28'}$$

Assuming that the long-run elasticity of exports supply with respect to global supply S_t^E is one, i.e. that in the long-run $\alpha_1 = 1$, then the ratio of the exports supply E_t^S to the potential exports supply S_t^E can be treated as an explained variable. This ratio will be dependent on the relative prices (the Armington model, Armington 1969). For the long-term, we have:

$$\ln\left(E_t^S / S_t^E\right) = \alpha_0 + \alpha_2 \ln\left(P_t^E / P_t\right) + \xi_t \tag{18.30}$$

where the elasticity α_2 stands for the competitiveness of the commodities to be exported. It is worth knowing that special computations are needed to estimate the volumes of the commodities intended for exports.

18.7.4 The Imports Functions

The import of commodities and services performs two major functions. Firstly, it is to complement the range of commodities and services that are either not produced domestically or produced in insufficient amounts (complementary imports). Secondly, imports compete against the domestically produced goods, thus contributing to higher efficiency of domestic supplies (competitive imports).

In the simplest case, the imports equation describing the first function can be specified as an identity, with imports defined as a residual. At the macro-scale this will be:

$$M_t = H_t + E_t - X_t \tag{18.31}$$

In the multisectoral models this identity holds for the distinguished commodity groups. In *these equations* imports play a balancing role. This becomes particularly clear, when output represents potential domestic output.

However, macroeconometric models use most frequently imports functions that represent mechanisms generating demand for imported commodities and include competitive issues.

The competitive intensity is usually represented by relative import prices defined as the ratios between import prices (in domestic currency) P_t^m and domestic producer prices P_t. Hence, at the macro scale the long-run equation will have the following form:

$$\ln M_t^* = \alpha_0 + \alpha_1 \ln X_t + \alpha_2 \ln\left(P_t^M / P_t\right) + \xi_t \tag{18.32}$$

Many authors assume that the share of imports in GDP is constant in the long-run. Therefore, imports elasticity with respect to GDP (calibrated) can be assumed at a level equal to one ($\alpha_1 = 1$). With this assumption, the explained variable can be redefined as the share of imports in GDP, i.e. M_t / X_t. In this case, the relative prices will be the only explanatory variable:

$$\ln(M_t / X_t) = \alpha_0 + \alpha_2 \ln\left(P_t^M / P_t\right) + \xi_t \tag{18.33}$$

It must be added that this relationship represents the result of optimizing the behaviour of economic agents that choose from among the domestic and imported commodities (the Armington model).

The acceptance of the above assumption is sometimes associated with an inclination to allow for structural changes, which leads to an increase in the growth rates of imports. An exponential trend is then introduced (Whitley 1994).

In the short-run, the shocks likely to come from unexpected changes in domestic demand represented by changes in the capacity utilization rate (WX_t) or from changes in foreign direct investment (B_t^J) are introduced. Then the short-term equation will have the following form:

$$\begin{aligned}\Delta \ln M_t = {} & \beta_0 + \beta_1\left(\ln M_{t-1} - \ln M_{t-1}^*\right) + \beta_2 \Delta \ln X_t + \beta_3 \Delta \ln\left(P_t^M / P_t\right) \\ & + \beta_4 \Delta \ln WX_t + \beta_5 \Delta \ln B_t^J + \xi_t\end{aligned} \tag{18.34}$$

In many macroeconometric models commodity imports are decomposed according to their use. There are distinguished imports of intermediate commodities (M_t^Z), investment imports (M_t^J) and consumer goods imports (M_t^C). The long-term equations explaining the demand for the distinguished groups of imported commodities will be specified below.

It is generally assumed that the imports of intermediate commodities depend directly on gross output (Q_t):

$$\ln M_t^Z = \alpha_0 + \alpha_1 \ln Q_t + \alpha_2 \ln\left(P_t^M / P_t\right) \tag{18.35}$$

where the relative prices are defined as the ratios of intermediate commodity prices, if the appropriate information is available. The multisectoral models, including the input-output models, use the data on the import-output ratios showing the unit use of materials and fuels.

The demand for imported, mostly complementary, investment goods mainly depends on the expansion of the domestic investment activities. In the case of the emerging economies a special channel for the imports of investment goods is distinguished, namely foreign direct investments (FDI).

Hence we have:

$$\ln M_t^J = \alpha_0 + \alpha_1 \ln J_t^V + \alpha_2 \ln FDI_t + \alpha_3 \ln\left(P_t^{MJ}/P_t^{JV}\right) + \xi_t \tag{18.36}$$

where

- J_t^V is investment expenditures on machinery and equipment (constant prices),
- P_t^{MJ} is the deflator of imported investment goods,
- P_t^{JV} is the deflator of domestic investment expenditures on machinery and equipment.

The demand for imported consumer goods is treated as competitive in most cases, which gives importance to the relative imports prices. The complementary imports help meet the domestic consumption. However, the major explanatory variable is the total consumption C_t, mainly because of restricted data availability. Hence we have:

$$\ln M_t^C = \alpha_0 + \alpha_1 \ln C_t + \alpha_2 \ln\left(P_t^{MC}/P_t^C\right) + \xi_t \tag{18.37}$$

where P_t^{MC} is the deflator of imported consumer goods and P_t^C is the deflator of individual consumption.

In the multisectoral models imports are decomposed further. The SITC classification is used most frequently to this end. Among the intermediate commodities the import of fuels is often distinguished.

As mentioned, the estimates of the long-term elasticities of imports with respect to GDP are either close to one or slightly exceed one. The long-term elasticities of consumer goods imports with respect to total private consumption are much higher, reaching 3 in the extreme cases. The price elasticities for the UK ranged between −0.5 and −0.8 (Whitley 1994) and for Poland between −0.3 and −0.8 (Welfe and Welfe 2004).

18.7.5 The Equations Explaining the Supply of Imported Commodities

The demand-determined models assume that the potential supply of commodities that importers may offer is unconstrained, hence the effective supply tends to match the demand for imported goods. The supply conditions are affected by import prices that will be discussed in the next section. Attempts have been made within the general equilibrium models and the DSGE models to estimate the supplies of imported commodities by looking at their sources.

18.8 Price Systems

18.8.1 Introduction

Prices perform many important functions in national economies. Changing, they clear the markets. They allow the flows of commodities and services to be transformed into the flows of incomes and expenditures (as deflators), i.e. they guarantee a transition from the real sphere to the sphere with financial flows (Courbis 1977; Tobin 1972). The most important role is played by the producer prices; the specification of the equations explaining these prices was shown in the previous chapter. In this section the discussion will be extended to include the equations explaining final goods prices and the prices of production factors that together constitute the price system functioning in the national economy. The discussion on how interest rates and exchange rates are generated will be postponed until the next chapters.

18.8.2 Equations Explaining the Prices of Domestic Final Goods

In specifying the equations explaining the prices of consumer or investment goods it is commonly assumed that the prices depend, as appropriate, on producer prices, import prices, trading mark-ups and indirect taxes (mainly VAT).

At the macro scale, the most frequent explanatory variables are GDP deflators; in the multisectoral models the deflators of gross output (or production sold) are applied in particular sections.

Like the retail price indices, the deflators of private consumption (CPI) are mostly computed as the functions of the geometric averages of GDP deflators, import deflators and indirect taxes. For the long-run, we have:

$$\ln P_t^{C^*} = \alpha_0 + \alpha_1 \ln P_t + (1 - \alpha_1) \ln P_t^{MC} + \alpha_2 \ln T_t^V + \varepsilon_t \tag{18.38}$$

where T_t^V is the rate of indirect taxes (VAT).

In the short-run, we have:

$$\begin{aligned} \Delta \ln P_t^C = {} & \beta_0 + \beta_1 \left(\ln P_{t-1}^C - \ln P_{t-1}^{C^*}\right) + \beta_2 \Delta \ln P_t \\ & + (1 - \beta_2) \Delta \ln P_t^{MC} + \beta_3 \ln \Delta T_t^V + \varepsilon_t \end{aligned} \tag{18.39}$$

The value of the parameter α_1 can be interpreted as the share of commodity flows of domestic origin. If the information on the decomposition of the consumer goods supplies is available, the value can be appropriately adjusted. Then Eq. (18.38) can be used to estimate price elasticity with respect to indirect taxes.

Equation (18.39) can be used to determine the rate of inflation and to make comparisons with the inflationary target assumed by the central bank.

Many macromodels decompose the deflators of consumer goods according to the groups of commodities and services, using the permanent components (core inflation) rather than the transitory components, such as the prices of agricultural products or fuel prices (those depend on the world business cycle).

The deflators of consumption of public institutions P_t^G are structured similarly to the deflators of private consumption. The deflators of inventories and their changes have a similar property.

The equations explaining the investment goods prices, i.e. the deflators of investment expenditures P_t^I, have a simple structure, because the prices usually do not include indirect taxes. Investment expenditures are frequently decomposed into amounts allocated to machines and equipment—here the deflators depend mainly on the prices in the electric machinery industry and on imports—and into amounts spent on buildings and constructions, where the deflators depend on the deflators in the building industry and transport. In general, the long-term equation reads:

$$\ln P_t^{J^*} = \alpha_0 + \alpha_1 \ln P_t^Q + (1 - \alpha_1) P_t^{MJ} + \varepsilon_t \tag{18.40}$$

and the short-term equation has the form:

$$\begin{aligned} \Delta \ln P_t^I = {} & \beta_0 + \beta_1 \left(P_{t-1}^I - P_{t-1}^{I^*}\right) + \alpha_2 \Delta \ln P_t^Q \\ & + (1 - \alpha_2) \Delta \ln P_t^{MJ} + \varepsilon_t \end{aligned} \tag{18.41}$$

where P_t^Q is the deflator of the appropriate section (or industry).

18.8.3 Equations Explaining Export and Import Prices

To specify the functions explaining export prices special assumptions about exporters' behaviour must be imposed. Following some Scandinavian models, it may be assumed that the exporters' role in price formation is meaningless, so the export prices are fully determined by the world prices:

$$P_t^{EW} = P_t^W \tag{18.42}$$

where

P_t^{EW} is the export prices of the exporting country in the world currency,
P_t^W is the world trade prices.

Many macroeconometric models assume, though, that exporters can efficiently compete in the world markets based on their export prices. Therefore, the effective prices of exported commodities will reflect the compromise between the world trade impacts (world prices) and exporters' activities (domestic prices converted into the world currency). When the homogeneity restriction is taken into account, the long-run equation will have the form:

$$\ln P_t^{EW*} = \alpha_0 + \alpha_1 \ln P_t^W + (1 - \alpha_1)\ln\bigl(P_t^E ER_t\bigr) + \varepsilon_t \qquad (18.43)$$

where ER_t is the exchange rate, P_t^E is export prices in domestic currency.

Allowing for the impacts of the likely shocks in exporters' behaviour, in the short-run the equation will be formed as shown below:

$$\Delta \ln P_t^{EW} = \beta_0 + \beta_1\bigl(\ln P_{t-1}^{EW} - \ln P_{t-1}^{EW*}\bigr) + \beta_2 \Delta \ln P^W + (1 - \beta_2)\Delta \ln(P_t / ER_t) + \beta_3 \Delta \ln U_t^E + \varepsilon_t \qquad (18.44)$$

where U_t^E is a variable representing the shocks in behaviour.

The estimate $\hat{\beta}_2 = 0.43$ was obtained for the stability period in Poland (Welfe 2002). Ignoring the likely impact of the 2008 shock would be misleading, though. The US dollar and euro exchange rates increasing suddenly by 20 % did not cause any decline in the export prices denominated in the two currencies, but contributed to higher domestic prices of domestic exports, i.e. to exporters' surplus. This effect could be captured by making an appropriate change to the variable $U_t^{E\cdot}$. Otherwise, assuming in this case (18.42), Eq. (18.44) could be solved for export prices in domestic currency P_t^E.

Imports deflators expressed in the world currency are commonly assumed exogenous. In the world multicountry models the deflators are computed as the weighted sums of the deflators of the exporting countries' exports to the country where the distinguished importers are based; the weights are obtained from the aforementioned world trade matrix. The deflators expressed in domestic currency are obtained from an identity that converts the world currency prices into domestic prices using the appropriate exchange rate:

$$P_t^M = P_t^{MW} \cdot ER_t \qquad (18.45)$$

where P^{MW} is the import prices in the world currency.

18.9 The Economic Mechanisms in the Real Sphere

In economic analyses based on macroeconometric models it frequently becomes necessary to treat jointly the relations in the real sphere. In such cases the multiplier analysis is commonly used. It helps identify the joint impacts of a single shock by treating the real sphere as a subsystem of the national economy (Klein et al. 1999).

In the demand-determined models three major economic mechanisms entering into the simultaneous feedbacks are typically distinguished (Welfe 2005). They are represented by the following multipliers: consumer multiplier, accelerator, and inflationary spiral.

The consumer multiplier summarizes the impacts of a single shock increasing household consumption. A rise in consumer demand (C_t) is followed by larger domestic output (X_t), modified by appropriately expanding competitive imports of

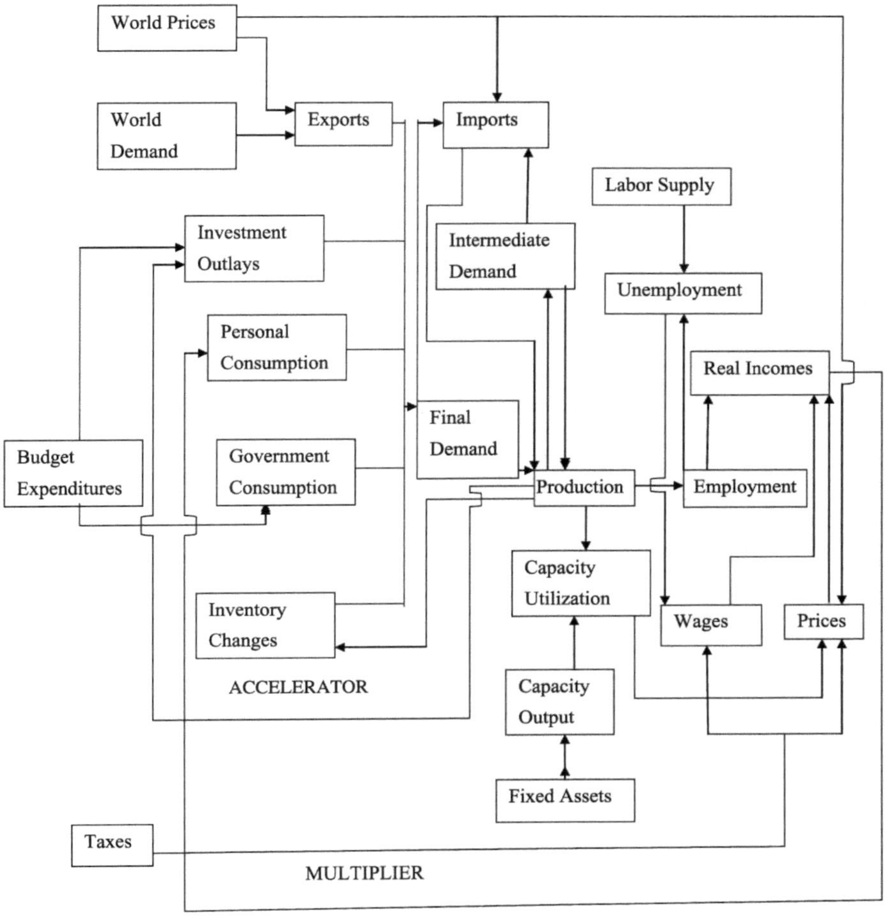

Fig. 18.1 Complete demand determined models. Real flows. Source: Klein et al. (1999, p. 113)

consumer goods. Growing output entails an increase in the number of working hours and, with a lag, in the number of employees (N_t) that is adjusted by the impact of technology changes. The wage bill and households' disposable incomes additionally grow and finally consumer demand expands.

The generalized consumer multiplier accounts for relationships containing financial flows. When the government budget increases its current expenditures the consumer expenditures of public institutions (G_t) also grow, as well as the wages bills related to public officers and old age and disability pensions, which increases households' incomes and stimulates their demand. The domestic output grows even more and the conditions for initiating the consumer multiplier are created.

The accelerator represents the relationships that control the investment process. An expected increase in production capacity triggers an rise in the demand for

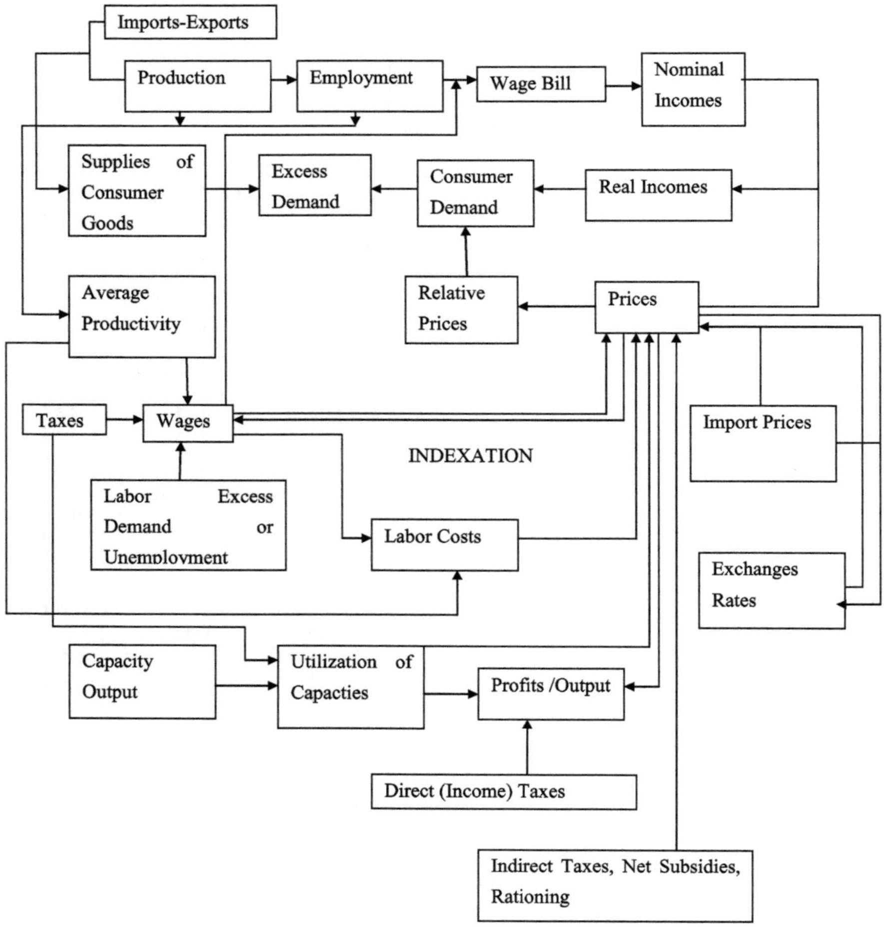

Fig. 18.2 Inflation. Source: Klein et al. (1999, p. 124)

investment goods. As a result, the domestic production of investment goods grows larger, constrained by additional imports of investment goods. Further increase takes place in the manufacturing industries. All this leads to stronger expectations of a production capacity increase.

These relationships are represented in Fig. 18.1.

The simplest version of the inflationary spiral is a summary of the relations between nominal wages and prices. An increase in the nominal average wages paid in the production sector will make labour costs grow, unless neutralized by increased labour productivity. The wage pressure will lead to an increase of producer prices and consequently of retail prices (CPI).

This has effect on the increase in nominal wages. It is worth noting that this feedback contains a mechanism reducing the rate of inflation, because the

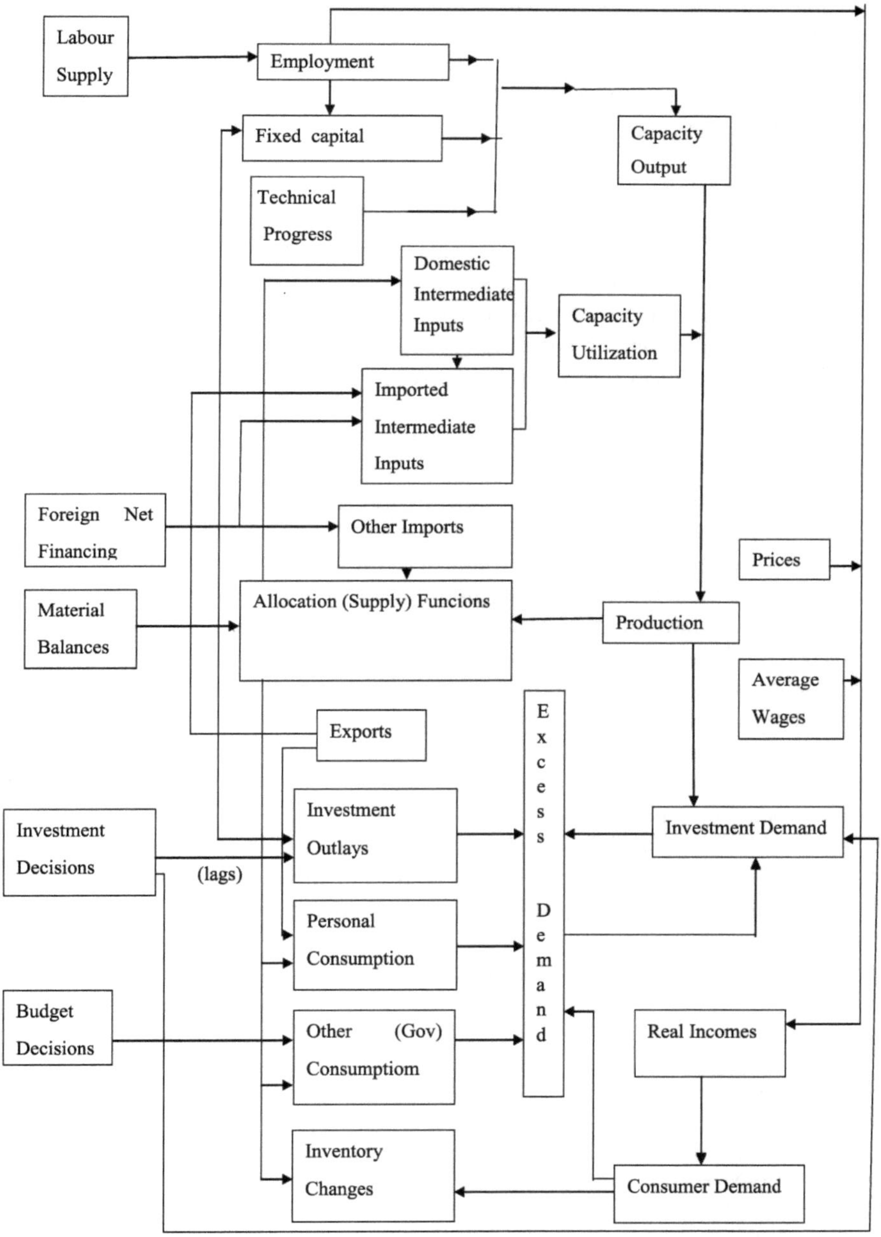

Fig. 18.3 Complete supply determined models. Real flows. Source: Klein et al. (1999, p. 112)

aforementioned increase in labour costs is generally unrelated to the increases in other items of production costs, such as import prices or interest rates. The likely impacts of changing interest rates and exchange rates will be discussed in the last chapter. Let us note that this system of wage and price equations, additionally enriched with the equation explaining the exchange rate, can be treated as a system of simultaneous equations. Hence it can be the subject of detailed cointegration analysis (Welfe 2002).

These relationships are shown in Fig. 18.2.

In the supply-determined models there is no room for the above economic mechanisms to function. Other feedbacks are entered instead.

In the case of consumer goods markets it is supply realized through market transactions. The changes in consumer demand are followed only by changes in excess demand. Hence, there are no conditions for changes related to the consumer multiplier to take place. A short-term supply-type consumer multiplier will appear instead (Barro and Grossman 1971). An increase in excess demand reduces labour force supply—this decreases output and commodity supply, so excess demand grows even larger. The decrease in the output can be partly offset, if the wage bill and household incomes become reduced at the same time. In the case of Poland this hypothesis was only partly confirmed, as the growing excess demand was followed by a declining number of hours worked, but not of employees (Welfe 1992).

In the above models the long-term supply accelerator plays an important role. An increase in the supply of investment goods augments (allowing for lags) the stock of fixed capital and subsequently of output and supplies, including the supplies of investment goods. This mechanism was a vital component of the policy of growth, particularly in the industrializing, developing countries (Welfe 2001).

In this class of models the short-term foreign trade multiplier can be additionally identified. An increase in exports determines the increase in aggregate imports and thus in production and exports.

The multisectoral models may have the bottleneck multiplier for chronic constraints in the supply of intermediate products. For instance, fuel delivery restrictions may constrain output as well as the energy industry, ultimately resulting to an energy crisis.

The above relationships are reproduced in Fig. 18.3.

References

Armington, P. (1969). *A theory of demand for products distinguished by place of production* (IMF Staff Papers XVI).

Bårdsen, G., Eitrheim, Ø., Jensen, E. S., & Nymoen, R. (2005). *The econometrics of macroeconomic modelling*. Oxford: Oxford University Press.

Barro, R. J., & Grossman, H. I. (1971). A general disequilibrium model of income and employment. *American Economic Review*, *61*, 82–93.

Blanchard, O., & Quah, D. (1989). The dynamic effects of aggregate supply and demand disturbances. *American Economic Review*, *79*, 635–673.

Bożyk, P., Maciejewski, W., & Piaszczyński, W. (1973). *Modele prognoz handlu międzynarodowego (Forecasting models of international trade)*. Warszawa: PWN.

Coen, R. M., & Hickman, B. G. (1976). Constrained joint estimation of factor demand and production function. *Review of Economics and Statistics, 52*, 287–300.

Courbis, R. (1977). *Les modeles de prix. Pour la prevision et la planification etude comparative*. Paris: Dunod.

Czyżewski, B., & Welfe, W. (1990). Handel zagraniczny w ekonometrycznych modelach gospodarki narodowej (Foreign trade in macroeconometric models). In J. Kotyński (Ed.), *Kraje Europy Środkowo-Wschodniej w handlu zagranicznym* (pp. 179–198). Warszawa: PWN.

Dreze, J. H., Bean, C., Lambert, J. P., Metha, F., & Snessens, H. R. (Eds.) (1990). *Europe's unemployment problem*. Cambridge: MIT Press.

Elmeskov, J. (1993). *High and persistent unemployment: assessment of the problem and its causes* (OECD Economics Department Working Paper No. 132).

Elmeskov, J., & MacFarland, M. (1993) Unemployment persistence. *OECD Economic Studies 21*, 59–88.

Fair, R. C. (2004). *Estimating how the macroeconomy works*. Cambridge: Harvard University Press.

Florczak, W., & Welfe, W. (2000). Wyznaczanie potencjalnego PKB, łącznej produktywności czynników produkcji (Potential GDP and TFP determination). *Gospodarka Narodowa, 11–12*, 10–55.

Friedman, M. (1968). The role of monetary policy. *American Economic Review, 58*, 1–17.

Grzęda Latocha, R. (2005). Ekonometryczna analiza koniunktury gospodarczej w krajach strefy Euro (Econometric analysis of business cycle in the Euro-area). *Ekonomista, 5*, 621–643.

Hickman, B. G., & Lau, L. J. (1973). Elasticities of substitution and export demand in a world trade model. *European Economic Review, 4*, 347–380.

Juszczak, W. (1982). Ekonometryczne modele zapasów w gospodarce socjalistycznej (Econometric inventory models in socialist economies). *Studia Prawno-Ekonomiczne, 29*, 131–146.

Klein, L. R., & Summers, R. (1966). *Studies in quantitative economics: Vol. 1. The Wharton index of capacity utilization*. Philadelphia: University of Pennsylvania.

Klein, L. R., Welfe, A., & Welfe, W. (1999). *Principles of macroeconometric modeling*. Amsterdam: North-Holland, Elsevier.

Layard, P. R. G., Nickell, S. J., & Jackman, R. (1991). *Unemployment: macroeconomic performance and the labour market*. Oxford: Oxford University Press.

Lovell, M. C. (1961). Manufactures inventories, sales, expectations and the acceleration principle. *Econometrica, 29*.

Lovell, M. C. (1962). Buffer stocks, sales expectations and stability. A multisector analysis of the inventory cycle. *Econometrica, 30*.

Nickell, S. J. (1984). The modelling of wages and employment. In D. H. Hendry & K. F. Wallis (Eds.), *Econometrics and quantitative economics*, Oxford: Basil Blackwell.

Phelps, E. S., et al. (1970). *Microeconomic foundations of employment and inflation theory*. New York: Norton.

Phillips, A. W. (1958). The relation between unemployment and the rate of change of money wage rates in the United Kingdom, 1961–1957. *Economica, 25*(4), 283–299.

Tobin, J. (1972). The wage-price mechanism: overview of the conference. In O. Eckstein (Ed.), *The econometrics of price determination conference*. Washington: Federal Reserve System.

Welfe, W. (1992). *Ekonometryczne modele gospodarki narodowej Polski (Econometric models of Poland's national economy)*. Warszawa: PWE.

Welfe, W. (2000). Empiryczne modele wzrostu gospodarczego (Empirical models of economic growth). *Ekonomista, 4*, 483–499.

Welfe, W. (Ed.) (2001). *Ekonometryczny model wzrostu gospodarczego (An econometric model of economic growth)*. Łódź: Wydawnictwo UŁ.

Welfe, W. (Ed.) (2002). Długookresowy, makroekonometryczny model W8D-2002 gospodarki polskiej (Long-term macroeconometric model W8D-2002 of the Polish economy). *Acta UŁ, Folia oeconomica, 172*. Łódź: Wydawnictwo UŁ.

Welfe, W. (2005). Stylized empirical model of economic growth. In W. Welfe & A. Welfe (Eds.), *Macromodels 2004, Problems of building and estimation of econometric models* (pp. 109–125). *Acta UŁ, Folia oeconomica, 190*. Łódź: Wydawnictwo UŁ.

Welfe, W. (Ed.) (2009). Makroekonometryczny model gospodarki polskiej opartej na wiedzy (A macroeconometric model of the knowledge economy for Poland). *Acta UŁ, Folia oeconomica, 229*. Łódź: Wydawnictwo UŁ.

Welfe, W., & Welfe, A. (2004). *Ekonometria stosowana (Applied econometrics)* (2nd ed.). Warszawa: PWE.

Welfe, W., Welfe, A., & Florczak, W. (1996). Makroekonomiczny roczny model gospodarki narodowej Polski (Macroeconomic annual model of the Polish economy) (Z Prac IRiSS No. 31). Warszawa.

Whitley, J. D. (1994). *A course in macroeconometric modelling and forecasting*. New York: Harvester Wheatsheaf.

Chapter 19
The Modelling of Financial Flows

19.1 Introduction

Transactions taking place between economic agents that have been characterized so far in quantitative terms have their financial counterparts. In macroeconometric models they are represented by financial flows, which are obtained by multiplying the quantity flows by the appropriate deflators. This concerns the flows of final demand components, the flows of production and the use of production factors.

On the other hand, autonomous financial flows take place within the secondary distribution of national income, which are related to income creation and the realization of economic agents' expenditures. The emerging savings (mark-up) ensure that the financial wealth of these agents will increase. The expenditures on commodities and services are one of the major channels through which financial flows are transmitted to the real sector (Klein et al. 1999). In general, the following institutional sectors of economic agents are distinguished: households, enterprises, public institutions (including financial ones), and foreign sector (Welfe 1992; Juszczak et al. 1993; Welfe and Welfe 2004).

19.2 The Models of Households' Incomes, Expenditures and Wealth Changes

Personal incomes are important component for households' economic activities. They are the main determinant of their consumption, which was referred to in Chap. 16. Because of this, macroeconometric models devote much attention to the generation of personal incomes, providing quite detailed descriptions of their range and particularly of their sources.

The national accounts distinguish the institutional accounts, which contain the nominal gross disposable incomes of households. The net incomes are obtained by deducting taxes and obligatory social contributions.

W. Welfe, *Macroeconometric Models*,
Advanced Studies in Theoretical and Applied Econometrics 47,
DOI 10.1007/978-3-642-34468-8_19, © Springer-Verlag Berlin Heidelberg 2013

In terms of their sources of origin, personal incomes are heterogeneous aggregates. The following components are usually distinguished: labour incomes, incomes of entrepreneurs, incomes from agriculture, transfers from public institutions (social services) and property incomes. Further decomposition of incomes depends on how complex a model is to be and on the availability of statistical data (Welfe 1993).

Gross labour incomes (i.e. incomes before tax) YP_t^{WB} are obtained when the average wages before tax WP_t^B are pre-multiplied by the number of employees N_t:

$$YP_t^{WB} = WP_t^B \cdot N_t \tag{19.1}$$

In the multisectoral models labour incomes are determined for particular sections and then added up.

The net labour incomes (i.e. after tax) YP_t^W are determined by deducting personal tax T_t^W:

$$YP_t^W = YP_t^{WB} - T_t^W \tag{19.2}$$

The incomes of entrepreneurs (excluding farmers) YP_t^{PR} represent special forms of participation in profits (dividends), which can be either direct or indirect. Hence, in the long-run we have:

$$YP_t^{PR*} = \alpha_0 + \alpha_1 ZP_t^N + \varepsilon_t \tag{19.3}$$

and in the short-run:

$$\Delta YP_t^{PR} = \beta_0 + \beta_1 \left(YP_{t-1}^{PR} - YP_{t-1}^{PR*}\right) + \beta_2 \Delta ZP_t^N + \varepsilon_t \tag{19.4}$$

where ZP_t^N is net profits (after tax).

If information on the level of profits is unavailable or unreliable, then symptomatic variables correlated with the level of profits are used, such as total receipts or value added.

Farmer incomes are estimated in a similar way. However, information on profits from agricultural activities is frequently not available. Then the most frequent symptomatic variable is value added generated by these activities, which is the source of consumer expenditures.

Transfers from mainly public financial institutions YP_t^S are an important component of personal incomes. The transfers primarily cover old-age and disability pensions YP_t^E, payments from social security funds, including payments to the unemployed, student scholarships and grants, payments from special funds and others YP_t^{SP}.

The level of old-age and disability pensions is regulated by law or agreements made with insurance companies. The implication of the applicable regulations having their main source in the concept of inter solidarity is that the insurance funds are to a large extent dependent on the subsidies from the government budget. In a few countries (for instance Chile and Poland) a different principle was implemented, which requires that obligatory direct payments be made from the wage bills to the funds and links the payments with employees' individual capital accounts. Regardless of which solution is used, the level of pensions depends on the level of wages paid over a predetermined period and on the duration of employment. It is reviewed (indexed) annually to address a living cost increase and, to some extent, the growth rate of average wages (cf. Welfe and Welfe 2004).

Therefore, the average level of pensions in a given year can be determined from average wages paid in that and the preceding years, adjusted for changes in the level of living costs. An equation approximating this relationship may have the following form:

$$WP_t^E = \alpha_0 \left(\sum_{s=1}^{S} h_s WP_{t-s}^B \right)^{\alpha_1} e^{\alpha_2 (P_t^i / P_{t-1}^i) + \varepsilon_t} \tag{19.5}$$

Other transfers are related to average wages more or less directly. The relationships can be approximated using the following equation:

$$WP_t^{SP} = \alpha_0 \left(WP_t^B \right)^{\alpha_1} e^{\xi_t} \tag{19.6}$$

Total transfer incomes can be obtained by pre-multiplying the average transfer income by the number of recipients. While the number of pensioners (L_t) is generally known, the number of other recipients fluctuates, hence an approximation relating the total value of the transfers to the wage bill is used. We then have:

$$YP^S = YP_t^E + YP_t^{SP} = WP_t^E L_t + \beta_0 \left(YP^{WB} \right)^{\beta_1} \tag{19.7}$$

The property incomes of households depend on the interest rates on bank deposits and securities, dividends from equities, etc. Therefore, these incomes are determined by the nominal interest rates on deposits (R_t^D) and average earnings on bonds and stocks (R_t^A), as well as by the volume of deposits BP_t^{DE} and stocks (BP_t^{AO}).

It can be assumed that in the long-run the total property incomes YP_t^{DA} will be represented by the following equation:

$$\ln YP_t^{DA^*} = \alpha_0 + \alpha_1 \ln\left(R_t^D BP_{t-1}^{DE} \right) + \alpha_2 \ln\left(R_t^A BP_{t-1}^{AQ} \right) + \varepsilon_t \tag{19.8}$$

In the short-run, unexpected shocks may occur, represented by variable U_t^{DA}:

$$\begin{aligned} \Delta \ln YP_t^{DA} &= \beta_0 + \beta_1 \left(\ln YP_{t-1}^{DA} - \ln YP_{t-1}^{DA^*} \right) + \beta_2 \Delta \ln\left(R_t^D BP_{t-1}^{DE} \right) \\ &\quad + \beta_3 \Delta \ln\left(R_t^A BP_{t-1}^{AO} \right) + \beta_4 U_t^{DA} + \varepsilon_t \end{aligned} \tag{19.9}$$

The nominal gross personal, disposable income of households (before tax) (YP_t^B) will be obtained by adding up its components:

$$YP_t^B = YP_t^{WB} + YP_t^{PR} + YP_t^S + YP_t^{DA} \tag{19.10}$$

The nominal net personal disposable income of households (after tax) YP_t will be derived by subtracting the personal income taxes and obligatory payments to the social security funds T_t^W:

$$YP_t = YP_t^B - T_t^W \tag{19.11}$$

Some macromodels use simplified relationships, mainly due to data scarcity. They relate personal disposable incomes to labour incomes and property incomes.

To determine the real disposable incomes (Y_t) the nominal incomes are divided by the appropriate deflator (PY_t):

$$Y_t = YP_t / PY_t \tag{19.12}$$

The nominal disposable incomes are the major source financing households' expenditures. An additional role as a source of households' ***receipts*** is played by consumer credits (KP_t^D). Both components constitute households' disposable funds. They are mainly used to purchase commodities and services (CP_t), then to repay credits (KP_t^{DS}) and to pay taxes (T_t^W). An increase in savings (ΔOP_t^G) is residual. This process is represented by the following balance equation:

$$YP_t^B + KP_t^D = CP_t + KP_t^{DS} + T_t^W + \Delta OP_t^G \tag{19.13}$$

Credit repayment is usually calculated by pre-multiplying the average rate of interest on consumer credits (R_t^G) by the lagged debt of households BKP_{t-1}^G.

Hence, we have:

$$KP_t^{DS} \cong R_t^G BKP_{t-1}^G \tag{19.14}$$

More detailed specifications of an increase in consumer credit and savings will be presented in the next chapter.

The saving increase is the basic component of the equation describing changes in the net wealth covering the financial assets owned by the households (WLP_t). In must be stressed that the notion of households' financial wealth can be understood differently (cf. also Chap. 16). The narrow definition contains, in addition to cash balances, also the value of possessed securities, including bonds and stocks. The broad definition which is sometimes used to specify the consumer demand functions takes account of the total debt of public institutions, the nominal value of domestic enterprises, and net foreign assets.

The net financial wealth of households is generated from the following equation:

$$WLP_t = WLP_{t-1} + \Delta OP_t^G \tag{19.15}$$

Their composition and changes will be shown in the next chapter.

19.3 The Modelling of Financial Flows of Enterprises

Financial flows in the enterprise sector are rather infrequently found in macroeconometric models. If they are present, then they are constrained to enterprises' receipts and the values of gross output and value added, on the one hand, and user costs and enterprises' profits, on the other, only exceptionally covering the capital flows of enterprises.

Enterprises' receipts are usually represented by the receipts from the sale of commodities and services, on which monthly information is available. As regards annual information, the data on gross output QP_t and value added XP_t are used. The gross output value is obtained by pre-multiplying the volume of output (Q_t) by producer prices (PQ_t) generated in the appropriate blocks of macromodels' equations:

$$QP_t = Q_t \cdot PQ_t \tag{19.16}$$

Value added is calculated as a difference between the value of gross output and the costs of material inputs (KP_t^m):

$$XP_t = QP_t - KP_t^m \tag{19.17}$$

The costs of material inputs can be decomposed to distinguish raw materials and materials of domestic and foreign origin:

$$KP_t^m = a_t^K Q_t^{KS} P_t + a_t^M M_t^S P_t^{MS} \tag{19.18}$$

where

Q_t^{KS} is domestic output of raw materials, materials and energy,
M_t^S is imports of raw materials, materials and energy,
a_t^K is average unit use of domestic inputs,
a_t^M is average unit use of imported inputs.

The information on the unit use of domestic and imported materials and energy is often scarce. To cope with this, a reduced form of Eq. (19.18) is used, which contains on the right-hand side the output values of the domestic raw-materials industries and imports of intermediate inputs, *respectively*.

The user costs are composed of production costs and producer taxes (T_t^A). The production costs (KP_t^P) contain labour costs (KP_t^W), the costs of use of raw-materials, materials and energy (KP_t^M), fixed capital depreciation (KP_t^A) and other costs (KP_t^i):

$$KP_t^P = KP_t^W + KP_t^M + KP_t^A + KP_t^C \tag{19.19}$$

Each of these components can be explained separately. The equations explaining particular cost components share the assumption that they should account for the real volume of inputs expressed in current prices. The real inputs can be obtained from the inverted production function, as presented in Chap. 17 (cf. also Walters 1963).

Labour costs can be determined by pre-multiplying the gross average wages (WP_t^B) by the number of employees (N_t), i.e. as a wage bill augmented by obligatory social insurance payments and wage taxes:

$$KP_t^W = WP_t^B \cdot N_t(1 + \mu_{1t} + \mu_{2t}) \tag{19.20}$$

where:

μ_1 is the rate of social insurance payments,
μ_2 is the rate of wage taxes.

The equations explaining average wages and employment were presented in Chaps. 17 and 18.

The costs of material inputs can be generally obtained by pre-multiplying the value of global output by the global unit material input coefficient A_t:

$$KP_t^M = A_t \cdot QP_t \tag{19.21}$$

They can be decomposed by input source as in the formula (19.19) and also according to different sections or groups of materials.

The depreciation costs are most frequently calculated as prescribed by the law. The rates of depreciation are commonly assumed to be predetermined and are applied to the gross value of fixed capital (KP_t):

$$KP_t^A = \delta KP_t \tag{19.22}$$

where δ is an average depreciation rate.

Other user costs are composed of different payments, which are frequently related to the wage bill, the payments to the insurance funds and the debt service. The equation explaining other costs can be approximated with the following formula:

$$\ln KP_t^i = \alpha_0 + \alpha_1 \ln KP_t^W + \alpha_2 \ln\left(R_t^P \cdot BP_{t-1}^P\right) + \xi_t \tag{19.23}$$

where

R_t^P is the interest rate on short-term credits,
BP_t^P is enterprises' debt at period end.

Because the information on particular cost components is scarce, the analysis is frequently limited to the approximate, reduced form of the relationship between the total costs and major cost components on which information is readily available. The components are the labour costs and the costs of use of imported raw materials, materials and energy. Then we have:

$$\ln KP_t^P = \alpha_0 + \alpha_1 \ln KP_t^W + \alpha_2 \ln M_t^S P_t^{mS} + \varepsilon_t \tag{19.24}$$

The unit costs in the production sector are obtained from dividing the total costs by the volume of output. It is worth noting that the unit labour costs may take the following form:

$$KP_t^W / Q = WP_t^B / (X_t / N_t) \cdot (X_t / Q_t)(1 + \mu_{1t} + \mu_{2t}) \tag{19.25}$$

They depend on the ratio between average wages and labour productivity (X_t/N_t) adjusted for the share of value added in gross output (X_t/Q_t), allowing for social contributions and taxes.

New research projects that came into being recently used the neoclassical approach to minimize production costs under imperfect competition. The shares of particular cost components in the total costs have to be explained. The shares are assumed to be the functions of relative prices in the respective and other cost components. The costs of particular components are obtained by inverting the production function, which allows determining the volumes of use of respective production factors (Jorgenson 1993).

The early investigations into enterprises' production costs that distinguished between constant and variable costs are also worth mentioning. They were mainly based on the cross-section data (Barczak 1971; Pawłowski 1965).

The gross profits of enterprises ZYP_t^B are obtained as a difference between their receipts and user costs. Given that prices are predetermined, they are residual. In general, they are derived from the following identity:

$$ZYP_t^B = QP_t - KP_t^P \tag{19.26}$$

The net profits of enterprises ZYP_t^N are calculated by deducting taxes T_t^Z and extraordinary profits and losses from gross profits:

$$ZYP_t^N \approx ZYP_t^B - T_t^Z \tag{19.27}$$

Therefore, the net profits are residual. This does not guarantee that their predicted values will show satisfactory accuracy. For this reason, several researchers attempted to explain the fluctuations in the net profits, using equations which regressed the profits on the major components of receipts and user costs. This approach is of utmost importance because net profits are a significant source financing investments, household incomes through dividends, as well as social services (Brzeszczyński and Kelm 2000).

19.4 The Modelling of Public Finances

Public finances involve the collection and distribution of monetary resources for the national government, local governments, social insurance systems, and other public institutions. In market economies public finances developed gradually, together with the widening scope of public tasks. This scope was country specific: moderate where the liberal state model prevailed, and extended in countries emphasising Keynesian concepts or in welfare states that paid much attention to government interference as an important instrument for reducing market inefficiencies (Brainard and Tobin 1968).

The alternative concepts concerning the role of the state and public finances gave rise to differences in fiscal policy. Briefly speaking, the neoclassical concepts required that the policy be neutral, accompanied by efforts to keep the government budget balanced and by measures preventing public debt from growing larger. The Keynesian views justified expansionary fiscal policy (i.e. one allowing the use of credit to finance public investments, etc.) and thus larger public debt.

Macroeconometric analysis is mainly interested in the government budget revenues and expenditures. Local government finances and special public funds are left aside, as the primary source of their revenues is the national budget.

The government budget revenues have domestic and foreign sources. Domestic revenues containing incomes from taxes and other incomes are the most important. Because the tax incomes play a key role, they are a common subject of macroeconometric analysis. The other revenues are either exogenous (e.g. central bank payments) or endogenous (for instance the dividends and profits of the state-owned firms), but these revenues are not easy to model.

The following taxes are commonly distinguished: direct taxes including the personal income tax (PIT) and the corporate income tax (CIT) and indirect taxes represented by the tax on the sale of commodities and services or the value added tax (VAT), the excise tax and custom duties.

For tax revenues to be estimated, their base, the tax scheme and the scope and amounts of tax exemptions must be determined. It is also necessary to allow for transfers made to local government budgets and special funds.

How much a person will pay in taxes is regulated by the law. This information can be obtained from the tax administration, but the actual amounts paid by particular individuals are usually not reported to the statistical offices. Consequently, the total flows of tax revenues must be approximated.

The personal income tax (PIT) is a progressive tax in most cases, although sometimes linear taxes involving a flat tax rate are used. The tax rates in the progressive systems are incremental, which means that a larger tax base (personal incomes) entails higher tax payments (Welfe and Welfe 2004).

The PIT revenues BYP_t^G can be estimated from the following approximate equation:

$$BYP_t^G = \alpha_0 + \alpha_1 YP_t^B + \varepsilon_t \tag{19.28}$$

where YP_t^B is the gross disposable personal income (the tax base).

The marginal effective tax rate α_1 can be substituted by its estimate t_t^G, calculated as a difference between the average tax rate and the fraction of tax exemptions:

$$BYP_t^G = \beta_0 + \beta_1 \ln\left(t_t^G YP_t^B\right) + \xi_t \tag{19.29}$$

The CIT revenues (BYP_t^P) are estimated in a similar way. These taxes are frequently linear, so they are easy to compute. We have then:

$$BYP_t^P = \alpha_0 + \alpha_1 ZYP_t^B + \varepsilon_t \tag{19.30}$$

where ZYB_t^B is the gross surplus (the tax base); if exemptions are not granted, then $\alpha_0 = 0$.

Today the paramount source of government revenues is VAT (BYP_t^V). Its main rate is accompanied by a range of reduced rates applied to particular foodstuffs, books and other cultural goods. Because detailed information on the sale of particular products is typically not available, the tax base is constrained to totals. This tax is most frequently levied on domestic and foreign consumer goods purchased by households and public institutions. The equation approximating these relationships treats all goods alike, assuming that the share of goods covered by preferential VAT rates is more or less constant. Hence, VAT revenues can be determined from the following equation:

$$\ln BYP_t^V = \alpha_0 + \alpha_1 \ln(CP_t + GP_t) + \alpha_2 \ln BYP_t^C + \xi_t \tag{19.31}$$

The revenues from custom duties (BYP_t^C) are generally determined as the functions of imports. The duty rates commonly differ depending on commodity and import sources. The detailed information is rather scarce. Hence, an approximation assuming that the average rates do not change significantly is used. The equation reads as follows:

$$\ln BYP_t^C = \alpha_0 + \alpha_1 \ln MP_t^C + \varepsilon_t \tag{19.32}$$

where MP_t^C is the value of imports from countries covered by import duties.

The total government revenues BYP_t are usually obtained from a bridge equation, assuming that the dynamics of other incomes and of tax revenues is similar. We have:

$$BYP_t = \alpha_0 + \alpha_1\left(BYP_t^G + BYP_t^P + BYP_t^V + BYP_t^C\right) + \xi_t \tag{19.33}$$

Macromodels initially treated government budget expenditures (BCP_t) as exogenous or as instruments of economic policy. Today only a few expenditure items are regarded as quasi-exogenous, i.e. the defence expenditures and the investment expenditures. Other expenditures are usually endogenized.

The budget expenditures can be generally decomposed into rigid expenditures, quasi-rigid expenditures and flexible expenditures that, being the instruments of economic policy, may show strong variations. Of practical importance is their decomposition into the current expenditures ($BCCP_t$) and the investment expenditures ($BCJP_t$), basically treated as an economic policy variable.

The current expenditures are further decomposed into expenditures on wage bills, pension funds, subsidies for local governments and state-owned firms, and on the service of domestic and foreign debt.

The wage-bill related expenditures depend on average wages and the number of employees in the public sector (cf. Chap. 18). How the allocations to pension funds are calculated has been explained earlier in this chapter. The expenditures on subsidies are regulated by law. Their amounts depend on the established unit norms and the respective base (Welfe and Welfe 2004).

The expenditures on debt service (BCP^I) are approximated using the average interest rates on credits (R_t) pre-multiplied by the lagged value of debt (BKP_t^I) as the explanatory variables:

$$\ln BCP_t^I = \alpha_0 + \alpha_1 \ln\left(R_t BKP_{t-1}^I\right) + \xi_t \tag{19.34}$$

The current budget expenditures can also be decomposed according to the sections and industries. The budget expenditures are allocated among institutions (ministries) representing particular sections following negotiations, where arguments accentuating the necessity to preserve infrastructure and employment play an important role. The outcomes of the negotiations can be represented by means of equations explaining the absolute values of the expenditures on an i-th section or the expenditures' shares in the total level of current expenditures.

In the first case we have:

$$BCCP_{it} = \alpha_0 + \alpha_1 BCCP_t + \varepsilon_t \tag{19.35}$$

In the second case:

$$BCCP_{it}/BCCP_t = \beta_0 + \beta_1(1/BCCP_t) + \varepsilon_t \tag{19.35$'$}$$

The results of the estimates for Poland are presented in Welfe and Welfe (2004).

19.5 Financial Flows—The Interim and Outside Links

In the demand-determined market economies financial flows between particular institutional sectors are interlinked. Financial flows taking place between households, enterprises and financial institutions are characterised by the following relationships that lead to important feedbacks.

The government budget expenditures on the public sector wages, old-age and disability pensions, and social benefits (some made through dedicated funds) increase the personal, disposable incomes of households. As a result, additional personal income tax amounts flow to the budget, thus increasing its revenues. With a budget deficit remaining constant even larger transfers can be made to households.

Owing to subsidies or tax exemptions enterprises increase their profits. Higher corporate taxes they pay then enlarge budget revenues. As before, if the budget deficit does not change, more subsidies can be paid to firms. These feedbacks are called fiscal multipliers.

The transmission to the real sector induces increases in budget expenditures, among which the increase in public investment expenditures plays a central role. This increase starts the accelerator mechanism. Growing expenditures on wages, pensions and social benefits that enlarge personal disposable incomes activate the consumer multiplier. These effects may be constrained, if larger expenditures are financed with credits (Welfe et al. 2002).

In the supply-determined economies, such as the former centrally-planned economies, the above transmission relationships are blocked. An increase in disposable incomes and consumer demand would result in an increase of excess demand, because of the prevalent scarcity of output and production factors. A comprehensive analysis of these relationships was provided in a study by Welfe (1990).

References

Barczak, A. (1971). *Ekonometryczne metody badania kosztów produkcji (Econometric methods in production costs analysis)*. Warszawa: PWN.

Brainard, W. C., & Tobin, J. (1968). Pitfalls in financial model building. *American Economic Review, 58*, 99–122.

Brzeszczyński, J., & Kelm, R. (2000). *Ekonometryczne modele rynków finansowych (Ecomometric models of financial markets)*. Warszawa: WIG.

Jorgenson, D. W. (1993). Econometric methods for modelling producer behavior. In Z. Griliches & M. D. Intriligator (Eds.), *Handbook of econometrics* (Vol. 3). Amsterdam: North-Holland.

Juszczak, G., Kaźmierska, M. M., Łapińska-Sobczak, N., & Welfe, W. (1993). Quarterly model of the Polish economy in transition (with special emphasis on financial flows). *Economic Modelling, 10*, 127–149.

Klein, L. R., Welfe, A., & Welfe, W. (1999). *Principles of macroeconometric modeling*. Amsterdam: North-Holland.

Pawłowski, Z. (1965). Modele ekonometrycznej analizy kosztów (Models of econometric cost analysis). *Ekonomista, 3*.

Walters, A. A. (1963). Production and cost functions. An econometric survey. *Econometrics, 31*, 1–66.

Welfe, A. (1990). State budget and inflation processes. Estimation for Poland. *Journal of Public Economics, 43*, 161–180.

Welfe, A. (1993). *Inflacja i rynek (Inflation and market)*. Warszawa: PWE.

Welfe, A., Karp, P., & Kelm, R. (2002). *Kwartalny makroekonometryczny model gospodarki Polski (Quarterly macroeconometric model of the Polish economy)*. Łódź: Wydawnictwo Uniwersytetu Łódzkiego.
Welfe, W. (1992). *Ekonometryczne modele gospodarki narodowej Polski (Econometric models of the Polish national economy)*. Warszawa: PWE.
Welfe, W., & Welfe, A. (2004). *Ekonometria stosowana (Applied econometrics)* (2nd ed.). Warszawa: PWE.

Chapter 20
Modelling the Money and Capital Markets

20.1 Introduction

In the last decades the modelling of money and capital markets has gained high status, following the increasing role of monetary policy. In the domestic financial markets where interest rates play a central role the basic interest rate has become the main policy instrument of central banks. In most developed countries central banks use the interest rate and other instruments to achieve the objective of keeping inflation at a prescribed level or within the prescribed limits, which in many instances is linked to a growth-supporting policy (Klein et al. 1999).

The exchange rate fluctuations are of general importance for national economies. In the past, central banks were regulating the exchange rate levels, but over time they relaxed their control, first by allowing the rates to fluctuate within the prescribed limits (a pegged rate) and more recently by floating them, leaving the likely interventions in the exchange rate markets to the banks.

The problems involved in the macromodelling of the above variables will be discussed in a broader context below. The discussion will cover the development of assets and liabilities in the banking system and changes in the balance of payments components. The relevant equations will be introduced. According to the new, modern tendencies in macromodelling, equations are combined into comprehensive subsystems, which are analysed by means of cointegration methods. This type of models for analysing the inflationary spiral has been constructed in Poland (Welfe et al. 2007) and Norway (Bårdsen et al. 2005), and a money market model for the United Kingdom has been presented in the monograph by Garratt et al. (2006).

20.2 The Modelling of Credit Activities

It is generally assumed that particularly in the long-run equations should explain the demand for credits (Bernanke and Blinder 1988; Fase 1995; Calza et al. 2003). The demand for real credits is positively related in the long-run to real GDP and negatively to the real short- and long-term interest rates, as it has been demonstrated for

W. Welfe, *Macroeconometric Models*,
Advanced Studies in Theoretical and Applied Econometrics 47,
DOI 10.1007/978-3-642-34468-8_20, © Springer-Verlag Berlin Heidelberg 2013

the Euro-area using the cointegration methodology (Calza et al. 2003). The short-run relationships are more complicated, as they allow for the impacts of supply. The additional variables are banks' domestic liabilities and net external assets, interest spread between the lending and deposit rates, as well as credit risk. The most common is the ECM and in the case of regime switches the Markov switching ECM is applied (Eller et al. 2010). Attempts have also been made to generate the equilibrium levels of credits and the deviations therefrom, the latter signalling either excessive credit expansion or deep decline in lending during financial crises, for instance in the years 2008–2009 (Psaradakis et al. 2004).

In modelling credits much interest is given to their decomposition into long- and short-term credits offered to enterprises and those available to households.

An important role is played by the long-term credits offered to enterprises, which are mainly used for investment purposes. They usually complement other sources of funding (profits, depreciation). Hence, the total value of an expected (planned) investment can be assumed to be the major explanatory variable. Because it takes a fairly long time to set up investment projects, the appropriate lags have to be introduced. The second important variable affecting entrepreneurs' readiness to apply for investment credits is the costs they will have to incur, represented by the long-term real interest rates (R_t^L). Therefore, the equation explaining the demand for long-term credits will have the following form:

$$\Delta BK_t^L = \alpha_0 + \alpha_1 J_t + \alpha_2 (1 + R_t^L)(P_t^J / P_{t-1}^J) + BK_{t-1} + \varepsilon_t \qquad (20.1)$$

where

BK_t^L is enterprises' debt due to long-term credits at period end (constant prices),
P_t^J is the investment deflator.

The estimated value of parameter α_1 points to the marginal share of credits in investment financing.

Short-term credits are used by enterprises to finance their current activities. Assuming that the shares of own and borrowed funds are approximately constant, the long-run level of debt ($BK_t^{O^*}$) can be explained as a function of the level of output (sales) or of total costs of enterprises' activities.

In the long-run, we have:

$$BK_t^{O^*} = \alpha_0 + \alpha_1 X_t + \varepsilon_t \qquad (20.2)$$

In the short-run, the borrowing costs should be acknowledged, i.e. the interest rate on short-term credits (R_t^S) must be introduced as an additional explanatory variable:

$$\Delta BK^O = \beta_0 + \beta_1 (BK_{t-1}^O - BK_{t-1}^{O^*}) + \beta_2 \Delta X_t + \beta_3 (1 + R_t^S)/(P_t / P_{t-1}) \qquad (20.3)$$

Consumer credits generally complement disposable personal incomes. Their volume that depends also on credit costs represented by the interest rate R_t^G can be affected by changes in the rate of economic stability that may be represented by a risk premium (or a trend).

In the long-run, we have:

$$\ln KR_t^{G^*} = \alpha_0 + \alpha_1 \ln Y_t + \alpha_2 \ln (1 + R_t^G)(P_t^C / P_{t-1}^C) \qquad (20.4)$$

in the short-run:

$$\begin{aligned}\Delta \ln KR_t^G = \beta_0 + \beta_1\left(KR_{t-1}^G - KR_{t-1}^{G^*}\right) - \beta_2 \Delta \ln Y_t \\ + \beta_3 \Delta \ln\left(1 + R_t^G\right)\left(P_t^C / P_{t-1}^C\right)\end{aligned} \tag{20.5}$$

where KR_t^G is the level of consumer credits (constant prices).

In some countries the money illusion phenomenon has been observed. In this case, the real interest rates have to be substituted by the nominal interest rates.

To determine the debt level of households the amount of debt repayments has to be estimated. Let us use δ_t to denote the average rate of repayments; then the level of repayments in constant prices will be (δKR_{t-1}^G) and the debt level will be given as:

$$BK_t^G = BK_{t-1}^G + KR_t^G - \delta KR_{t-1}^G \tag{20.6}$$

20.3 Equations Explaining the Demand for Money and Financial Assets

Financial flows going into the money and capital markets form the assets of a broadly understood banking system. They consist of cash and bank deposits of economic agents (households, enterprises, public and financial institutions) and of other financial assets. The deposits finance the national economy via the credit system (Welfe and Welfe 2004).

Households' monetary assets are composed of (a) cash and call deposits at banks (OPL_t^G), which are treated as disposable cash, and (b) other bank deposits representing households' savings (ODP_t^G). It can be therefore assumed that:

$$OLP_t^G = \alpha_0 + \alpha_1 CP_t + \varepsilon_t \tag{20.7}$$

and

$$\Delta OLP_t^G = \beta_0 + \beta_1 YP_t + \beta_2 R_t^D + \zeta_t \tag{20.8}$$

where R_t^D is the average interest rate on deposits.

The monetary assets are important components of households' financial wealth $SAVP_t$. Other components are securities, stocks and shares. In a broadly understood notion it is assumed that the households control total value of securities and shares together with their claims toward the government that are equal to the budget deficit.

The equation explaining households' real financial wealth has the following form in the long-run:

$$\ln SAV_t^* = \alpha_0 + \alpha_1 \ln Y_t + \alpha_2\left(R_t^D - \Delta PY_t / PY_t\right) + \zeta_t \tag{20.9}$$

in the short run:

$$\begin{aligned}\Delta \ln SAV_t = \beta_0 + \beta_1\left(\ln SAV_{t-1} - \ln SAV_{t-1}^*\right) + \beta_2 \Delta \ln Y_t \\ + \beta_3 \Delta\left(R_t^D - \Delta PY_t / PY_{t-1}\right) + \beta_4 \ln PY_t + \zeta_t\end{aligned} \tag{20.10}$$

The bank deposits of enterprises (ODP_t^P) are mainly used in handling their current transactions. Hence, the level of enterprises' bank deposits will depend on the value of the enterprises' lagged receipts:

$$\ln ODP_t^{P^*} = \alpha_0 + \alpha_1 \ln QP_{t-1} + \varepsilon_t \tag{20.11}$$

and their increase:

$$\Delta \ln ODP_t^P = \beta_0 + \beta_1\left(\ln ODP_{t-1}^P - \ln ODP_{t-1}^{P^*}\right) + \beta_2 \Delta \ln QP_{t-1} + \varepsilon_t \tag{20.12}$$

Likewise, the bank deposits of financial institutions, including the government budget (ODP_t^B), depend on the value of the institutions' revenues or expenditures.

By adding up the money deposits of the above groups of economic agents, we arrive at the total value of assets (ODP_t):

$$ODP_t = ODP_t^G + ODP_t^P + ODP_t^B \tag{20.13}$$

These total assets represent the total demand for money of the domestic economic agents. When the demand of foreign agents is added, the total demand for money in the national economy is obtained. The reduced form of the equation explaining global money demand will have the following form:

$$\begin{aligned} \Delta ODP_t = {} & \beta_0 + \beta_1 \Delta YP_t + \beta_2 (YP_t - CP_t) + \beta_3 XP_t \\ & + \beta_4 ER_t\left[(R_t - \Delta PX_t / PX_{t-1}) - \left(R_t^W - \Delta PX_t^W / PX_{t-1}^W\right)\right] \end{aligned} \tag{20.14}$$

where the first component stands for changes in cash balances, the second is the difference between households' disposable income and the value of their consumption, the third is enterprises' deposits and the last component denotes the demand for domestic money that results from an inflow of foreign capital being dependent on the differences between real domestic and foreign interest rates.

The total assets will be obtained from the identity:

$$ODP_t = ODP_{t-1} + \Delta ODP_t \tag{20.15}$$

Money supply, for instance M2, can be obtained assuming that at a predetermined interest rate it will be equal to the demand for money as determined from (20.14).

20.4 Interest Rates

The most important role is played by the basic interest rates of central banks, which are treated as a major instrument of monetary policy. Regarding them as an exogenous variable would be misleading, though, despite the fact that their levels are set by the central bank or the national monetary board. In most cases the institutions take their decisions attempting to keep inflation rates within the prescribed limits, so they analyse the macroeconomic conditions beforehand. Besides the lagged rate of inflation, other analysed aspects are the impacts of budget deficit, the rate of current account balance, etc. Assuming that the rate of inflation is a crucial factor, the following equation will approximate this relation:

$$R_t = \alpha_0 + \alpha_1 \left(\Delta P_t^C / P_{t-1}^C \right) + \xi_t \tag{20.16}$$

In several macromodels special rules for determining central banks' interest rates were used. One of the most known is the Taylor rule (Taylor 1993). In its simplest version the current nominal interest rate depends on the rate of inflation and the lagged interest rate:

$$R_t = (1 - \lambda)\left[\left(P^* - P_{t-1} \right) / P_{t-1} \right] + \lambda_t R_{t-1} \tag{20.17}$$

where

R_t is the basic (long-term) interest rate,
P^* is the desirable price level.

In countries where central banks are responsible for keeping prices as well as the rates of economic growth at the desired levels, the problem of how to determine a basic interest rate being the best compromise in terms of fulfilling the two objectives has to be solved. The solution can be found by minimization of the loss function of the central bank (Garratt et al. 2006).

Generally, the interest rates that commercial banks pay to deposit owners or charge on credit borrowers (R_t^K) depend on the basic refinancing interest rates of the central bank. In the short-run interest rate adjustments may be delayed. This relationship is usually described using the bridge equations. In the long-run, we have:

$$\ln R_t^{K^*} = \alpha_0 + \alpha_1 \ln R_t + \varepsilon_t \tag{20.18}$$

and in the short-run:

$$\Delta \ln R_t^K = \beta_0 + \beta_1 \left(R_{t-1}^K - R_{t-1}^{K^*} \right) + \beta_2 \Delta \ln R_{t-1}^K \tag{20.19}$$

In some markets the main cause of interest rate fluctuations is changes in the relation between money demand and supply. Then the equilibrium interest rates have to be determined. The situation can be illustrated by an interest rate determination process conducted by investors who may choose to allocate their disposable funds either to machinery and equipment or to securities. To take a decision, they must contrast the real returns they expect from their fixed capital investment projects with the real interest rates (yields) they expect from securities. A similar problem must be solved by investors choosing between domestic and foreign securities (Garratt et al. 2006, pp. 71–73).

20.5 The Balance of Payments

Market transactions carried out with foreign economic agents are presented in the respective components of the balance of payments. Current, capital and financial transactions are distinguished.

Current transactions account in the first place for the receipts from exports, for the expenditures on imported commodities and services, and for the foreign trade balance treated as residual. In most cases the exports and imports equations are bridge equations that link the deflated values of exports and imports with their counterparts that rest on foreign trade commodity flows. Additionally, the balances of incomes from the property of production factors utilised abroad and the costs of foreign production factors are distinguished, as well as the balance of transfers and the balance of unclassified current transactions. The equations of the above components are specified individually. To obtain the total balance of current transactions the components are added up.

Capital transactions cover non-refundable receipts, debt remissions, the purchase and sale of patents, copyright, etc. These are commonly treated as exogenous.

Financial transactions encompass foreign direct investments, foreign portfolio investments, derivatives of financial instruments, transactions involving the use and repayment of credits.

Because foreign direct investments have special significance in investment processes, equations are built to explain their fluctuations that mostly arise from the rate of GDP growth and risks premium.

Portfolio investments are important for balancing the current account, as they narrow the gap caused by a negative trade balance. Their volume depends on the profitability of capital inflows. Hence the major explanatory variables are the differences between the domestic and foreign interest rates and the risk premium.

20.6 The Exchange Rate Functions

The exchange rate plays a crucial role not only as a means of balancing foreign trade transactions in the balance of payments, but also in explaining inflationary processes and export and import changes affecting output and employment. In the past it was one of the major instruments of economic policy. Its control function has been relaxed over the last thirty years; the central banks first allowed it to vary within the prescribed bands (pegged rates) and finally to float freely, its fluctuations being dependent on the relation between demand and supply in the foreign currency markets.

In modelling the exchange rates the most frequent point of departure is the conception of purchasing power parity (PPP). According to this approach, the exchange rate (EX_t) depends on the relation between domestic prices (P_t^D) and foreign prices (P_t^F):

$$ex_t^* = \alpha_0 + \alpha_1 \left(p_t^D - P_t^F\right) + \varepsilon_t \tag{20.20}$$

where the variables are expressed in logs.

The indices of domestic and foreign prices should cover the tradables only.

The above equation represents the long-term relationship, assuming arbitrage in the exchange rate markets that equalizes the domestic and foreign prices (recalculated into the same currency).

However, considerable deviations from values generated by the equation are observed in both the short- and long-run. Their main cause is capital flows, particularly the transfer of portfolio investments to countries where interest rates are relatively high. Their inflow leads to the appreciation of the domestic currency, while outflow results in its depreciation. More generally, the expected exchange rate in the terminal markets is assumed to differ from the effective exchange rate, because the domestic and foreign interest rates are different. This approach, known as the uncovered interest parity (UIP), is most frequently applied in macroeconometric models. In empirical research, it leads to the following equation

$$ex_t = \beta_0 + \beta_1\left(r_t^D - r_t^F\right) + \xi_t \tag{20.21}$$

where r^D, r^F are domestic and foreign interest rates, respectively.

The above approach does not include the "fundamental" macroeconomic variables, such as GDP and money supply. To fill the gap the monetary theory of exchange rates is used. Within the approach we distinguish (a) the flexible-price monetary model (FPMM) presented by the Chicago school (Frenkel 1976; Bilson 1978), (b) the sticky-price monetary model (SPMM) of a Keynesian origin, where sticky prices are assumed (Dornbusch 1976; Frankel 1979; Fleming 1962; Mundell 1963; Hooper and Morton 1982).

The approach treats exchange rates as asset prices. All models mentioned above use the PPP approach. In general, in the monetary models the equation explaining the exchange rate can be written as:

$$\begin{aligned} ex_t = {} & \alpha_0 + \alpha_1\left(m_t - m_t^F\right) + \alpha_2\left(x_t - x_t^F\right) + \alpha_3\left(r_t - r_t^F\right) + \alpha_4\left(p_t^e - p_t^{eF}\right) \\ & + \alpha_5 TB_t - \alpha_6 TB_t^F + \nu_t \end{aligned} \tag{20.22}$$

where

m_t, m_t^F is money supply, respectively domestic and foreign,
x_t, x_t^F is GDP, respectively domestic and foreign,
p_t^e, p_t^{eF} is the expected rate of inflation, respectively domestic and foreign,
TB_t, TB_t^F is cumulative balance of trade, respectively domestic and foreign.

In the Frenkel-Bilson model the parameters α_4, α_5 and α_6 are zero; in the Dornbusch-Frankel model that allows for long-term deviations from the PPP the parameters α_1 and α_6 are the only ones to take 0. The Hooper-Morton model assuming that all variables are significant takes account also of the impacts of fluctuations in the foreign trade balance.

The monetary approach did not dominate the manner of entering the exchange rate equations into macroeconometric models. Particularly, in the short-run the exchange rate volatility was only infrequently associated with the changes in the GDP growth rates and money supply. Besides, the monthly data on the GDP were not available. On the other hand, proposals were put forward to define explicitly the equilibrium levels of the exchange rates and to search for factors capable of explaining deviations from the levels.

The most natural equilibrium level seemed to be one ensuring equality between the current account balance (CA) and the financial account balance (FA) in the balance of payments, i.e. CA = FA (MacDonald 2007).

A simplified approximation of the current account balance is a sum of foreign trade balance and net receipts from foreign assets (NFA):

$$CA_t = (E_t - M_t) + \nu NFA_t \tag{20.23}$$

where ν is the rate of profit from foreign assets.

Let us remind that the standard exports equation has the form:

$$E_t = X_t^{F\alpha}\left(EX_t P_t^F / P_t\right)^{\beta} \tag{20.24}$$

and the imports equation is:

$$M_t = X_t^{\gamma}\left(EX_t P_t^F / P_t\right)^{\lambda} \tag{20.25}$$

Substituting these equations into (20.23), we have:

$$CA_t = X^{F\alpha}\left(EX_t P_t^F / P_t\right)^{\beta} - X_t^{\gamma}\left(EX_t P_t^F / P_t\right)^{\lambda} - \nu NFA_t \tag{20.26}$$

The financial account balance can be approximated by allowing for factors affecting capital investments, i.e. for the differences between the domestic and foreign interest rates and for the expectations of exchange rate changes in the future. This yields the following equation:

$$FA_t = \Theta\left(R_t / R_t^F\right) / EX_t + EX_t^e \quad 0 < \Theta < \infty \tag{20.27}$$

After equalizing the current account and financial account balances given by Eqs. (20.26) and (20.27), the equilibrium position is reached. Solving for the exchange rate, we obtain the equilibrium exchange rate. The solution (the variables have been transformed into logs) is as follows:

$$ex_t^* = \left(p_t - p_t^F\right) - \frac{1}{\Phi}\left(\gamma X_t - \alpha X_t^F\right) - \Theta\left(r_t - r_t^F + ex_t^e\right) \quad \Phi = \beta + \lambda < 0 \tag{20.28}$$

the net foreign assets are ignored and the expected exchange rate ex_t^e is retained to be additionally estimated.

This long-term equation includes the PPP and the UIP exchange rate parities, as well as the fundamental variable GDP which determines exports and imports. This equation allows obtaining the real exchange rate (REX):

$$rex_t = ex_t + \left(p_t^F - p_t\right) \tag{20.29}$$

The short-term equation may include additional variables, e.g. currency reserves. With exports and imports inserted explicitly (instead of their determinants, i.e. domestic and foreign GDP), we obtain:

$$\begin{aligned}\Delta ex_t = \beta_0 + \beta_1\left(ex_{t-1} - ex_{t-1}^*\right) + \beta_2\Delta\left(r_t - r_t^F\right) + \beta_3\Delta(e_t - m_t) \\ + \beta_4\left(s_t - s_t^*\right) + \xi_t\end{aligned} \tag{20.30}$$

where s_t, s_t^* are ratios between currency reserves and imports, respectively effective and demanded.

The estimation of the equation parameters poses some difficulties. If the price indices of tradables are not available, then the deflators of consumer goods are used instead. Secondly, the determination of the optimal level of currency reserves is not unique, so this variable is often skipped (Brzeszczyński and Kelm 2002).

Several other models of the equilibrium exchange rate were developed in the past, accentuating different combinations of factors having effect on exchange rates. Their parameters were the most frequently estimated using the cointegration methodology and regarding the models composed of several equations—the optimal control models (cf. Whitley 1994; Wdowiński 2010).

The most widespread is the capital enhanced equilibrium exchange rate (CHEER) model, which was developed and empirically verified by Juselius and MacDonald (2004, 2007), see also Kębłowski and Welfe (2010) for Poland. This hybrid model combines the purchasing power parity (PPP) with the impacts of capital flows represented by the uncovered interest parity (UIP). Hence, we have:

$$ex_t = \left(p_1 - p_t^F\right) - \frac{1}{\Theta}\left(r_t - r_t^F\right) \tag{20.31}$$

The behavioural equilibrium exchange rate (BEER) model that Clark and MacDonald (1998) proposed represents a more general approach. It stresses the role of specific foreign trade-related factors in determining the long-term exchange rate. The explained variable is the real exchange rate REX. It is assumed that in the long-run it depends on the net foreign assets (NFA), on the ratio between the prices of tradables and non-tradables (TNT) and on the terms of trade index (TOT) (all in logs):

$$rex_t^* = \alpha_0 + \alpha_1 nfa_t + \alpha_2 tnt_t + \alpha_3 tot \tag{20.32}$$

and in the short-run we have:

$$rex_t = rex_t^* - \left(r_t^F - r_t\right) \tag{20.33}$$

The subsequent models use more complicated small systems of equations. They are solved using the optimal control methods and yield the equilibrium level of the real exchange rate. These are the fundamental equilibrium exchange rate (FEER) model developed by Wren-Lewis (1992) and the natural real exchange rate (NATREX) model formulated by Stein et al. (1995).

The FEER model uses a small macroeconometric model that ensures either domestic and foreign equilibria, or, in its simplified version, equality between the current balance and the financial balance of the balance of payments. In this case the respective equation can be written as:

$$REX_t X_t^{*F\alpha} / X_t^{*\gamma} NFA_t^{*\nu} = FA_t \tag{20.34}$$

where (*) is the long-term (equilibrium) value.

J. Whitley demonstrated how domestic and external equilibria can be combined. He assumed that domestic equilibrium can be obtained by choosing the appropriate value of NAIRU and external equilibrium by using a real exchange rate (REX) at which CA is offset by the structural capital flows, i.e. CA is equal to FA (cf. Whitley 1994, pp. 124–127).

The NATREX model generates an equilibrium exchange rate that ensures equality between the current account and ex-ante savings reduced by investment. It is composed of three equations that explain the real exchange rate, foreign debt and capital. The exogenous variables are represented by a savings-to-GDP ratio, productivity of capital, long-term real interest rate and terms of trade in small open economies. The model has been verified in many research projects.

20.7 Equations Systems

Processes taking place in the money and capital markets can be jointly analysed within separate submodels, where the markets are linked with the rest of economy. The examples of these submodels are the models stressing the role of exchange rates, which were mentioned in the previous section. Other models show the links between the exchange rates, interest rates, the prices of commodities and services and wages, entering the inflationary spiral mentioned in the previous chapters. Interest rates that represent the debt service costs indirectly contribute to price changes, like the exchange rates. On the other hand, they depend on the rates of inflation. These relationships are simultaneous and represent important feedbacks.

They can be jointly analysed using the cointegration methodology. Let us mention at this point numerous empirical contributions from the Łódź Institute of Econometrics. The most representative among them is the model presented in the paper by Welfe et al. (2007). In this small system of equations for Poland the producer prices depend on labour costs and import prices. The average wages are determined by prices and labour productivity. The import prices depend on the world prices and exchange rates, with the latter being determined by the ratio between domestic and world prices.

The model distinguishes four long-term stochastic equations and one identity. Because most variables are non-stationary, the equation parameters are estimated using the maximum likelihood method.

The links between prices and wages were a subject of thorough analysis conducted for the Norwegian economy, which utilised the cointegration approach (Bårdsen et al. 2005).

At the beginning of the 21st c. a system of equations linking the monetary and capital markets in the United Kingdom was built, summarized then in the monograph by Garratt et al. (2006). Its major long-term relationships explained the purchasing power parity of money, the desired interest rate parity ultimately determining resource allocation between financial and real investments, and the interest rate parity determining the allocation of resources between domestic and foreign assets.

This submodel is special in that the UK variables are directly linked to the respective variables for the world economy, following the common trends hypothesis.

References

Bårdsen, G., Eitrheim, Ø., Jansen, E. S., & Nymoen, R. (2005). *The econometrics of macroeconomic modelling*. Oxford: Oxford University Press.

Bernanke, B. S., & Blinder, A. S. (1988). Credit, money and aggregate demand. *American Economic Review, Papers and Proceedings*, *78/88*, 435–439.

Bilson, J. F. O. (1978). The current experience with floating exchange rates: an appraisal of the monetary approach. *Exchange Rates and Open Economies*, *68*, 392–397.

Brzeszczyński, J., & Kelm, R. (2002). *Ekonometryczne modele rynków finansowych (Econometric models of financial markets)*. Warszawa: WIG.

Calza, A., Gartner, C., & Sousa, S. M. (2003). Modelling the demand for loans to the private sector in the Euro area. *Applied Economics*, *35*, 107–117.

Clark, P. B., & MacDonald, R. (1998). *Exchange rates and economic fundamentals: a methodological comparison of BEERs and FEERs* (IMF Working Paper No. 98).

Dornbusch, R. (1976). Expectations and exchange rate dynamics. *Journal of Political Economy*, *84*, 1161–1176.

Eller, M., Froemmel, M., & Srzentic, N. (2010). Private sector credit in CESEE: long-run relationships and short-run dynamics. *Focus on European Economic Integration*, *Q2*, 50–78.

Fase, M. M. G. (1995). The demand for commercial bank loans and the lending rate. *European Economic Review*, *39*, 99–115.

Fleming, J. M. (1962). Domestic financial policies under fixed and under floating exchange rates. *IMF Staff Papers*, *12*, 369–380.

Frankel, J. A. (1979). On the mark: theory of floating exchange rates based on real interest rate differentials. *American Economic Review*, *69*, 610–622.

Frenkel, J. A. (1976). A monetary approach to the exchange rate: doctrinal aspects and empirical evidence. *Scandinavian Journal of Economics*, *78*, 200–224.

Garratt, A., Lee, K., Pesaran, M. H., & Shin, Y. (2006). *Global and national macroecomometric modelling: a long-run structural approach*. Oxford: Oxford University Press.

Hooper, P., & Morton, J. E. (1982). Fluctuation in the dollar: a model of nominal and real exchange rate determination. *Journal of International Money and Finance*, *1*, 39–56.

Juselius, K., & MacDonald, R. (2004). The international parities between Japan and the USA. *Japan and the World Economy*, *16*, 17–34.

Juselius, K., & MacDonald, R. (2007). International parity relationships and a nonstationary real exchange rate. Germany versus the US in the post Bretton Woods period. In V. Morales (Ed.), *Recent issues in international macroeconomics*. New York: Nova Publications.

Kębłowski, P., & Welfe, A. (2010). Estimation of the equilibrium exchange rate: the CHEER approach. *Journal of International Money and Finance*.

Klein, L. R., Welfe, A., & Welfe, W. (1999). *Principles of macroeconometric modeling*. Amsterdam: North-Holland.

MacDonald, R. (2007). *Exchange rate economics: theories and evidence*, New York: Routledge.

Mundell, R. A. (1963). Capital mobility and stabilization policy under fixed and flexible exchange rates. *Canadian Journal of Economics and Political Science*, *29*, 475–485.

Psaradakis, Z., Sola, M., & Spagnolo, F. (2004). On Markov error correction models, with an application to stock prices and dividends. *Journal of Applied Econometrics*, *19*, 69–88.

Stein, J. L., Allen, P. R., et al. (Eds.) (1995). *Fundamental determinants of exchange rates*. Oxford: Oxford University Press.

Taylor, J. B. (1993). Discretion versus policy rules in practice. *Carnegie-Rochester Conference Series on Public Policy*, *39*, 195–214.

Wdowiński, P. (2010). *Modele kursów walutowych (Models of exchange rates)*. Łódź: Wydawnictwo Uniwersytetu Łódzkiego.

Welfe, A., Kelm, R., & Majsterek, M. (2007). Agregatowy model inflacji (Aggregate model of inflation). *Przegląd Statystyczny*, *49*, 15–31.

Welfe, W., & Welfe, A. (2004). *Ekonometria stosowana (Applied econometrics)* (2nd ed.). Warszawa: PWE.

Whitley, J. D. (1994). *A course in macroeconomic modelling and forecasting*. New York: Harvester, Wheatsheaf.

Wren-Lewis, S. (1992). On the analytical foundations of the fundamental equilibrium exchange rate. In C. P. Hargreaves (Ed.), *Macroeconomic modelling of the long-run*. London: E. Elgar.

Chapter 21
Prospects of Macroeconometric Modelling

Macroeconometric modelling has become a well-established field of applied macroeconomics and a worldwide activity involving empirical research into the functioning and growth of national economies. The use of macroeconometric models in forecasting and simulation analyses of the likely economic policy outcomes has expanded to the majority of countries. The models have become an important instrument of worldwide analyses and forecasts conducted by international organizations and renowned research institutions, as well as central banks of many countries.

The above factors have stimulated the expansion of a vast "industry" of macroeconometric modelling. Its centre has shifted from academic institutions to public authorities—governments, central banks and in many industrialized countries to commercial organizations. This tendency is likely to continue into the future. The academic institutions seem to be constrained in handling small macromodels that require further development of efficient estimation techniques based on the cointegration methodology.

The development of the models' structure has been characterized by rivalry between the "old" mainstream structural models of macro-Keynesian orientation and the new models, i.e. the dynamic stochastic equilibrium models of neo-Keynesian orientation. The first type still predominates among the large macroeconometric models operated in most countries. The DSGE models founded on solid theoretical microeconomic foundations are more or less experimental and at some point in the future they may perhaps substitute the structural models.

The neoclassical pressure gave microeconomic orientation to the internal structure of many macroeconometric models after the household and enterprise sectors were distinguished. However, the question about how the sectors should be positioned within the whole economy has not been successfully answered so far. The attempt at finding a solution is presented in Chap. 18 of this book.

Two different lines in the development of macromodelling activities are worth stressing. One is the tendency towards building large macroeconometric models, frequently counting in excess of 30,000 equations in the case of multicountry models. This tendency extends to the one-country models, if sectoral or regional dis-

W. Welfe, *Macroeconometric Models*,
Advanced Studies in Theoretical and Applied Econometrics 47,
DOI 10.1007/978-3-642-34468-8_21,

aggregation is of practical importance. The other line involving the construction of one-sectoral models being either replicas of large models or representing theoretical issues is mostly seen among academic centres. These models may be applied to particular areas of economy, such as inflation or money markets, where the cointegration techniques can be applied.

There are important questions that still need to be discussed. They concern the problem of dealing with a structural change. The information that is increasingly available from the cross-section data may occur helpful in this case. The role of rational expectations and of the learning process still needs to be clarified, taking into account the achievements of social psychology. The role of stock variables (e.g. wealth in consumption) has to be stressed.

Macromodellers tend to reduce exogeneity by extending the scopes of the models. The use of the engineering functions allows explaining the impact of technical progress. This has opened up the door to broad analyses of the impacts of innovation and technological progress in determining total factor productivity, breaking the path to endogenize the growth theory empirically.

A fruitful and fast-developing field of research is the modelling of money and capital markets and particularly of exchange rates. This research is promoted by central banks, which are interested in having models capable of monitoring monetary policy impacts.

Macromodels are extended in cooperation with the I-O modellers to cover the sectoral and environmental issues. The demographic sector is frequently included incorporating the health sector. The impacts of human capital and, more recently, of social capital are a new promising area of research.

Most macromodels employ new estimation techniques developed at academic centres, which mainly involve the use of the cointegration approach. The two-step Engle-Granger procedure that allows distinguishing between long-run relationships and short-run adjustments is broadly applied. More complex cointegration techniques are used to estimate the small-equation subsystems. The question about how to combine small subsystems into a large system covering a whole national economy still remains to be answered. The applications of the Bayesian estimation methods have recently developed.

Author Index

A
Abbey, J. L. S., 231, *236*
Abrahamsen, Y., 118, *118*
Adams, F. G., 22, *42*, 244, *270*
Addison, D., 247, *270*
Adelman, F., 18, *42*
Adelman, I., 18, *42*, 214, *238*
Adeniyi, J., 234, *239*
Adolfson, M., 97, *98*
Agarwala, R., 215, *237*
Agenor, P., 270, *272*
Aglietta, M., 63, *71*
Ahmed, A. M., 225, *239*
Al-Din, N. A. K., 207, *237*
Alexander, S. L., 23, *44*
Allard, D., 69, *72*
Allard-Prigent, C., 71, *71*, 305, *312*
Allen, C., 54, *60*
Allen, P. R., 391, *393*
Amano, A., 205, *237*
Amer, R., 228, *237*
Anders, H. D., 126, *181*
Andersen, F. M., 93, *98*
Andersen, L. C., 26, *42*
Anderson, R. G., 32, *42*, 278, *278*
Ando, A., 24, 25, *42*, 191, *201*, 242, *273*, 308, *312*, *313*
Andre, C., 246, *271*
Armington, P., 359, *368*
Artus, P., 69, 70, *72*, *73*
Arya, G. N., 210, *237*
Audenis, C., 71, *71*
Auderis, C., 305, *312*
Audric, S., 149, *188*

B
Bachetta, P., 117, *120*
Bachman, D., 28, *43*
Bakaev, O., 177, *186*
Bakaev, O. O., 178, *181*
Ball, R. J., 48, 50, *60*, *61*, 241, 243, *270*, *272*
Bank of England, 59, *61*
Ban, K., 205, *237*
Barczak, A., 124, 128, *181*, 376, *380*
Bårdsen, G., 6, *7*, 21, *43*, 95, *98*, 278, *278*, 356, *368*, 383, 392, *393*
Barker, T. S., 49, *61*
Barrel, R., 150, 179, *181*
Barrell, R., 242, 263, *270*, 306, *312*
Barro, R. J., 297, *301*, 336, *343*, 348, 368, *368*
Barteczko, K., 135, 151, *181*
Barten, A., 102, *119*
Barten, A. P., 77, *83*
Basdevant, O., 172, *181*, 308, *312*
Bayoumi, T., 242, 250, *270*, 334, *343*
Bean, C., 107, *119*, 297, *302*, 348, 349, *369*
Beck, M., 107, *119*
Becker, B., 242, 263, *270*, 306, *312*
Beckerhoff, D., 104, *120*
Beenstock, M., 51, *61*
Beffy, P.-O., 197, *200*, 329, *343*
Belfiori, L., 109, *119*
Beltran-del-Rio, A., 235, 236, *237*
Benabou, R., 337, *343*
Benĉik, M., 176, *181*
Benčik, M., 176, *186*
Beneš, J., 149, *181*
Beneš, L., 149, *188*
Beng, G. W., 208, *239*
Benhabib, J., 336, *343*
Berg, L., 97, *98*
Berger, K., 71, *71*, 305, *312*
Bergman, L., 85, 86, 88, 94, 96, *98*, *99*
Bergsman, J., 214, *238*

W. Welfe, *Macroeconometric Models*,
Advanced Studies in Theoretical and Applied Econometrics 47,
DOI 10.1007/978-3-642-34468-8, © Springer-Verlag Berlin Heidelberg 2013

Bergström, R., 85, 96, 97, *98*, *99*
Bergström, V., 85, 96, 97, *98*, *99*
Bernanke, B. S., 383, *393*
Berner, R., 261, *273*
Bessarab, S., 137, *182*
Beynet, P., 149, *188*
Bhattacharya, B. B., 215, *237*
Bielkin, W. D., 136, *182*
Billaudot, B., 64, *72*
Bilson, J. F. O., 389, *393*
Biorn, E., 95, *98*
Bjerholt, O., 88, *98*
Bjerkholt, O., 85, 94, 95, *98*
Bjornstad, R., 94, *98*
Black, R., 35, *43*
Blake, A. P., 303, *313*
Blanchard, O., 308, *312*, 354, *368*
Blanchard, O. J., 35, *43*
Blinder, A. S., 383, *393*
Bloch, L., 69, *72*
Boatwright, B. D., 50, *60*
Bobińska, K., 124, *182*
Bocian, A., 135, 151, *181*, *182*
Bodin, L., 131, *182*
Bodkin, R. G., 3, 4, *7*, 9, 13, *13*, 18, 41, *43*, 47, 50, *60*, *61*, 70, *72*, 77, *83*, 117, *119*, 203, 205, 206, 213, 225, 235, 236, *237*, *239*, 241, 242, 244, *271*, 285, *301*
Boeschoten, W. C., 192, *201*
Boeuf, P., 69, *72*
Bogaert, H., 102, *119*
Bognai, A., 195, *200*
Bondarenko, G., 177, *186*
Bonnet, X., 197, *200*, 329, *343*
Borensztein, E., 335, *343*
Bossier, F., 102, *119*
Boudier, M.-A., 70, *73*
Boulle, J., 64, *72*
Bournay, J., 69, *73*
Boutillier, M., 69, 70, *72*, *73*
Boyer, R., 64, *72*
Bożyk, P., 129, *182*, 356, *369*
Brada, J. C., 126, *182*
Bradley, J., 108, 109, 116, *119*, 155, 170, 179, *182*, 194, *200*
Bradley, T., 247, *273*, 307, *313*
Brainard, W. C., 325, *343*, 377, *380*
Braithwaite, C., 41, *45*, *46*
Brautzsch, H.-U., 107, *119*
Brayton, F., 27, 31, *43*, 262, *270*
Brillet, J. L., 65, *72*, 149, *188*, 192, *201*
Broer, P., 77, *83*
Brooks, R., 6, *7*, 211, 212, *237*
Brown, T. M., 34, *43*, 104, *119*, 304, *312*
Brumberg, R. H., 308, *313*
Brumer, K., 5, *7*
Brunner, K., 27, *45*
Brzeszczyński, J., 377, *380*, 391, *393*
Budd, A., 54, *61*
Budnik, K., 154, *182*
Buraczas, A. I., 136, *182*
Burgraeve, K., 103, *120*
Burns, T., 50, *60*
Burridge, M., 54, *61*, 267, *270*
Burża, T., 135, *182*
Buscher, H. S., 194, *201*
Buslei, H., 194, *201*
Bussery, H., 63, *72*
Byrne, J., 242, 263, *270*, 306, *312*
Byron, R., 50, *61*

C

Cagas, M. A., 208, *237*
Cain, B., 41, *45*, *46*
Calza, A., 383, 384, *393*
Campbell, J. Y., 308, *312*
Campero, L.-F., 117, *120*
Cappelen, Å., 94, *98*
Capros, P., 170, *182*, 194, *201*
Carlson, K. M., 26, *42*
Carlsson, E., 97, *99*
Carlsson, S., 85, 96, *99*
Carlucci, F., 195, *200*
Carnot, N., 71, *71*, *72*
Carr, J. L., 41, *46*
Carrot, N., 305, *312*
Celov, D., 93, *98*
Chamberlin, G., 270, *270*
Charemza, W., 128, 134, 135, 179, *182*, *183*, *189*, 297, *301*
Charpin, F., 70, *73*
Charpin, J. M., 65, *72*, *73*
Chen, K., 213, *237*
Chenery, H. B., 215, *237*
Chew, K., 208, *239*
Chips, B., 118, *118*
Chirinko, R. S., 321, 325, *343*
Choudhry, N. K., 41, *43*, *46*
Christensen, A. M., 87, *98*
Chung, A., 42, *44*
Church, K. B., 60, *61*
Ciepielewska, B., 124, 128, *181*
Cincera, M., 335, *343*
Ciupagea, C., 170, 179, *182*
Clark, C. S., 231, *236*
Clark, P., 261, *273*
Clark, P. B., 391, *393*
Coe, D. T., 334, *343*

Coen, R. M., 25, *43*, *44*, 312, *312*, 348, 352, *369*
Coenen, G., 197, *201*
Cohen, D., 336, *343*
Coletti, D., 35, *43*
Collechia, A., 332, *343*
Conrad, K., 194, *201*
Costa, A. M., 123, 124, *182*, *186*
Côté, D., 42, *43*
Courbis, R., 63, 65, 70, *71*, *72*, 101, *119*, 151, 153, 175, *182*, *186*, *189*, 337, *343*, 362, *369*
Cripps, T. F., 49, *61*
Crispolti, V., 335, *343*
Cufer, U., 150, 179, *181*
Curdia, V., 97, *98*
Czerwiński, Z., 133, 135, *182*, *183*
Czyżewski, B., 131, *183*, 356, *369*

D
Dadajan, V., 137, *182*
D'Adda, C., 109, *119*, 267, *271*
D'Alcantara, G., 102, *119*
Dalsgaard, T., 246, *271*
Dalziel, A., 51, *61*
Dam, P. U., 86, *98*
Darby, J., 55, *61*, *62*
David, J.-H., 69, *72*
Davidson, J. E. H., 5, *7*, 306, *313*
Davis, C. M., 128, 135, *183*, *189*, 297, *301*
de Bever, L., 34, *43*
de Biolley, T., 102, *119*
de Bondt, G. J., 192, *201*, 306, *313*
De Gregorio, J., 335, *343*
de la Fuente, A., 336, *343*
de Leeuw, F., 24, *43*
de Menil, G., 69, *73*, 105, *119*
de Monchy, G., 64, *72*
de Mooij, R., 77, 82, *83*, *84*
De Wet, G. L., 234, *237*
Deaton, A. S., 305, *313*
Dębski, W., 131, 133, *183*
Dehove, M., 64, *72*
Deleau, M., 69, *72*
Delessy, H., 264, 265, *271*, 329, 342, *344*
den Haan, R. J. A., 76, *84*
Den Hartog, H., 76, *83*, 117, *119*
Desai, M. J., 213, 214, *237*
Desaulniers, D., 41, *45*, *46*
Devulis, G. S., 136, *182*
Dhaene, G., 102, *119*
Dhar, S., 54, *61*, 267, *270*
Dicks, G., 54, *61*
Dieppe, A., 196, *201*
Dinenis, E., 54, *61*
Dixon, P. D., 211, *237*
Dlouhy, V., 126, *183*
Dobrescu, E., 170, *183*
Domenech, R., 336, *343*
Donders, J., 82, *84*
Dooley, M., 248, *272*
Dornbusch, R., 389, *393*
Douven, R. C., 191, *201*
Dramais, A., 191, 194, *201*, 242, 251, *271*, *273*, 325, *344*
Dreger, C., 107, *119*, 196, 200, *201*, 322, 329, *344*
Drehuis, W., 117, *119*
Dreze, J. H., 107, *119*, 297, *302*, 348, 349, *369*
Driehuis, W., 76, 83, *84*
Dubey, A., 336, *345*
Ducanes, G., 208, *237*
Duchene, S., 71, *71*, 305, *312*
Duessenberry, J. G., 4, *7*, 18–21, 23, *43–45*, 322, *344*
Duggal, V. G., 22, *42*, *43*
Duguay, P., 35, *44*
Dumazet, D., 70, *72*
Dungan, P., 41, *44*
Duong, M. H., 108, *119*
Durand, B., 69, 70, *72*, *73*
Ďuraš, J., 176, *183*
Dutta, M., 30, *44*, 210, 213, *238*, *240*

E
Eaton, J., 335, *344*
Eckstein, O., 18, 23, 24, 27, 28, *43*, *44*, 341, *345*, 362, *369*
Economie, 70, *73*
Edison, H., 261, *271*
Eisner, R., 320, *344*
Eitrheim, Ø., 6, *7*, 21, *43*, 95, *98*, 278, *278*, 356, *368*, 383, 392, *393*
El-Sheikh, S., 229, *237*
Eller, M., 384, *393*
Elliot, D. A., 26, *44*
Elliot, J., 231, *237*
Elmeskov, J., 354–356, *369*
Engelbrecht, H. J., 334, 337, *344*
Engle, R. F., 6, *7*, 295, 301, *302*
Erceg, C. J., 262, *271*
Erikson, I., 97, *98*
Ernsäter, L., 96, *98*
Eskin, V., 30, *45*
Estrada, A., 117, *119*, *121*, 196, *201*
Ettlin, F. A., 97, *98*
Evans, H. P., 50, *62*
Evans, M. K., 21, *44*, 324, 326, *344*

Exposito, P., 230, *237*
Eyford, B. L., 41, *45*, *46*
Ezaki, M., 207, *238*

F
Fagan, G., 196, *201*, 329, *344*
Fair, R. C., 5, *7*, 23, 25–27, 32, 33, *44*, 268, 269, *271*, 276, 278, *278*, 281, *283*, 287, 294, 300, *302*, 304, 310, 312, *313*, 316, 317, 323, 338, 339, *344*, 353, *369*
Fanning, C., 108, *119*
Faruqee, H., 248, *272*, 294, *302*, 325, *345*
Fase, M. M. G., 82, *84*, 117, *119*, 192, *201*, 383, *393*
Faubry, E., 69, *73*
Fauqueur, A. Q., 64, *72*
Fazio, A., 69, *72*
Fazzari, S. M., 325, *343*
Fernandez, J. L., 117, *119*
Fernandez-Corugedo, E., 303, *313*
Ferrari, M., 109, *119*
Fic, T., 154, *183*
Fidrmuc, J., 101, *119*
Finocchiaro, D., 97, *98*
Fischer, J., 3, *7*
Fisher, F. M., 21, *44*
Fisher, P. G., 59, 60, *61*, *62*, 329, *345*
Fitz Gerald, J., 109, *119*, 194, *200*
Fleming, J. M., 389, *393*
Florczak, W., 118, *118*, 149, 152, 153, 174, *183*, *188–190*, 333, 337, *344*, *346*, 348, 349, *369*, *370*
Fomasari, C., 109, *119*
Fomin, J., 124, *183*
Fonteneau, A., 64, 65, *72*, *73*
Foot, D. K., 34, *43*
Fouquet, D., 65, *72*
Fouquet, D. J., 65, *73*
Frączek, R., 154, *188*
Fraichot, J.-P., 69, *72*
Frandsen, S. E., 86, *98*
Frankel, J. A., 389, *393*
Franz, W., 107, *120*
Frenkel, J. A., 389, *393*
Friedman, M., 10, *13*, 308, *313*, 354, *369*
Frisch, R., 10, *13*
Froemmel, M., 384, *393*
Fromm, G., 4, *7*, 18, 19, 21, 23, *43*, *44*
Fujiwara, H., 205, *239*
Fuss, H., 104, *120*

G
Gagnon, J., 261, *271*
Gahlen, B., 103, *120*
Gajda, J., 135, 149, *188*, *189*
Gajda, J. B., 132, 137, *183*
Gali, J., 117, *120*
Galli, G., 109, 116, *120*, *121*
Gana, J. L., 242, *271*
Ganges, B., 244, *270*
Garbicz, M., 135, *183*
Garganas, N. C., 108, *120*
Garratt, A., 54, 60, *61*, 278, *278*, 383, 387, 392, *393*
Garrett, P., 245, 248, *271*
Gartner, C., 383, 384, *393*
Gaudemet, J.-P., 64, *72*
Gavrilenkov, E. E., 137, 172, *183*
Gavura, M., 176, *184*
Gelauff, G. M. M., 77, *84*
Georgakopulos, P., 170, *182*, 194, *201*
Gergelyi, K., 125, *187*
Ghali, H. S., 228, *237*
Ghartey, E. E., 231, *237*
Gibbs, D., 6, *7*, 211, 212, *237*
Gilby, S., 41, *45*, *46*
Ginsburgh, V., 288, *302*
Gjelsvik, M. L., 94, *98*
Global Insight Co, 304, *313*
Goday, A., 94, *98*
Godley, W. A. H., 49, *61*
Goeggelmann, K., 107, *120*
Göggelmann, K., 194, *201*
Goldberger, A. S., 4, *7*, 16, *44*, *45*, 337, 340, *344*
Goslinga, S. S., 75, *84*
Gottschalk, S., 242, 263, *270*, 306, *312*
Graafland, J., 82, *84*
Grabek, G., 154, *184*
Gradzewicz, M., 154, *183*, *186*
Grajek, M., 151, *184*
Gramlich, E., 24, *43*
Granger, C. N. J., 6, *7*
Granger, W. J., 295, 301, *302*
Green, D., 136, *184*
Green, E. W., 28, *44*
Green, G., 24, *45*
Greszta, M., 151, 154, *182*, *184*
Griffin, P., 154, *186*
Griliches, Z., 376, *380*
Grimm, B. T., 24, *44*
Grinderslev, D., 93, *98*
Gronicki, M., 135, *182*
Grossman, H. I., 297, *301*, 348, 368, *368*
Gruber, J., 134, *187*
Gręda Latocha, R., 331, *344*, 352, *369*
Gu, Q., 213, *237*
Guerrieri, L., 262, *271*

Guillaume, H., 65, *73*
Guisan, C., 230, *237*
Gust, C., 262, *271*
Guzik, B., 135, *183*

H
Haas, R., 248, *271*, *272*
Haavelmo, T., 11, *13*
Haber, G., 176, *184*, *190*
Halabuk, L., 5, *7*, 124, 127, *184*, *187*
Hall, R. E., 308, *313*
Hall, S. G., 54, *60*, *61*, 147, *184*
Haluška, J., 149, 174, 175, *184*, *186*, *188*
Hamada, F., 206, *238*
Hansen, C. B., 86, *98*
Hansen, H.-J., 106, *120*
Hansen, J., 86, *98*
Hansen, J. V., 86, *98*
Haque, N. U., 203, *238*
Hargreaves, C. P., 391, *394*
Harnett, I., 51, 55, *61*, *62*
Harrison, R., 59, *61*, 304, 310, *313*, 324, 327, 329, 330, *344*
Hart, P. E., 5, *7*
Hasselman, B. H., 76, *84*
Hazelwood, A., 48, *61*
Heathfield, D., 50, *61*
Heidbrink, W., 107, *120*
Heilemann, U., 4, *7*, 103, 105, 107, *120*
Hejl, A., 125, *186*
Helliwell, J., 245, 248, *271*
Helliwell, J. F., 34, 35, 42, *43*, *44*
Helpman, E., 334, *343*
Hendry, D. H., 5, *7*, 281, *283*, 293, 300, *302*, 305, 306, *313*, 340, *345*, 354, *369*
Henry, J., 196, *201*, 329, *344*
Henry, S., 42, *44*
Henry, S. G. B., 54, *61*, 172, *183*, 270, *270*
Hernandez-Cata, E., 261, *273*
Herring, R., 191, *201*, 242, *273*
Herzog, P., 64, *73*
Hetemäki, M., 88, *98*
Heyets, V., 177, 178, *184*
Hickman, B. G., 24, 25, *42–44*, 241–244, *271*, 312, *312*, 348, 352, 357, *369*
Hicks, J. R., 12, *13*
Higgins, C., 136, *184*
Hildenbrand, W., 133, *182*
Hilton, K., 50, *61*
Hirsch, A., 24, *45*
Hirsch, A. A., 24, *44*, *45*
Hlédik, T., 149, *181*
Ho, M. S., 332, 337, *344*
Ho, W. M., 42, *44*
Hoffmaister, A. W., 334, *343*
Hoffman, D. L., 32, *42*, 278, *278*
Holden, K., 54, *61*
Holland, P., 150, 179, *181*
Holly, S., 50, 54, *60*, *61*
Holm, I., 94, *98*
Holsen, J. G., 247, *272*
Hong, I., 209, *239*
Hooper, P., 389, *393*
Howe, C. W., 231, *238*
Howe, H., 261, *273*
Howrey, E. P., 30, *44*
Huff, L. A., 230, *238*
Hughes-Hallet, A., 270, *272*
Huizinga, F., 77, *84*
Hulej, M., 154, *182*
Hulyak, K., 127, 128, *184*
Hunt, B., 35, *43*, 248, 249, *272*
Hurst, I., 242, 263, *270*, 306, *312*
Husson, M., 64, *72*
Huth, C., 118, *118*
Hwa, E. C., 26, *45*
Hymans, S. H., 23, 30, *44*
Hyndman, R. M., 42, *44*

I
Ichimura, S., 204, 207, 213, *238*
in't Veld, J., 251, 261, *271*, *273*, 325, *344*, *345*
Intriligator, M. D., 376, *380*
Ireland, J., 55, *61*, *62*
Isard, P., 248, *272*, 294, *302*, 325, *345*
Issing, O., 195, *201*
Iyaniwura, J., 234, *239*

J
Jackman, R., 340, *345*, 354, *369*
Jacobs, J., 83, *84*
Jacobs, J. P. A., 60, *61*, 278, *278*
Jacobson, L. R., 242, *271*
Jaffe, A. B., 335, *344*
Jakab, Z. M., 180, *184*
Jakob, Z., 150, 179, *181*
Jakubczyc, T., 124, 128, *181*
Jansen, E. S., 6, *7*, 95, *98*, 278, *278*, 383, 392, *393*
Janssen, M., 132, *188*
Jaquette, P., 28, *43*
Jaremienko, J. W., 136, *184*
Järnhäll, B., 85, 96, 97, *98*, *99*
Jean, C. S., 41, *45*, *46*
Jeanfils, P., 102, 103, *120*, 196, *201*
Jemielianov, A. S., 136, *184*
Jensen, E. S., 21, *43*, 356, *368*
Jensen, M., 95, *98*

Jiapei, W., 213, *238*
Johansen, S., 6, *7*
Johansson, S., 85, 96, 97, *98*, *99*
Johnson, K., 243, *272*
Jongbloed, B. W. A., 83, *84*
Jonkergouw, E., 234, *237*
Jorgenson, D. W., 20, *45*, 322, 332, 336, 337, *344*, 376, *380*
Jump, G. V., 34, 41, *43*, *45*, *46*
Jurek, W., 135, *183*
Juselius, K., 281, *283*, 391, *393*
Juszczak, G., 134, 151, *185*, 371, *380*
Juszczak, W., 132, *185*, 351, *369*

K
Kadjar, A., 69, *72*
Kaergärd, N., 86, *98*
Kalecki, M., 10, *13*, 124, *185*
Kalinin, D., 130, *185*
Kalisiak, Z., 129, *185*
Karani, H., 231, *238*
Karl, K., 28, *43*
Karp, P., 151, 152, *188*, 380, *381*
Kaski, E.-L., 87, *99*
Kawokami, A., 206, 207, *240*
Kazlauskas, A., 93, *98*
Kazmi, N. A., 215, *238*
Kaźmierska, M. M., 134, 151, *185*, 371, *380*
Keating, G., 54, *61*
Kębłowski, P., 151, 152, *188*, 391, *393*
Keller, W., 334, *344*
Kelm, R., 107, *121*, 147, 151, 152, *184*, *188*, 227, 234, *237*, *239*, 377, 380, *380*, *381*, 383, 391, 392, *393*
Kendall, M. G., 76, *84*
Kenen, P., 243, *272*
Kenessey, Z., 124, 127, *184*
Keynes, J. M., 11, 12, *13*
Keyzer, M., 288, *302*
Khan, A. H., 225, *239*
Khong, V., 70, *72*
Kiedrowski, R., 135, *183*
King, A. S., 126, *182*
Klas, A., 126, *185*
Klein, L. R., 3, 4, *7*, 9, 13, *13*, 15, 16, 18, 19, 21, 22, 30, *43–45*, 47, 48, 50, *60*, *61*, 70, *72*, 77, *83*, 101, *120*, 124, 129, 132, 133, 135, 148, *185*, *188*, 203, 204, 213, 230, *237*, *238*, 241–244, *271*, *272*, 281, *283*, 285, 286, 288, 291, *301*, *302*, 316, 321, 324, 326, 331, 337, 340, *344*, 352, 357, 364–367, *369*, 371, *380*, 383, *393*
Klein, T., 174, 175, *186*
Kłos, B., 135, 152, 154, *184*, *185*
Kmenta, J., 21, *45*, 275, *278*
Knudsen, D., 87, *98*
Kobayashi, K., 207, *238*
Koekemoer, R., 234, *237*
Kohno, A., 205, *239*
Koizumi, S., 204, *238*
Kokoszczyński, R., 152, 154, *185*
Kolasa, M., 154, *182*, *183*
Kolek, J., 124–127, 136, *185*, *187*
Komiya, R., 204, *239*
Koning, M. A., 83, *84*
Konopczyński, M., 135, *183*
Kornai, J., 124, 128, *185*
Kortelainen, M., 88, *99*
Kortum, S., 335, *344*
Koschel, H., 194, *201*
Koster, M., 234, *238*
Kot, A., 154, *183*
Kotasz, G., 124, 127, *184*
Kotowitz, Y., 41, *43*, *46*
Kotyński, J., 134, *185*, 356, *369*
Kovacs, M. A., 150, 179, *181*
Kovács, M. A., 180, *184*
Koyck, L. M., 321, *345*
Kponsihoin, K., 235, *238*
Kramer, P., 192, *201*
Kratena, K., 102, *120*
Krelle, W., 104, *120*, *121*
Kretan, V., 176, *183*
Krishnamurty, K., 215, 225, *238*
Krueger, A., 337, *345*
Kudrycka, I., 129, 135, *185*
Kuh, E., 4, *7*, 19, *43*
Kuiper, J., 34, *45*
Kuipers, S. H., 83, *84*
Kukkonen, P., 85, 96, *99*
Kulikowski, R., 135, *185*
Kumhof, M., 149, *181*, 250, *272*
Kuper, G. H., 83, *84*
Kurihara, K. K., 308, *313*
Kurokawa, M., 206, 207, *240*
Kurosawa, Y., 205, *239*
Kushnirskij, F. I., 136, *184*
Kuszczak, J., 42, *43*
Kuszewski, T., 135, *183*
Kwack, S., 261, *273*
Kwack, S. Y., 231, *237*
Kydland, F. E., 6, *7*, 11, *13*, 282, *283*
Kyn, O., 125, *186*

L
Lahiri, K., 203, *238*
Lam, J.-P., 42, *43*
Lambelet, J.-C., 117, *120*

Lambert, J. P., 107, *119*, 297, *302*, 348, 349, *369*
Lane, P. R., 250, *272*, 277, *278*
Langer, H. G., 104, *120*
Łapińska-Sobczak, N., 133, *186*, 371, *380*
Laroque, G., 70, *72*
Laséen, S., 97, *98*
Lau, L. J., 242, 243, *271*, 357, *369*
Laursen, T., 154, *186*
Lawson, T., 13, *13*
Laxton, D., 35, *43*, 248–250, *272*, 294, *302*, 325, *345*
Layard, P. R. G., 53, *61*, 298, *302*, 329, 340, *345*, 354, *369*
Le Fouler, L., 246, *272*
Le Van, C., 65, *72*
Lee, G., 335, *345*
Lee, J. W., 335, 336, *343*
Lee, K., 60, *61*, 278, *278*, 383, 387, 392, *393*
Leith, C., 55, *61*
Leppä, A., 87, 88, *98*, *99*
Lettau, M., 309, *313*
Levin, A. T., 31, *43*, 262, *270*, *272*, 305, *313*
Levine, P., 54, *61*
Levington, P., 51, *61*
Levinson, M., 172, *186*
Li, B., 213, *239*, *240*
Li, H.-Y., 210, *238*
Li, S., 213, *239*, *240*
Liang, Y., 213, *238*, *240*
Lichtenberg, F. R., 335, *345*
Liebenberg, M., 24, *45*
Lindahl, M., 337, *345*
Linde, J., 97, *98*
Lipinski, C., 106, *120*
Liu, T. C., 19, 26, *45*
Liu, Y., 42, *43*
Lo, J. C., 210, *238*
Lobban, P. W. M., 50, *60*
Lodh, B. K., 41, *45*, *46*
Logeay, C., 108, *119*
London, A., 42, *45*
Longbottom, J. A., 60, *62*, 329, *345*
Longva, S., 94, *98*
Longworth, D., 35, *44*
Lopez-Salido, A., 117, *120*
Loue, J.-F., 69, *72*
Lovell, M. C., 351, *369*
Lucas Jr., R. E., 27, *45*, 324, *345*
Lucas Jr., R. J., 5, *7*
Ludvigson, S., 309, *313*
Łukinov, I., 177, *186*
Lunsing, H., 82, *84*
Luttimore, R., 308, *313*
Lybeck, J. A., 85, 96, 97, *98*, *99*
Lynn, R. A., 247, *273*, 307, *313*
Łyziak, T., 152, 154, *185*

M

Mac Garvie, M., 334, *345*
MacDonald, R., 390, 391, *393*
MacFarland, M., 355, 356, *369*
MacGregor, M. E., 42, *44*
Maciąg, W., 135, *183*
Maciejewski, W., 129, 130, 136, *182*, *185*, *186*, 356, *369*
Magtibay-Ramos, N., 208, *237*
Mahalanobis, P. C., 214, *238*
Mahmud, I. K., 228, *238*
Majsterek, M., 87, *99*, 172, *186*, 281, *283*, 383, 392, *393*
Mak, W., 30, *45*
Makarova, S., 179, *182*
Malgrange, P., 64, 69, *72*, *73*
Malinvaud, E., 290, *302*
Mankiw, N. G., 308, *312*
Manne, A. S., 214, *238*
Mannermaa, K., 87, *99*
Männistö, H.-L., 88, *99*
Marcellino, M., 200, *201*, 322, 329, *344*
Marconi, D., 335, *343*
Marjanek, K., 128, *188*
Markowski, K., 118, *121*, 132, 134, *185*, *189*
Marquez, J., 261, *271*
Marston, R., 191, *201*, 242, *273*
Martens, B., 102, *119*
Martihhina, A., 87, *99*
Marwah, K., 3, 4, *7*, 9, 13, *13*, 18, 41, *43*, *45*, 47, 50, *60*, *61*, 70, *72*, 77, *83*, 203, 213, 214, 225, *237–239*, 241, 242, 244, *271*, 285, *301*
Marwaha, S. S., 51, *62*
Marzouk, M. S., 235, *239*
Masera, F., 69, *72*
Masoleh, A. S. S., 229, *239*
Masson, P., 248, *271*, *272*, 309, *313*
Mathis, J., 64, *72*
Matsuura, S., 205, *239*
Matthews, K. G. P., 51, *62*
Maxwell, T., 34, 35, *43*, *45*
May, S., 34, *45*
Mayes, D., 54, *61*, 267, *270*
Mazier, J., 64, *72*
Mazur, K., 135, *183*
Mc Guire, M., 196, *201*
McAdam, P., 270, *272*

McCarthy, M. D., 22, *43*, *45*
McCoy, D., 109, *119*, 194, *200*
McKibben, W., 270, *272*
McRae, R. N., 42, *44*
Meen, G., 54, *61*, 267, *270*
Melliss, C. L., 50, *62*
Meltzer, A. H., 5, *7*, 27, *45*
Meredith, G., 248, *272*, 309, *313*
Merikull, J., 87, *99*
Mestre, R., 196, *201*, 329, *344*
Metha, F., 107, *119*, 297, *302*, 348, 349, *369*
Michalevskij, W., 123, 136, *186*
Michaudon, H., 70, *73*
Michel, G., 70, *72*
Mienszikow, S. M., 123, 124, *182*, *186*
Mikhaylenko, K., 173, *186*
Miller, G. W., 50, *60*
Mills, G., 5, *7*
Milo, W., 126, 128, 131, 132, 134, 137, *182–184*, *187*, *188*
Minassian, G., 148, *186*
Minford, A. P. L., 51, *62*, 270, *272*
Miszczyński, M., 126, 128, 131, 132, 134, 137, *182–184*, *187*, *188*
Modesto, L., 109, 116, *119*, *120*, 170, *182*, 194, *200*
Modigliani, F., 24, 25, *42*, 308, *312*, *313*
Moncoble, J.-E., 69, *73*
Montfort, B., 197, *200*, 329, *343*
Montiel, P. J., 203, *238*
Moral, E., 117, *119*
Morales, V., 391, *393*
Moran, K., 42, *44*
Morana, C., 200, *201*
Mori, K., 204, *239*
Moriguchi, C., 243, *272*
Morigushi, C., 205, *237*
Morin, P., 69, *73*
Morshuk, A., 206, 207, *240*
Morton, J. E., 389, *393*
Mot, E. S., 83, *84*
Movchauk, A., 244, *270*
Muellbauer, J., 308, *313*
Muet, P.-A., 65, 69, *72*, *73*
Mukminova, T. A., 178, *181*
Müller, S., 104, *120*
Mundell, R. A., 389, *393*
Murawski, K., 154, *182*, *183*
Murchison, S., 42, *45*
Musila, J. W., 231–233, *239*

N
Nagaya, S., 205, *239*
Najim, T. H., 228, *239*
Nakamura, S., 104, *120*
Naqui, S. N., 225, *239*
Narasimham, G. L. V., 24, *44*
Nasse, P., 69, *73*
Nataf, A., 64, *73*
National Institute for Economic Research, 155, *186*
Neal, D., 54, *61*
Neal, E., 267, *270*
Neck, R., 176, *184*, *190*
Nehru, V., 336, *345*
Nerlove, M., 3, *7*
Neves, P., 116, *120*
Nevin, P., 41, *45*, *46*
Nickell, S. J., 53, *61*, 298, *302*, 329, 340, *345*, 354, *369*
Nikolov, K., 59, *61*, 304, 310, *313*, 324, 327, 329, 330, *344*
Nilsson, C., 97, *98*, *99*
Nixon, J., 54, *60*, 172, *183*
Nobay, A. R., 308, *313*
Nordström, T., 96, *98*
Nowell, E., 270, *272*
Nyari, Z., 124, 127, *184*
Nymoen, R., 6, *7*, 95, *98*, 278, *278*, 356, *368*, 383, 392, *393*
Nynoen, R., 21, *43*

O
Officer, L. H., 34, 35, *44*, *45*
OECD, 245, *272*, 305, *313*
Ogwang, T., 225, 226, *239*
Okker, V. R., 76, 77, *83*, *84*
Okólski, M., 288, *302*
Olayide, S. O., 234, *239*
Olexa, M., 149, 174, 175, *184*, *186–188*
Olive, G., 64, *72*, *73*
Olofin, S., 234, *239*
Olsen, K., 95, *99*
Olsen, Ø., 85, 86, 88, 94, 96, *98*, *99*
Ondko, P., 176, *183*
Onishi, A., 266, *272*
Onofri, P., 109, *119*
Opara, W., 129, *186*
Oplotnik, Z., 150, *187*
Orłowski, W., 288, *302*
Orságová, J., 149, 174, *184*, *188*
Orságovă, M., 175, *186*
Oser, U., 107, *120*
Oshikoya, T. W., 234, *239*
Oteafy, R. A., 229, *239*
Ozaki, T., 207, *239*
Özatay, F., 118, *120*

P
Pace, L. F., 132, *188*
Padoa-Schioppa, T., 69, *72*
Pagan, A., 6, *7*
Pain, N., 150, 179, *181*
Pajestka, J., 123, *186*
Palčákova, H., 175, *184*
Paldam, M., 86, *98*
Pàlenik, V., 176, *181*, *183*, *186*
Pandit, V., 215, 225, *238*
Panek, E., 135, *183*
Pani, P. K., 215, *239*
Paries, M. D., 197, *200*, 329, *343*
Parkhomenko, V. W., 178, *186*
Parkin, M., 308, *313*
Parmenter, B. R., 211, *237*
Patterson, K., 51, 55, *61*, *62*
Pawilno-Pacewicz, J., 135, *183*
Pawłowski, Z., 124, 128, *181*, 376, *380*
Peaucelle, L., 69, *73*
Pedersen, K. N., 247, 248, *272*, *273*, 307, *313*
Peel, D. A., 54, *61*
Peetersen, A., 243, *272*
Peeterssen, A., 243, *272*
Perera, N., 225, *239*
Pesaran, M. H., 13, *13*, 60, *61*, 278, *278*, 308, *313*, 383, 387, 392, *393*
Pesin, F., 71, *71*
Petersen, C. E., 247, *273*, 307, *313*
Petit, P., 69, *73*
Phelps, E. S., 354, *369*
Philip, J.-M., 235, *238*
Phillips, A. W., 17, *45*, 340, *345*, 353, *369*
Piaszczyński, W., 129, *182*, *185*, 356, *369*
Pichelmann, K., 101, *119*
Plasmans, J., 65, 69, *72*
Plasmans, J. E. J., 191, *201*
Plich, M., 105, *121*, 153, 154, *187*
Plourde, A., 42, *44*
Polenske, K. R., 41, *43*
Poli, G., 267, *271*
Poloz, S., 35, *45*
Popkin, J., 24, *45*
Porcher, R., 69, *73*
Portes, R., 134, *187*
Post, J. J., 75, *84*
Powell, A. A., 211, *237*
Prasad, E., 248, *272*, 294, *302*, 325, *345*
Prescott, E. C., 6, *7*, 11, *13*, 282, *283*
Presin, F., 305, *312*
Preston, R. S., 22, 41, *45*, *46*
Price, S., 303, *313*
Prigent, C., 70, *73*
Proost, S., 194, *201*
Przystupa, J., 152, 154, *185*
Psaradakis, Z., 384, *393*
Pytlarczyk, E., 106, *120*

Q
Qin, D., 208, *237*
Quah, D., 354, *368*
Quandt, R. E., 134, *182*, *187*, 300, *302*
Quinn, M., 59, *61*, 304, 310, *313*, 324, 327, 329, 330, *344*
Quising, P., 208, *237*

R
Rabenanarjara, R., 69, *72*
Rae, D., 196, *201*, 246, *273*
Ralle, P., 69, *72*
Ramin, J. E., 41, *45*, *46*
Ramsay, G., 59, *61*, 304, 310, *313*, 324, 327, 329, 330, *344*
Ramsey, J. B., 21, *45*, 275, *278*
Rankaduwa, W., 225, 226, *239*
Rao, P. S., 41, *45*, *46*
Rao, U. L. G., 225, 226, 231, 232, *237*, *239*
Rasche, R. H., 24, 25, 32, *42*, *46*
Rasche, R. M., 278, *278*
Regal, A., 117, *119*
Reiffers, J. L., 65, *72*
Remka, H., 135, *183*
Renton, G. A., 50, *60*, *62*
Reyment, M., 95, *98*
Rhomberg, R. R., 34, *46*
Ricchi, O., 55, *62*
Richards, R. G., 334, *345*
Richardson, P., 245, 246, *271*, *273*
Riley, C. J., 50, *62*
Rimmer, R. J., 211, *237*
Riordan, E. J., 247, *273*, 307, *313*
Robidoux, B., 42, *46*
Robinson, B., 54, *61*
Robinson, G., 51, *62*
Rocco, P., 28, *43*
Roeger, W., 251, 261, *271*, *273*, 325, *344*, *345*
Rogers, J. H., 262, *272*, 305, *313*
Romański, J., 132, 134, *187*
Rose, D., 35, *43*, *45*
Rosted, J., 85, 88, 94, *98*
Rot, M., 154, *182*
Rubaszek, M., 154, *183*
Ruffing, K. G., 244, *271*
RWI, 105, *121*
Ryan, M., 196, *201*
Rybarczyk, B., 154, *182*
Ryding, J., 51, *62*

S

Sabanty, L., 153, *189*, *190*, 337, *346*
Sachs, J. D., 270, *272*
Saggi, K., 334, *345*
Sahn, D., 232, *240*
Saibaba, P., 215, *238*
Sailer, M., 103, *120*
Saillard, Y., 69, *73*
Saiyed, H. M., 41, *45*, *46*
Salman, A. K., 227, *239*
Salmon, P., 70, *72*
Salou, G., 245, 248, *271*
Sapocinski, T., 135, *181*
Sargan, J. D., 5, *7*
Sarrazin, H. T., 104, *121*
Sasinek, M., 174, 175, *186*
Satchi, M., 270, *270*
Sato, K., 204–206, *238*, *239*
Satunovskij, L. M., 136, *182*
Sault, J. E., 60, *61*
Sava, G., 171, *187*
Sawyer, J. A., 34, 41, *43*, *46*, 123, 124, *182*, *186*, 242–244, *271–273*
Schellhorn, M., 107, *121*
Scheufele, R., 107, *121*
Schmidt, P., 20, *46*
Schmidt, T., 194, *201*
Schmidt, T. F. N., 194, *201*
Schreyer, P., 332, *343*
Schweitzer, T. T., 41, *45*, *46*
Scobie, H. M., 229, *237*, 248, *272*
Scott, A., 59, *61*, 304, 310, *313*, 324, 327, 329, 330, *344*
Seibel, C., 63, *71*, *72*
Sekerka, B. L., 125, *186*
Semudram, M., 208, *239*
Sepp, U., 150, 179, *181*
Sgherri, S., 60, *61*
Shapiro, H. T., 5, *7*, 23, 24, 35, *44*, *46*, 124, *187*
Shen, L., 213, *239*
Shepherd, J. F., 50, *62*
Shim, S., 209, *239*
Shin, Y., 60, *61*, 278, *278*, 383, 387, 392, *393*
Shinkai, Y., 204, *238*
Shishido, S., 205–207, *239*, *240*, 244, 269, *270*, *273*
Simon, A., 128, *187*
Sims, C. A., 5, *7*, 27, *46*, 301, *302*
Sinai, A., 28, *44*
Sisak-Fekete, Z., 128, *188*
Skolka, J. V., 41, *43*
Skripnichenko, M., 177, *184*
Śledziński, W., 135, *183*
Smets, F., 197, *201*, 310, *313*, 329, *345*
Smidkova, K., 150, 179, *181*
Smidt, J., 86, *98*
Smith, K., 333, *345*
Smith, P., 54, *61*
Smith, R., 13, *13*
Smolny, W., 107, *120*
Snessens, H. R., 107, *119*, 297, *302*, 348, 349, *369*
Snyker, W., 246, *272*
Soekarni, M., 208, *240*
Sok, H., 65, *72*
Sola, M., 384, *393*
Soloviev, J. P., 123, *186*
Solow, R., 333, *345*
Sonbaty, A., 228, *237*
Sosvilla-Ribero, S., 109, 116, *119*
Sosvilla-Rivero, S., 170, *182*, 194, *200*
Soto, M., 336, *343*, *345*
Sousa, S. M., 383, 384, *393*
Spagnolo, F., 384, *393*
Spiegel, M., 336, *343*
Sprague, A., 51, *62*
Srba, E., 5, *7*, 306, *313*
Srinivasan, T. N., 211, *237*
Srzentic, N., 384, *393*
St. Amant, P., 42, *43*
Stäglin, R., 105, *121*
Stalder, P., 117, *121*
Stancu, S., 171, *187*
Staszewska-Bystrova, A., 301, *302*
Stein, J. L., 391, *393*
Steiner, V., 194, *201*
Stepanov, Y., 171, *187*
Stephan, S., 108, *119*
Sterdyniak, H., 70, *73*, 264, 265, *271*, 329, 342, *344*
Sterken, E., 83, *84*
Stevens, G., 261, *273*
Stewart, I. A., 35, *44*
Štiblar, F., 150, *187*
Stiroh, K. J., 332, 336, 337, *344*
Stokes, E., 42, *45*
Stokman, A. C. J., 192, *201*
Stolen, N. M., 94, *98*
Stone, J. R. N., 303, *313*
Straszak, A., 132, *188*
Štrauch, D., 126, *187*
Strotz, R. H., 320, *344*
Strout, A. M., 215, *237*
Strumelle, P., 102, *119*
Studeman, A. C. J., 306, *313*
Sturm, P., 245, 248, *271*
Suchecki, B., 48, *62*, 303, *313*

Suits, D. B., 18, *46*
Šujan, I., 124–127, 136, 149, 174, *184*, *185*, *187*, *188*
Šujanova, M., 149, 174, *184*, *188*
Summers, R., 352, *369*
Swanson, E., 336, *345*
Symansky, S. A., 248, *272*, 309, *313*
Szafrański, J., 154, *188*
Szakolczai, G., 128, *188*

T
Tabeau, A., 154, *188*
Tampubobn, E., 207, *238*
Tanaka, S., 205, *239*
Tanna, S. K., 59, *61*
Tanny, S. M., 41, *43*
Tarkka, J., 85, 88, 96, *99*
Tarnicka, M., 154, *182*
Tatemoto, M., 205, *240*
Tavlas, G. S., 231, *237*
Taylor, J. B., 27, *46*, 59, *62*, 294, 300, *302*, 387, *393*
Terlizzese, D., 109, 116, *120*, *121*
Tetlow, R., 35, *43*, *45*
Theiss, E., 124, 127, *184*
The Michigan Model of World Production, 245, *273*
Thomas, R., 59, *61*, 304, 310, *313*, 324, 327, 329, 330, *344*
Thonet, P., 64, *73*
Thorbecke, E., 214, *238*
Timofiejuk, I., 288, *302*
Tinbergen, J., 10, 11, *13*, 75, *84*
Tinsley, P., 27, *43*
Tjan, H. S., 76, *83*
Tkăč, M., 176, *184*
Tkravczenko, T. T., 178, *181*
Tobin, J., 325, 341, *343*, *345*, 362, *369*, 377, *380*
Tomashiro, K., 206, 207, *240*
Tomasini, S., 109, *119*
Tomaszewicz, Ł., 132, 133, *186*, *188*, 286, 288, *302*
Tomaszewski, W., 124, *183*
Tomczyk, P., 126, 135, *185*, *187*
Tong, W., 212, *240*
Treadway, A. B., 324, *345*
Trier, P., 86, *98*
Tryon, R. W., 31, *43*, 261, 262, *270–272*, 305, *313*
Tsurumi, H., 41, *46*
Tujula, M., 88, *99*
Turlelboom, B., 248, *272*, 294, *302*, 325, *345*
Turner, D., 196, *201*, 246, *272*, *273*
Turner, D. S., 59, 60, *61*, *62*, 329, *345*
Tyrell, N., 54, *61*, 267, *270*

U
Uchida, T., 205, *240*
Uebe, G., 3, *7*
Ueno, H., 204, *240*
Utzig-Lenarczyk, G., 154, *184*

V
Valavanis-Vail, S., 18, *46*
Vallet, D., 64, 65, *72*, *73*
Van den Beld, C. A., 76, *84*
Van den Berg, P. J. C. M., 77, *84*
van den Noord, P. J., 83, *84*
van der Baken, W., 102, *119*
van der Stelt-Scheele, D. D., 83, *84*
Van der Wiel, H., 332, *345*
van Els, P. J. A., 192, *201*, 306, *313*
Van Frausum, Y., 232, *240*
van Ierland, E. C., 83, *84*
Van Leeuwen, G., 332, *345*
van Pottelsberghe de la Potterie, B., 335
Van Pottelsberghe de la Potterie, B., 335
van Pottelsberghe de la Potterie, B., *343*, *345*
Van Pottelsberghe de la Potterie, B., *345*
Van Regemorter, D., 194, *201*
van Welsan, D., 306, *312*
van Welsum, D., 242, 263, *270*
Vandome, P., 48, *61*
Vàvra, D., 149, *181*, *188*
Ventura, J., 247, *273*
Verbruggen, J., 278, *278*
Verdoorn, P. J., 75, *84*
Verlinden, J., 102, *119*
Vidal de la Blache, O., 69, *73*
Vilares, M. J., 70, *73*
Villa, P., 70, *73*
Villani, M., 97, *98*
Vincenzi, L., 109, *121*, 267, *271*
Viren, M., 154, *188*
Visco, I., 109, 116, *120*, *121*
Visholm, T., 85, 96, *99*
Voisin, P., 65, *72*
Voleu, A., 170, *182*
Volle, M., 69, *72*
von Haberler, G., 10, *13*
von Ungern-Sternberg, T., 305, *313*
Voorhoeve, W., 83, *84*
Vromans, M., 77, *84*
Vukotič, V., 150, *187*

W
Waelbroeck, J., 22, 41, *46*, 191, *201*, 242, 243, *273*

Wago, H., 205, *239*
Walker, J., 54, *61*, 267, *270*
Wallis, K. F., 3, *7*, 10, *13*, 59, 60, *61*, *62*, 196, *201*, 278, *278*, 329, 340, *345*, 354, *369*
Walters, A. A., 375, *380*
Wang, H., 213, *239*, *240*
Wang, J., 335, *346*
Warburton, P., 51, *61*
Waslander, H. E. L., 34, 41, *43*, *46*
Watanabe, T., 205, *240*
Waud, R. N., 20, *46*
Wdowiński, P., 87, *99*, 108, *121*, 151, 153, 171, 174–177, *183*, *184*, *186*, *187*, 391, *393*
Welfe, A., 4–6, *7*, 101, *120*, 134, 148, 151–153, 170, *183*, *185*, *188–190*, 230, *238*, 243, 244, *272*, 281, *283*, 286, 288, 291, *302*, 308, *313*, *314*, 316, 317, 321, 341, 343, *344–346*, 348, 349, 357, 361, 364–367, *369*, *370*, 371–373, 378–380, *380*, *381*, 383, 385, 391, 392, *393*
Welfe, W., 4, 5, *7*, 49, *62*, 70, *72*, 87, *99*, 101, 105–108, 118, *118–121*, 124, 126, 129–135, 146–153, 155, 170–172, 174–180, *181–190*, 227–230, 234, *237–240*, 243, 244, 267, *271*, *272*, 281, *283*, 286, 288, 291, 297, 300, *302*, 304, 308, 312, *313*, *314*, 315–317, 321, 324, 325, 327, 331–335, 337, 338, 341, 343, *344*, *346*, 348–353, 356–358, 361, 364–368, *369*, *370*, 371, 373, 378, 379, *380*, *381*, 383, 385, *393*
Westphal, U., 105, *119*
Weyerstrass, K., 176, *184*, *190*
Whalley, J., 211, *237*
Whelan, K., 109, *119*
Whillans, J. D., 41, *45*, *46*
Whitaker, J. K., 5, *7*
Whitley, J., 3, *7*, 9, *13*, 47, 48, 50, 51, *62*, 242, 263, 269, *273*
Whitley, J. D., 59, 60, *61*, *62*, 85, *99*, 288, 291, 298, *302*, 329, *345*, 354, 357, 358, 360, 361, *370*, 391, 392, *394*
Wickens, M., 308, *313*
Wieland, V., 197, *201*
Williams, J. C., 31, *43*, 262, *270*
Willis, M., 41, *45*, *46*
Willman, A., 85, 88, 96, *99*, 117, *121*, 196, *201*
Winder, J. W. L., 41, *43*, *46*
Winker, P., 107, *119–121*, 194, *201*
Winter, D., 134, *187*
Woelfling, M., 127, *190*
Wold, H., 21, *46*
Wolters, J., 4, *7*, 103, 105, 107, *120*
Wong, B.-S., 42, *46*
Wouters, R., 197, *201*, 310, *313*, 329, *345*
Wren-Lewis, S., 52, 55, *61*, *62*, 391, *394*
Wright, J., 109, *119*
Wróbel, E., 152, 154, *185*
Wu, J., 213, *240*
Wuger, M., 102, *120*
Wulfsberg, F., 95, *99*

X

Xu, B., 335, *346*

Y

Yaari, M. E., 35, *46*, 308, *314*
Yagci, F., 118, *121*
Yap, J. T., 208, *240*
Yaxiong, Z., 213, *238*
Yeo, S., 5, *7*, 134, *187*, 306, *313*
Yohn, F., 69, *73*
Youcai, L., 213, *238*
Yu, Q.-L., 212, *240*
Yu, T., 210, *238*, *240*

Z

Zachłod-Jelec, M., 309, *314*
Zagame, P., 69, *73*
Zaidman, C., 69, *73*
Zajchowski, J., 129, 130, *185*, *186*
Zaleski, J., 155, 179, *182*
Zatoń, W., 228, 229, *237*, *239*
Zayid, E., 235, *240*
Zhang, Y., 213, *240*
Zhu, B., 213, *240*
Zienkowski, L., 286, *302*
Żółkiewski, Z., 154, *186*, *188*, 288, *302*
Żółtowska, E., 135, *189*
Zonzilos, N. G., 108, *121*
Zwiener, R., 108, *119*

Subject Index

A
Accelerator, 11, 12, 15, 17, 29, 75, 129, 131, 132, 136, 141, 170, 175, 195, 196, 208, 210, 226, 246, 248, 265, 267, 268, 292–294, 321, 364, 365, 368, 380
Accounting balances, 387, 390
ADAM model, 85, 86, 89
Adaptive expectations, 5, 33, 53, 54, 57, 81, 89, 90, 96, 276, 294, 317, 341
Adjustments, 31, 35, 41, 51, 55, 58, 64, 65, 77, 97, 102, 103, 107, 117, 123, 128, 131, 134, 146, 149, 153, 154, 176, 177, 181, 192, 194, 197, 198, 227, 232, 249–251, 255, 258, 259, 262, 263, 270, 276, 277, 282, 285, 286, 289, 290, 292, 294, 296, 300, 301, 316, 320, 322, 326, 327, 330, 332, 338, 347, 351, 357, 387, 396
  costs, 103, 110, 249, 277
  short-run adjustments, 97, 153, 294, 301, 396
Aggregates, 12, 372
Asian Link System, 207
Assets, 16, 17, 20, 35, 75, 76, 173, 192, 246, 263, 268, 270, 286, 287, 296, 303, 309, 374, 383–386, 390–392
ATLAS model, 269
Augmented Phillips curve, 341
Autoregressive, 234
AWM model, 196, 198, 329

B
Balance
  equations, 332, 335, 374
  of payments, 25, 152, 173, 195, 197, 212, 225, 251, 266, 283, 348, 349, 356, 383, 387, 388, 390, 391
Bank
  of Greece models, 108
  of Italy models, ix, 109
  of England models, 53, 55
Bank-Gem, 247, 306
Banking loans, 24, 25, 249
Bargaining models, 53, 86, 96, 107, 192, 197, 251, 263, 270, 340
BEA model, 19, 21, 23
BEA/CEF model, 163, 172
Behaviour relations, 55, 251, 291
BOF models, 88, 90, 92, 93, 97, 310
BOJ model, 216
Bond finance, 20, 23, 54, 96, 171, 373, 374
Bonn models, 104
Bottleneck multiplier, 368
Bottom-up, 198
Brookings model, 4, 18–21, 23, 49
Budget
  deficit (government deficit), 155, 385
  multiplier, 293, 380
Business cycle, 6, 9–11, 19, 24, 75, 104, 105, 228, 277, 282, 322, 331, 351, 354, 363

C
Calibration, 59, 79, 80, 110, 157, 160–162, 223, 255, 256, 258, 277
Cambridge Growth Project, 48
Candide Model, 41
Capacity
  output, 17, 19, 22, 25, 28, 29, 31, 52, 63, 76, 82, 86, 94, 96, 97, 104–106, 192, 209, 211, 214, 246, 251, 263,

W. Welfe, *Macroeconometric Models*,
Advanced Studies in Theoretical and Applied Econometrics 47,
DOI 10.1007/978-3-642-34468-8, © Springer-Verlag Berlin Heidelberg 2013

Capacity (*cont.*)
265, 267, 268, 296, 331, 333, 351, 352, 359, 366
utilization rate, 52, 82, 209, 211, 246, 249, 263, 282, 298, 324, 327, 328, 331–333, 338, 339, 352, 354, 360
Capital
productivity, 150, 318, 331
stock, 320
Capital accumulation, 64, 249
Capital-labour ratio, 76, 104
Capital-output ratio, 318, 319, 327
CEM models, 125, 138, 139
CEMI-RUN model, 172
CES production function, 83, 93, 103, 107, 150, 194, 211, 251
CIEMEX model, 224, 236
CMEA models, 125, 126, 132, 137, 146
CMEM model, 212, 218
Cobb-Douglas production function, 22, 23, 25, 29, 42, 52, 55, 94, 103, 117, 153, 170, 173, 174, 179, 195–197, 200, 209, 214, 217, 231, 232, 249, 251, 297, 318, 327, 331, 333, 334, 348
Cointegration, 6, 21, 27, 29, 54, 57, 60, 95, 199, 200, 211, 232, 270, 275, 278, 281, 283, 295, 301, 368, 383, 384, 391, 392, 395, 396
cointegration
space, 281
vector, 301
Combined forecast, 86, 93, 103, 128, 206, 226, 231, 245
COMLINK model, 244
Commodity
models, 12, 29, 32, 49, 54, 59, 64, 86, 94, 104, 107, 117, 127, 130, 131, 133, 135, 156, 158, 159, 164, 165, 175, 214, 243–246, 252, 256–258, 264, 268, 285, 286, 288, 290, 296, 297, 303, 304, 315, 338, 347, 356, 360–362
prices, 215, 361
Communication lines, 332, 335
COMPACT model, 55, 60, 191, 193–195, 278
Compensation, *see* wages
Competitiveness, 171, 265, 315, 356, 357, 359
Complete models, 95, 133, 304, 365, 367
Computable general equilibrium (CGE) models, 6, 10, 31, 35, 42, 48, 152, 154, 160, 170, 194, 211, 245, 253, 270, 277, 282, 288, 358
Computerization, 332
Consistent, 26, 27, 38, 40, 41, 53, 55, 87, 129, 153, 155, 242, 294, 308
Constant prices, 19, 291, 292, 315, 316, 318, 347, 361, 384, 385
Consumers' demand function, 5, 11, 12, 15–17, 19, 20, 22–24, 27–29, 31, 32, 48, 49, 52, 53, 55, 71, 75, 82, 86–88, 93, 94, 96, 104, 106, 109, 117, 124–126, 130, 134, 147, 150, 153, 155, 172, 174–177, 180, 195–197, 208, 210, 211, 213, 226, 228, 229, 231, 232, 247, 263, 264, 266, 267, 276, 281, 282, 296–298, 303–309, 350, 361, 364, 365, 368, 374, 380
Keynesian, 16, 22, 24, 29, 71, 86, 87, 96, 197, 303
long-run consumers' demand, 24, 31, 197, 264, 304, 309, 350
short-run consumers' demand function, 24, 29, 31, 69, 88, 196, 208, 209, 211, 232, 248, 264, 306, 350, 360
with expected long life income, 55, 296
with permanent income, 29, 55, 88, 117, 276, 296, 308
Consumption, 4, 5, 12, 15, 16, 20, 24, 28, 31, 34–36, 41, 48, 54–57, 64, 75, 97, 103, 104, 125–127, 130, 131, 134, 136, 148, 159, 170, 174–176, 178, 195, 198, 209, 211, 213, 215, 226, 227, 232–234, 245, 248–251, 262–264, 266, 268, 276, 277, 286, 287, 291, 292, 294, 295, 299, 303–311, 315, 347, 349, 350, 354, 355, 361–364, 371, 386, 396
government consumption, *see* public institution demand, 213, 350
intermediate, 226, 277, 349, 361
multiplier, 15, 131, 170, 175, 305, 364
private consumption, *see* consumers demand, 226, 232, 303, 349, 361–363
Convergence, 130, 149, 196, 233, 242, 278, 383, 386
Cooperation, 39, 106, 118, 124, 127, 129, 149, 150, 154, 244, 264, 396
tax, 371
Costs, 11, 19, 20, 22, 24, 27–29, 31, 33, 41, 52, 59, 65, 71, 76, 82, 86–88, 93, 94, 96, 97, 103, 104, 109, 110, 116, 149, 150, 153, 173, 192, 195–197, 206, 209, 210, 212, 215, 226, 232, 233, 246, 248–251, 259, 263, 265, 267, 270, 277, 282, 286, 289, 293, 295, 296, 299, 322–327, 329, 334,

337–341, 355, 357, 366, 368, 373–377, 384, 388, 392
cost minimizing, 282
function, 20, 384
labour costs, 20, 27, 41, 52, 71, 82, 93, 96, 104, 149, 150, 173, 195, 197, 206, 210, 212, 232, 233, 246, 248, 251, 263, 265, 267, 293, 299, 329, 334, 337–339, 355, 357, 366, 368, 375, 376, 392
minimizing, 248, 282
of adjustment, 277
of capital, 88, 195
production costs, 93, 109, 338, 340, 368, 375, 376
total, 376, 384
unit costs of fabrication, 20, 22, 24, 28, 33, 52, 65, 82, 86, 88, 93, 94, 96, 97, 104, 116, 153, 192, 212, 215, 226, 232, 233, 246, 248, 251, 263, 265, 267, 270, 289, 293, 329, 337, 338, 357, 376
Counterfactual, 23
CREA model, 115, 117
Credits, *see* loans
Crowling pag, 51, 299, 358, 364, 391, 392
CUBS model, 51
Current account, *see* balance of payments
Current prices, 175, 341, 375
Customs, 377, 378

D

Data, x, 10, 12, 15, 16, 18–20, 22, 24–26, 30, 34, 35, 39, 41, 42, 49, 50, 56, 75, 76, 82, 83, 86, 87, 94, 104, 107, 108, 118, 124, 125, 129–135, 147, 149, 151, 152, 154, 170, 172–174, 179, 180, 192, 194, 195, 197, 208, 211, 212, 214, 215, 225–229, 231–233, 235, 247, 265, 268, 269, 288, 294, 310, 311, 326, 331, 332, 334, 335, 337, 361, 372–374, 376, 389, 396
cross-section, 334, 335, 376, 396
time series data (sample), 334, 335
panel, 49
Decomposition, 11, 23, 36–40, 56–58, 66–68, 78–82, 89–92, 110–115, 138–146, 152, 156–169, 193, 198, 199, 216–224, 252–260, 264, 286, 287, 304, 334, 362, 372, 379, 384
of forecast, 27, 50, 56, 108, 148, 242, 244, 269
of models, 11, 23, 36–38, 40, 58, 67, 68, 78–82, 92, 110, 111, 113–116, 138–146, 152, 153, 156–160, 163, 164, 166–170, 193, 198, 199, 216–219, 221, 223, 224, 252–256, 258, 259, 264, 287, 288, 303, 304, 334, 371, 372, 384
of multipliers, 36–38, 40
Deflators, 11, 17, 18, 20, 28, 32, 33, 93, 148, 152, 155, 209, 234, 249, 265, 266, 269, 299, 304, 362–364, 371, 391
see prices
DEM models, 126, 140
gross domestic product (GDP) deflators
investment deflators, 28, 32, 148, 155, 209, 234, 249, 266, 269, 299, 362
investment deflator, 20, 384
Demand, x, xi, 4, 5, 10–12, 15–17, 19, 20, 22–33, 47–49, 52–55, 63–65, 68–71, 75–77, 79, 81–83, 85–88, 91–94, 96, 97, 101–104, 106–110, 112–114, 116–118, 124–137, 140–143, 145, 147, 149–156, 158–168, 170–181, 192–197, 199, 200, 206, 208–218, 220, 221, 223–229, 231–236, 241, 243, 245–248, 251–253, 256–270, 275–277, 281–283, 286–293, 296–300, 303–310, 315–321, 323–332, 337–339, 342, 343, 347–354, 356–358, 360, 361, 364, 365, 368, 371, 374, 379, 380, 383–388
aggregate, 94, 106
constraints, 97, 326, 349, 357
consumption demand (demand for consumer goods, households demand), 4, 5, 12, 16, 20, 24, 27, 28, 31, 47, 48, 54, 55, 75, 88, 97, 103, 104, 106, 124, 125, 128, 131, 134, 152, 170, 174–177, 209, 210, 215, 226, 227, 232, 234, 248, 251, 262–264, 266, 268, 276, 277, 282, 286, 303–308, 347, 349, 350, 356, 361, 364, 365, 368, 374, 379, 385, 386
determined system, 19, 20, 22, 24, 26, 28, 29, 31, 33, 49, 54, 55, 63, 65, 69–71, 75–77, 82, 83, 85–88, 92–94, 96, 97, 103, 104, 106–109, 112–114, 116–118, 124, 126, 127, 129–135, 137, 140, 142, 143, 145, 147, 149, 150, 152, 153, 155, 156, 158, 160–168, 170–176, 180, 181, 192–197, 200, 206, 208–218, 220,

Demand (*cont.*)
221, 223–229, 231–234, 236, 241, 243, 245, 246, 248, 251–253, 257–259, 261–270, 277, 281, 282, 289, 292, 293, 297–299, 315–319, 321, 330, 331, 342, 348–354, 356, 358, 361, 364, 365, 368, 379, 380, 386, 387
effective, 31, 33, 54, 65, 107, 118, 134, 172, 266, 326, 330, 338, 358
excess, 5, 12, 26, 33, 48, 118, 125, 126, 128, 131, 133, 134, 289, 290, 297, 307, 308, 342, 343, 352, 353, 368, 380
final, 4, 11, 12, 16, 17, 20, 22–24, 28–30, 33, 55, 64, 71, 86–88, 93, 96, 97, 101, 102, 104, 106–109, 116, 117, 131, 133, 134, 149, 150, 152, 153, 155, 163, 167, 170–176, 178, 179, 192, 194, 196, 206, 208, 209, 211, 213–215, 226, 228, 232, 234, 247, 262, 266, 267, 277, 282, 283, 286, 288, 296, 297, 315, 316, 330, 348–351, 371
final domestic, 171, 172, 175, 209
for exported commodities function, 130, 357
for imports function, 291
for increase of inventory, 22, 33, 213, 299, 315, 316, 350, 351
for money, 25, 33, 171, 179, 208–210, 212, 213, 215, 229, 231–233, 246, 263, 266, 267, 291, 293, 299, 386, 387
function, 15, 20, 88, 104, 176, 208, 213–215, 235, 246
gap, x, 33, 163, 173, 197, 212, 307, 308, 338, 339
investment, 4, 11, 12, 17, 19, 20, 24, 27–29, 31, 32, 49, 52, 71, 86–88, 93, 103, 104, 109, 117, 118, 133, 134, 150, 152, 153, 170, 172–178, 180, 192, 197, 208–210, 215, 228, 229, 232, 248, 251, 263–268, 281, 282, 292, 297, 299, 303, 310, 321, 323–326, 350
Deposits of households, 75, 385
Depreciation, 17, 31–33, 96, 196, 292, 310, 320, 322, 324–326, 332, 334, 335, 338, 375, 376, 384, 389
knowledge capital, 153, 297, 333–335, 337
DESMOS model, 191, 193
Deterministic trend, 268
Devaluation, 172, 195, 214
Disaggregation of models, 4, 6, 12, 20, 28, 29, 36, 41, 49, 65, 67, 69, 102, 159, 164, 180, 218, 223, 252, 264, 275, 276, 287, 356
by branches, 23, 65, 66
regional disaggregation of models, 180
Discouraged worker, 33, 311
Disequilibrium models, 5, 70, 106, 124–126, 128, 134, 147, 289, 297, 300, 348, 350
Disposable income, 16, 20, 24, 31, 55, 59, 87, 88, 150, 153, 172, 173, 175, 195–197, 209, 210, 226, 228, 229, 231–234, 246, 248, 249, 263, 264, 266, 267, 293, 303–305, 307–312, 347, 373, 386
Distribution, 9, 11, 20, 32, 118, 127, 132, 133, 137, 155, 178, 249, 286, 292, 304, 332, 341, 371, 377
asymptotic (limit) distribution, 130
Disturbance term, 297, 318, 322
DIW models, 105, 199, 250
DMS model, 64, 65, 69, 192
Dobrescu models, 155
Domestic final demand, 149, 155, 266, 267, 350
DRI model, 21, 22, 27–29, 39–42, 269, 334
Dummy variables, 232
DUTCH models, 50, 75, 78, 82, 83
Dynamic, 6, 10, 16, 18, 22, 30, 31, 33, 35, 37, 39, 49, 58, 59, 65–67, 77, 97, 102, 103, 117, 125, 127, 129–133, 135, 149, 154, 158, 160, 194, 211, 212, 243, 249, 250, 255, 256, 258, 270, 277, 282, 288, 289, 300, 305, 306, 317, 325, 338, 349, 395
adjustments, 31, 102, 103, 117, 154, 194, 249, 258, 270, 300, 338
forecasts, 30
multipliers, 243
simulations, 16, 18, 35, 37, 135, 154
Dynamic stochastic general equilibrium (DSGE) model, 97, 149

E

Economic fluctuations, 15
Economic growth, 9, 31, 51, 116, 132, 153, 180, 194, 215, 219, 225, 228, 230, 235, 333, 387
theories, 9, 31
Economic regimes, 147, 288, 290, 297, 348
Economy, ix, x, 4, 6, 9, 10, 12, 15, 16, 18, 21–23, 25–28, 30–35, 38, 39, 41, 42, 47–52, 55, 60, 65, 66, 69–71, 75,

78, 82, 83, 85, 87, 88, 90, 92–95, 97, 101–103, 105, 107–109, 116–118, 123–128, 130, 132–137, 147–155, 160–162, 169–174, 176–180, 191–195, 200, 204–216, 225–236, 241–244, 250, 252, 261–270, 276–278, 281, 285–288, 291, 294, 297, 304, 306, 308, 331–335, 338, 342, 343, 351, 352, 356, 358, 362, 364, 385, 386, 392, 395, 396
centrally planned, 47, 103, 123, 129, 212, 243, 286, 297, 343
knowledge-based economy, 151, 276
market economy, 47, 133, 134, 147, 170, 172, 173, 178, 180, 213, 308, 342
new economy, 33, 277
transformed to market economy, 147
Education, 79, 230, 324, 336
costs, 324
effective demand, 79
EITF model, 244, 245, 253
ELSA models, 206, 207
Empirical growth models, 3, 31, 85, 96, 116, 153, 396
Mankiw-Romer-Weil models, 265
models *a la* Barro, 297, 336, 348, 368
Employment, x, 4, 5, 9, 12, 15, 19, 22–26, 29, 31, 33, 49, 52, 65, 69, 71, 76, 82, 83, 85–88, 93, 96, 97, 102, 103, 105–109, 116, 117, 125–127, 130, 133, 149, 150, 152, 153, 155, 170, 172, 174–178, 192, 194, 195, 197, 206, 208–211, 214, 215, 225, 227–229, 231–234, 245, 246, 251, 262, 263, 265–268, 282, 286, 287, 290–293, 296, 297, 299, 316–321, 326–329, 331, 332, 334, 337, 343, 347, 348, 352–354, 373, 375, 379, 388
EMSE model, 165, 174, 175
EMU model, 179, 192, 193, 195, 196, 200
Endogenous, 20, 29, 33, 48, 51, 60, 96, 131, 155, 194, 222, 268, 291, 297, 300, 377
Energy, 5, 23–25, 28, 29, 39, 40, 42, 49, 56, 69, 94, 155, 161, 170, 173, 179, 194, 246, 251, 262, 266, 290, 317, 331, 338, 348, 349, 368, 375, 376
Environmental, 154, 396
EPA models, 204, 205, 216, 269
Equilibrium, 6, 9, 10, 12, 27, 29, 59, 70, 77, 79, 80, 82, 86, 97, 149, 154, 170, 172, 195, 211, 245, 249, 250, 269, 276, 277, 282, 288, 289, 294, 297, 323, 349, 351, 354, 356, 362, 384, 387, 389–392, 395
long-run, 27, 77, 86, 97, 294, 387, 391
steady state, 59, 294
Equities, 373
Error correction mechanism (ECM), 5, 6, 27, 29, 31, 35, 55, 56, 68, 77, 79, 81, 83, 87–89, 91, 93, 96, 103, 107, 108, 110, 113, 115, 117, 148–150, 152, 157–160, 163, 165, 166, 175–177, 180, 192–194, 197–200, 208, 211, 217, 218, 222, 232, 246, 247, 254, 258, 259, 261, 276, 277, 295, 301, 305, 306, 317, 323, 327, 328, 384
ESRC model, 59
Estimation
methods for nonstationary time series, 301
two-step Engle-Granger estimation method, 211, 396
of equation parameters, 15, 16, 49, 131, 261, 282, 300, 391
equation parameter estimation based on error correction models (ECM), 5, 6, 29, 31, 35, 56, 68, 77, 79, 81, 83, 87–89, 91, 93, 96, 103, 107, 108, 110, 113, 115, 117, 148–150, 152, 157–160, 163, 165, 166, 175–177, 180, 192–194, 197–200, 208, 211, 217, 218, 222, 232, 246, 247, 254, 258, 259, 261, 276, 277, 295, 301, 305, 306, 317, 323, 327, 328, 384
ordinary least squares (OLS) estimation, 4, 6, 11, 15, 18, 19, 24–26, 34–39, 41, 55, 56, 66, 67, 78–80, 86, 88–94, 104, 106, 107, 110, 111, 113, 127, 131, 133, 138–146, 155–162, 164–168, 175, 177, 193, 207, 211, 214–224, 226, 228, 231, 233, 252, 253, 256–259, 264, 266, 300, 301
system parameter estimation, 300
two stage least squares (TSLS), 4, 15, 18, 19, 25, 32, 34, 36, 37, 39, 42, 78, 80, 83, 89, 92, 107, 113, 127, 131, 138, 141–144, 157–159, 162, 167, 171, 214, 215, 219, 260, 268, 300, 323, 339
ESTMOD model, 87, 89
EUROMON model, 191–193, 306
European Commission models, 69, 250
European evaluation, 4, 101, 102, 116, 123, 135, 154, 178–180, 191–193, 195, 267
Ex-ante prediction, 392

Ex-post forecasting, 118, 128, 233
Excess demand, 5, 26, 118, 125, 128, 131, 133, 134, 289, 290, 297, 307, 308, 342, 343, 353, 368, 380
Exchange rate, 25, 37, 51, 109, 148, 149, 173, 197, 212, 233, 243, 248, 249, 261, 283, 298, 299, 338, 355, 358, 364, 368, 383, 388–392
balance of payments, 25, 173, 212, 283, 383, 388, 390, 391
equilibrium model, 6, 10, 27, 59, 70, 82, 86, 97, 149, 154, 170, 172, 211, 245, 249, 250, 269, 289, 391
models of, 148, 150, 174, 391
monetary models, 389
PPP and UIP models, 298, 388–391
Exogeneity, 396
Expectations, xi, 5, 11, 19, 22, 26–29, 31–33, 38, 40, 41, 47, 51, 53–59, 68, 69, 71, 81, 87–93, 96, 102, 103, 105, 106, 110, 154, 173, 194, 195, 197, 198, 206, 232, 242, 246, 248, 249, 255, 256, 258–263, 268–270, 276, 277, 293, 294, 296, 300, 301, 306, 317, 321, 322, 326, 329, 330, 339–341, 366, 390, 396
adaptative, 256
inflationary, 92, 154, 197, 198, 206, 232, 246, 293, 366
learning, 54, 56, 259, 270, 294, 396
rational, 5, 27, 31, 32, 47, 51, 53–59, 88, 90, 93, 102, 103, 106, 110, 197, 198, 242, 248, 249, 255, 256, 258, 259, 261–263, 268–270, 276, 293, 294, 300, 306, 341, 396
Exports, 15, 18, 65, 82, 86, 93, 96, 97, 125, 126, 128, 130–132, 148, 149, 152, 155, 172–175, 177, 192, 208–211, 213, 214, 217, 220, 221, 223, 225–229, 231–235, 241, 243, 247, 249, 252, 263, 265–267, 269, 282, 289, 291, 292, 296–299, 315, 347, 348, 356–359, 364, 368, 388, 390
demand, 65, 93, 128, 130, 131, 149, 152, 155, 172, 173, 175, 192, 208–210, 214, 220, 221, 223, 226–229, 232, 247, 252, 263, 266, 267, 282, 291, 298, 357
functions, 86, 128, 130, 131, 220, 297, 356, 358
prices, 82, 86, 97, 149, 192, 208–211, 232, 233, 241, 243, 263, 267, 269, 291, 292, 299, 358, 364
supply, 130, 132, 148, 177, 208, 211, 232, 297, 356, 358, 359
External shocks, 4, 11, 18, 23, 24, 33, 42, 51, 64, 69, 149, 196, 200, 208, 244, 246, 360, 364
Extrapolation, 195, 196

F

Fair Mc model, 32
Fair model, 25, 27, 287, 312
Federal Reserve Board (FRB) models, 25, 27, 31, 258, 261, 262, 305–307, 309, 338, 339
Feedback variables, 18, 173, 338
FEER exchange rate, 391
FIFI model, 63, 65, 66
Final capital goods, 11, 76, 109, 125, 126, 128, 226, 227, 233–235, 243, 277, 282, 310, 316, 320, 322, 335, 350, 356, 357, 361–363, 366, 368
Financial flows, 4, 28, 29, 31, 32, 64, 69, 76, 77, 82, 83, 86, 88, 93, 94, 96, 103, 104, 106, 116–118, 124, 126, 133, 134, 148–150, 152, 154, 170, 171, 174, 178–180, 197, 204–206, 209, 212, 213, 225, 227, 231, 245–247, 251, 262, 263, 265, 276, 281, 282, 285–287, 293, 298, 303, 362, 365, 371, 374, 375, 379, 385
Financial sector, 5, 11, 21, 23, 24, 26, 29, 32, 34, 64, 76, 82, 86–88, 93, 104–106, 117, 133, 167, 176, 192, 193, 195, 209, 228, 229, 231, 234, 247, 287, 305
First differences, 18, 76, 253, 301, 306, 322, 336, 337
Fiscal policy, 11, 22, 24, 42, 103, 148, 195, 200, 210, 261, 277, 322, 377
Fixed capital, *see* fixed assets
fixed capital-labour ratio, 76, 104
total fixed capital, 332
Flexible accelerator, 75, 246, 292, 321
Flow of funds model, 23
FOCUS model, 41
Forecasts, 5, 6, 18, 19, 22, 24–27, 30, 32, 36–42, 48–51, 54, 56–59, 65–68, 70, 78–81, 83, 89–93, 96, 102, 103, 106, 109–116, 118, 124, 126, 127, 132, 134, 137–143, 145–149, 151–154, 156–165, 167, 168, 170–172, 174, 176–178, 181, 191, 193, 196–200, 205, 208, 209, 212, 216–218, 220–225, 232, 234–236, 242–247, 252–255, 257–260,

262–265, 281, 294, 311, 329, 333, 341, 351, 395
Foreign direct investment (FDI), 150, 169, 170, 180, 246, 335, 360, 361
Foreign speculative investment, 51
Foreign trade, 19, 28, 55, 63, 64, 85, 87, 93, 97, 101, 104, 107, 117, 118, 126, 127, 130–132, 135, 137, 149, 156, 164, 165, 174, 175, 177, 179, 180, 192, 208, 210, 211, 213–215, 225–227, 229, 231, 233–235, 242, 243, 245, 249, 264, 282, 286, 287, 292, 299, 347, 352, 356, 368, 388–391
  multiplier, 368
Forward consistent, *see* rational
FRB MCM model, 261
FRB/US model, 25, 27, 31, 309, 339
FRB/WORLD model, 261
FREIA model, 76–79
French models, 65, 69–71
FUGI model, 241, 257, 265, 266
Fully
  demand determined model, 234
  supply determined model, 126, 127, 132, 155, 225, 233
Functioning of the economy, 3, 85, 123, 130, 181, 297, 362, 395

G
GAMA models, 64, 65, 67, 70, 270
Games, 6, 9, 10, 97, 149, 154, 170, 172, 195, 211, 245, 250, 269, 277, 282, 288, 289, 297, 349, 351, 354, 356, 362, 384, 387, 389–392, 395
Gauss-Seidel algorithm, 21, 300
GDP deflator, 173, 175, 197, 209, 213, 226, 231–234, 249, 268, 322
GEM models, 194, 242, 247, 250, 254, 256, 258, 262, 263, 265
Germany models, 4, 101, 103–105, 107, 111, 112
GLOBAL INSIGHT model, 29
Government expenditures, 178, 350, 365, 372, 377, 379, 380, 386
  current expenditures, 20, 208, 251, 286, 334, 335, 350, 365, 379, 386
  expenditure multiplier, 365
  investment expenditures, 24, 29, 51, 233, 234, 310, 311, 315, 321, 325, 332, 347, 361, 363, 379, 380
Gross domestic product (GDP), 12, 28–30, 32, 59, 87, 95, 102, 148–150, 153, 155, 162, 164, 170, 172–179, 195–197, 199, 200, 208–210, 212, 213, 226, 228, 229, 231–235, 246, 247, 249, 254, 266–269, 286, 290, 291, 293, 299, 315, 322–325, 328, 330, 331, 334, 338, 339, 347–349, 352, 354, 357–362, 383, 388–390, 392
Gross domestic product per capita, 24, 127, 178, 264, 268, 269, 304
Growth
  model, 153, 205
    Keynesian growth models, 16, 22, 24, 28–30, 41, 42, 52, 53, 63–65, 71, 83, 85–88, 96, 97, 108, 116, 154, 192, 205, 206, 262, 281, 292, 303, 389, 395
  models, 9, 17, 24, 29, 30, 48, 56, 80, 116, 153, 193, 205, 231, 297, 396
    endogenous growth models, 29, 297
    neoclassical growth models, 258
  rates, 175, 178, 269, 360, 389

H
Her Majestýs Treasury (HMT) models, 52–54, 57
HERMES model, 64, 67, 69, 102, 109
HERMIN models, 108, 109, 114, 116, 155, 161, 162, 170, 171, 178, 179, 191, 193, 194, 200
Heteroskedasticity, 372
Hickman-Coen models, 21, 25
Homogeneity, 212, 305, 306, 318, 339, 364
Hours, 17, 33, 96, 251, 287, 311, 326, 327, 330, 365, 368
  hours worked, 17, 33, 287, 311, 326, 327, 330, 368
Household incomes, 134, 292, 368, 377
Housing, 305
Human capital, 153, 297, 334–337, 396
  databases, 336
  measurement methods, 153, 297, 334–337, 396
    income method of human capital measurement, 153, 335–337
  per employee, 335, 336
  quality, 337
Human resources, *see* labour force
HWWA models, 105

I
Identities, 11, 15, 41, 125, 173, 235, 245, 247, 293, 330, 348, 353

Immigration, 53, 118
Impact multiplier, 365
IMPEC model, 151, 153, 159
Import functions, 127, 197
Imports, 17, 18, 33, 52, 63–65, 93, 94, 96, 97, 104–106, 125, 128, 130–132, 149, 152, 173–175, 177, 192, 195, 197, 206, 208–211, 220, 225–227, 230, 232–235, 241, 243, 247–252, 254, 262, 263, 265, 282, 289, 291–293, 296–299, 315, 317, 331, 334, 335, 347–349, 356–361, 363, 364, 366, 368, 375, 378, 388, 390, 391
  imports of investment goods, 233, 335, 361, 366
Income tax, 322, 355, 377, 378, 380
Independence, 3, 31, 85, 127, 171, 210, 230, 235, 332
Industry, 22, 23, 56, 57, 64, 66–68, 77, 80, 83, 88, 93, 94, 96, 125, 129, 131, 133, 136, 148, 195, 204, 209, 210, 225, 227–229, 231, 245, 264, 285, 287, 297, 326, 336, 352, 363, 368, 395
Inflation, *see* deflators, prices
Inflationary task, 196, 283, 337, 339, 341, 366, 383
INFOSTAT models, 125, 148, 156, 164, 165, 174, 175, 180
Innovations, 77
Innovative activity, 77, 132, 153, 396
Input-output (I-O)
  efficients, 135
  matrix, 140, 207, 241, 357
  model, 41, 49, 56, 65, 85, 89–91, 94, 104, 105, 114, 126, 132, 145, 159, 212, 223, 233, 234, 245, 275, 276
Input-output system, 136
INSEE models, 63–70, 149, 157, 192, 193, 196, 197, 199, 200
Institutional equations, 312
Integration order, 64
Interactive, 261
Interest rates, 11, 12, 16, 17, 20, 23, 24, 26, 28, 32, 51, 55, 60, 64, 76, 82, 86–88, 93–97, 104, 105, 108, 118, 153, 170, 172, 173, 177, 195, 196, 211, 212, 230, 232, 234, 246, 247, 249, 261–264, 266, 267, 269, 283, 286, 294, 298, 299, 304–306, 310, 322–324, 326, 338, 362, 368, 373, 379, 383–390, 392
INTERLINK model, 245, 246, 253, 254, 305
Intermediate goods, 226, 227, 277, 288, 331, 359
International and external balance, 391, 392
Interpolation, 332
Inventories, 133, 175, 211, 296, 316, 317, 347, 350–352, 363
Investment risk, 248
Investments, 5, 10–12, 15, 17, 18, 24, 33, 34, 54, 63, 65, 75, 94, 96, 106, 125–131, 134, 149, 153, 170, 195, 209, 211, 213, 226, 229, 231, 233, 247, 249, 250, 286, 296, 317, 320, 361, 377, 388–390, 392
  investment
    demand function, 213, 320
    in fixed capital, 10, 15, 17, 33, 96, 125–127, 129, 131, 170, 211, 229, 231, 249, 250, 286, 317, 320
    in human capital, 153
    projects, 12, 17, 324, 384, 387
ISI model, 126
Italy models, ix, 109, 114, 267

J

JADE model, 77

K

Kalecki's theory of growth, 9, 123, 124
KDI model, 217
KEIO-UNIV. Model, 216
KESSU models, 87, 88, 90
Keynesian models, 281, 338
Klein Model, 4, 9, 12, 15, 17–20
Klein-Goldberger Model, x, 4, 16–18, 34, 50, 337, 340
Klein-Shinkai model, 204
Knowledge
  absorption, 347
  capital, 153, 297, 333–335, 337
    foreign knowledge capital, 335
    knowledge capital embodied in fixed capital, 334
    transfer of foreign knowledge capital, 335
KOF model, 115, 117, 118
KOMPAS model, 76, 77, 79
KONSTANZ model, *see* MDM model
Kornai's theory of shortage models, 227, 290
KOSMOS model, 92, 95, 96
Koyck transformation, 77
KP models, 124, 129, 130, 142
KVARTS model, 91, 95
KYOTO Univers. model, 205, 207, 216

L
Labour
force, 12, 33, 102, 127, 130, 133, 153, 175, 211, 227, 231, 246, 268, 270, 276, 282, 286, 287, 292, 293, 295, 296, 303, 311, 312, 315, 335, 347, 348, 353, 368
intensity, 318
productivity function, 155, 319
Lag distributions, 20, 22, 24, 53, 54, 77, 321, 322
LAM model, 178
LATIN AMERICA models, 4, 203, 204, 235, 236, 244
Learning by doing, ix
Life expectancy, 309, 310, 337
Life-cycle hypothesis, 31, 308
LIMA model, 101
Linear, 11, 12, 48, 124, 127, 129–131, 136, 138, 142, 155, 158, 177, 212, 229, 234, 304, 378
Linear expenditure system (LES), 48, 56, 94, 104, 243, 304
LINK project, 22, 37, 41, 86, 97, 101, 104, 105, 108, 116, 117, 128, 133, 134, 143, 153, 158, 162, 163, 174, 180, 191, 203, 205, 207, 208, 210, 213, 230, 241–244, 252, 269, 288
Liquidity, 12, 17, 305, 308, 309
Literacy ratio, 79, 230, 324, 336
LITMOD model, 90, 93
LIVERPOOL models, 51, 53, 58, 270
Loans, 24, 25, 109, 175, 208, 212, 213, 229, 231, 233, 235, 247, 249, 254, 266, 283, 304, 310, 315, 326, 374, 376, 379, 380, 383–385, 388
London Business School (LBS) models, 50, 52–56, 262
Long run multiplier, 195, 305
Long-life, 55, 296
Long-linear, 264
Long-term supply accelerator, 368
Lucas critique, xi, 5, 53, 276, 294

M
MAC models, 193, 223, 233
MAC-SIM model, 192
Machinery and equipment, 23, 31, 76, 96, 104, 132, 227, 229, 325, 332, 349, 352, 361, 387
Macroeconomic modelling Bureau models, 59
Macroeconomics, 3, 270, 395
Keynesian macroeconomics, 41, 42, 52, 262, 281, 290, 303, 395
Marginal propensity, 12, 16, 17, 24, 309
to consume, 12, 16, 17, 24, 309
to investment, 12, 16, 17, 24
Market, 19, 30, 32, 33, 47, 51, 52, 54, 69, 82, 93, 94, 101, 104, 106, 107, 117, 118, 124–126, 128, 131, 133, 134, 141, 147–149, 152, 154, 156, 162, 170–173, 175, 178, 180, 181, 195, 208–210, 212, 213, 215, 225–228, 234, 270, 282, 287, 289, 290, 293, 296, 298, 307–309, 325, 326, 330, 331, 336, 337, 339, 342, 351–354, 357, 368, 377, 379, 383, 387
Marwah models, 41, 214
Marxian system, 9
Material balance, 126, 171, 286
Material Product System (MPS), 124, 286
Matrix, 48, 80, 207, 241–243, 246, 252, 265, 266, 269, 286, 315, 316, 357, 364
MCM model, 196, 198, 258, 261, 262
MDM, *see* Konstanz model
MEDIG model, 135, 143
MEMMOD model, 106, 262, 323, 328, 334, 339
METRIC model, 67–69
MIMOSA models, 257, 264, 265, 304, 305, 311, 329, 339, 342
MIRABEL model, 102
Misspecification, 17, 33
MODAG model, 94
Models, ix–xi, 3–6, 9–12, 15, 16, 18, 19, 21–23, 25–29, 31, 32, 34–37, 41, 42, 47–56, 59, 60, 64–67, 69–71, 75–79, 82, 83, 85–89, 92–97, 99, 101–111, 116–118, 123–139, 141–143, 147–156, 158, 159, 169–181, 191–198, 203–216, 219, 223, 225–231, 233–236, 241–256, 258, 259, 261–270, 275–278, 281, 282, 285–291, 293–298, 300, 301, 303–308, 310–312, 315–317, 319–324, 326–329, 331, 332, 334, 337–342, 347–353, 356–365, 367, 368, 371, 372, 374, 383, 389, 391, 392, 395, 396
block-rescursive model, 21, 125, 301
business cycle model, 6, 9, 10, 75, 105, 277, 282, 331, 351, 369
computational general equilibrium (CGE) models, 6, 10, 31, 35, 42, 48, 152, 154, 160, 170, 194, 211, 245, 253, 270, 277, 282, 288, 358

Models (*cont.*)
disequilibrium models, 5, 70, 124–126, 128, 134, 147, 289, 297, 300, 348, 350
dynamic stochastic general equilibrium (DSGE) models, 250, 282, 349
empirical model of growth, 3, 153, 357, 395
error correction model (ECM), 5, 6, 27, 29, 31, 35, 55, 56, 68, 77, 79, 81, 83, 87–89, 91, 93, 96, 103, 107, 108, 110, 113, 115, 117, 148–150, 152, 157–160, 163, 165, 166, 175–177, 180, 192–194, 197–200, 208, 211, 217, 218, 222, 232, 246, 247, 254, 258, 259, 261, 276, 277, 295, 301, 305, 306, 317, 323, 327, 328, 384
knowledge-based economy models, 151, 181, 276
long-run W8D model, 159, 333
macroeconometric model, x, 3–6, 9, 10, 12, 15, 16, 18, 21, 26, 27, 30, 31, 33, 35, 41, 42, 47–50, 54, 60, 69, 70, 75, 82, 95, 96, 101, 103, 104, 106–109, 116, 117, 123–130, 132, 134–137, 147–151, 154, 155, 170–181, 191, 192, 194–197, 203–207, 209–215, 225–236, 241, 242, 247, 250, 261, 267, 275, 276, 278, 281, 282, 285, 286, 288, 291, 293, 298, 300, 301, 303, 305, 307, 311, 316, 317, 321–324, 326–328, 331, 332, 337, 338, 340, 342, 348, 351–353, 356, 360, 363, 364, 371, 377, 380, 389, 391, 395
neo-Keynesian, 69, 109, 197, 207, 242, 298, 395
neoclassical, 6, 42, 47, 75, 177, 194, 206, 242, 245, 270, 276, 282, 295, 316, 322, 327, 377, 395
model specification, 3, 5, 16, 17, 24, 29, 30, 32, 33, 51–54, 57, 60, 64, 67, 69, 77, 78, 80, 93, 96, 109, 127, 129, 133–135, 149, 150, 155, 173, 175, 177, 179, 195, 196, 207–210, 213, 215, 217, 226, 241, 243, 246, 247, 249–251, 254, 261, 265, 268, 270, 277, 282, 283, 292, 295–299, 304, 306, 308, 310–312, 321, 322, 324, 329–332, 335, 338, 340–342, 351, 358, 362
models of the
EURO Area, 102, 191, 195–197, 200, 246, 248, 270, 304
European Union, 109, 155, 161, 170, 179, 191–195, 251, 256, 259, 262, 264
Hungary, 124, 125, 128, 140, 141, 146, 150, 157, 168, 169, 179, 194
Polish economy, 128, 130, 132, 151
world economy, 47, 241, 264, 276
multiplier model, 12, 15–18, 22, 24, 36, 38, 131, 152, 170, 175, 281, 293, 294, 305, 364, 365, 368, 380
real business cycle model, vii, 6, 9–11, 19, 24, 75, 104, 105, 228, 277, 282, 322, 331, 351, 354, 363
regional models, 212, 288
structural models, 5, 268, 395
structural vector equilibrium correction model (SVEqCM), 199, 200
vector
autoregression model (VAR), 5, 6, 27, 60, 103, 167, 198, 301
equilibrium correction model (VEqCM), 199, 200, 259, 278
MODIS model, 91, 94
MOGLI model, 64, 65, 70
MONA model, 87, 89
Money, x, 5, 12, 16, 25, 26, 30, 32, 33, 51, 57, 60, 69, 76, 93, 116, 152, 155, 171, 174, 176, 179, 206, 208–210, 212, 213, 215, 225, 226, 228, 229, 231–234, 246, 249, 263, 266, 267, 269, 283, 285, 287, 289, 291, 293, 299, 301, 304, 383, 385–387, 389, 392, 396
demand, 25, 26, 33, 171, 179, 208–210, 212, 213, 215, 229, 231–233, 246, 263, 266, 267, 283, 291, 293, 299, 385–387
illusion, 304, 385
policy, 25, 51, 69, 93, 152, 176, 209, 210, 212, 234, 263, 383, 386, 396
supply, 26, 155, 176, 206, 208–210, 226, 232, 234, 249, 266, 269, 386, 387, 389
MORKMON model, 82
MPS model, 21, 24, 25
MSM, RMSM World Bank models, 223, 235, 247, 254
Multicollinearity, 318, 319
Multicountry models, 4, 47, 178, 242, 264, 269, 276, 288, 310, 329, 342, 357, 364, 395
MULTIMACI model, 102

MULTIMOD model, 191, 196, 248, 249, 255, 294, 309, 310, 325
Multipliers, 11, 22, 37, 195, 243, 282, 305, 364, 380
MZE model, 196, 197, 199, 329, 339

N
NAIRU, 53, 107, 116, 196, 197, 200, 269, 282, 298, 329, 354, 355, 392
National Income and Product Accounts (NIPA), 28, 30
National Institute of Economic and Social Research (NIESR) models, 50, 52–54, 57, 150, 169, 241, 262, 263, 329, 332
Natural rate of increase, 311, 319, 327, 340, 354, 360
NAWRU, 265, 282, 355
NBS model, 165, 176
NECMOD model, 154
Neoclassical income, 276, 295, 327
Net business income, 371
Newton method, 21, 300
NIGEM model, 47, 150, 155, 157, 169, 178–180, 196, 242, 259, 262–264, 306
NIRA System, 207, 244, 252
NIRA-LINK model, 244, 252
Nominal GNP, 26
NONAME model, 103, 110
Nonlinear equation system, 234, 300
Nordic countries models, 47, 85, 89, 363

O
OECD type model, 196, 241, 245, 246, 253, 254, 258, 259, 261–264, 267, 270, 337, 356
OEF models, 52, 54, 57, 259, 267
OFCE models, 64, 67–69, 257, 264
Oil price, 11, 249
Openness of economy, 34, 229, 334
Optimal control, 176, 391
ORANI model, 211
Ordering of equations, 291
Ordinary least squares (OLS), 4, 6, 11, 15, 18, 19, 24–26, 34–39, 41, 55, 56, 66, 67, 78–80, 86, 88–94, 104, 106, 107, 110, 111, 113, 127, 131, 133, 138–146, 155–162, 164–168, 175, 177, 193, 207, 211, 214–224, 226, 228, 231, 233, 252, 253, 256–259, 264, 266, 300, 301
  dynamic ordinary least squares, 35, 39, 110
  ordinary least squares with changing intercept, 6, 18, 55, 88, 89, 91, 93, 158, 165, 166, 211
  truncated ordinary least squares, 36
Output, 4, 5, 9, 15, 17, 19–22, 24–26, 28, 29, 31, 33, 48, 52–56, 63–65, 68, 69, 75, 82, 83, 86–88, 92–94, 96, 97, 101, 104–106, 108, 109, 116, 117, 123, 125–127, 129–137, 145, 147, 149, 150, 152–155, 165, 166, 170, 172–175, 177–179, 192, 194–200, 205, 206, 208–211, 213–215, 217, 218, 225, 227–233, 235, 246, 248, 251, 262, 263, 265–268, 276, 277, 282, 286, 288, 290, 292, 293, 296, 297, 303, 307, 308, 315–319, 321, 322, 326–333, 336–338, 348, 349, 351, 352, 354, 358–362, 364, 365, 368, 374–376, 380, 384, 388
  see production
  effective output, 54, 118, 209, 297, 331, 338, 352
  gross output, 21, 93, 94, 123, 125–127, 130, 132, 136, 137, 145, 165, 213, 229, 235, 286, 315, 316, 360, 362, 374–376
  observed output, 123, 129–133, 315, 331
  potential output, 19, 22, 25, 26, 28, 29, 52, 83, 86, 94, 97, 101, 104–106, 116, 131, 152, 153, 173, 174, 177, 179, 192, 197, 200, 209, 210, 229, 251, 262, 266–268, 282, 297, 317, 330, 331, 338, 348, 349, 352
Overseas assets, 215, 335, 388

P
PANI model, 219
Parameter changes, 59, 147
Patents, 335, 388
Pawlowski's model, 124, 128
Permanent personal income, 29, 55, 117, 276, 308
Personal income, 127, 305, 378, 380
Philadelphia model, 21, 235, 242
Phillips curve, 24, 41, 93, 94, 109, 116, 153, 246, 249, 265, 340, 341
Planning models, 130, 135, 137, 214, 233
  central planning, 75, 79, 80, 126, 151, 207
  developing countries, 203, 215, 226, 227, 229, 230, 236, 241, 243, 244, 247, 254, 255, 266, 276, 290, 297, 326, 368
  indicative planning, 63

Policy, xi, 3–5, 11, 12, 19, 21–27, 29, 30, 33–37, 39–42, 49–51, 54–59, 64, 65, 67–69, 77, 78, 83, 85–87, 89, 90, 92–97, 101–104, 106–118, 125–130, 132–135, 137, 139, 141, 142, 146–155, 157–159, 161, 164–166, 170–174, 176–181, 191–193, 195–200, 205, 206, 209–212, 215–218, 220, 222–226, 228–236, 242, 244, 245, 247–250, 252–256, 258–267, 269, 270, 275–277, 281, 293, 322, 350, 354, 368, 377, 379, 383, 386, 388, 395, 396
  economic, xi, 3–5, 12, 22, 26, 27, 29, 35, 40, 42, 49–51, 54, 59, 64, 68, 69, 77, 85, 86, 95, 102, 107, 108, 113, 116, 125, 127, 128, 135, 137, 149–153, 155, 172–174, 180, 181, 193, 196, 200, 206, 209, 212, 224, 226, 228, 234, 235, 242, 244, 249, 256, 259, 261, 262, 265, 267, 269, 270, 275, 293, 350, 354, 379, 388, 395
  fiscal, 11, 19, 22, 24, 27, 42, 50, 59, 93, 103, 109, 118, 148, 149, 152, 166, 176, 179, 180, 195, 200, 206, 210, 250, 261, 277, 322, 377
  monetary, xi, 11, 19, 24–27, 29, 35, 42, 50, 51, 55, 58, 59, 93, 95, 97, 102–104, 106, 108, 109, 113, 114, 116–118, 148–150, 152, 157, 165, 166, 170, 176, 179, 180, 193, 195–198, 200, 205, 206, 210–212, 217, 224, 229, 232, 234, 247, 248, 250, 261, 277, 383, 386, 396
  scenario analyses, 154
Population, 16, 33, 103, 210, 212, 213, 222, 231, 233, 264, 291–293, 309, 311, 336
Post recursive block, 21, 26, 300
Potential employment, 83, 268
Potential output, 19, 22, 25, 26, 28, 29, 52, 83, 86, 94, 97, 101, 104–106, 116, 131, 152, 153, 173, 174, 177, 179, 192, 197, 200, 209, 210, 229, 251, 262, 266–268, 282, 297, 317, 330, 331, 338, 348, 349, 352
PPP exchange rate, 246, 298, 388, 390
PRIAMO model, 259, 267
Prices, *see* deflators
  producer prices, 21, 24, 28, 33, 54, 71, 109, 176, 212, 251, 263, 276, 277, 282, 287, 292, 293, 296, 299, 316, 319, 337–339, 348, 355, 360, 362, 366, 375, 392
  retail prices, 71, 87, 96, 175, 338, 341, 366
  world prices, 63, 65, 88, 94, 96, 97, 109, 212, 233, 268, 269, 292, 338, 355, 363, 364, 392
Principal components, 138, 211
Production
  capacity, 12, 17, 63, 76, 105, 106, 365, 366
  elasticity with respect to, 297, 327, 333
    employment, 327
    fixed capital, 297, 333
  factor productivity, 153, 297
  factors, 5, 12, 17, 24, 29, 41, 54, 71, 82, 86, 93, 94, 116, 118, 125, 133, 150, 171, 177, 179, 194, 196, 200, 227, 229, 242, 246, 248, 251, 262, 263, 276, 277, 282, 286, 288–290, 296, 299, 316–319, 322, 330, 331, 342, 347, 348, 352, 362, 371, 376, 380, 388
  function, 15, 17, 20, 22–25, 29, 41, 42, 52, 55, 76, 78, 80, 82, 83, 93, 94, 103, 105–107, 110, 117, 126, 133, 148–150, 153, 155, 167, 170, 173–177, 179, 194–197, 200, 209–211, 214, 215, 217, 225, 231, 232, 246, 249, 251, 256, 276, 277, 292, 297, 317–319, 326, 327, 330, 331, 333, 334, 348, 349, 352, 375, 376
    Cobb-Douglas production function, 22, 23, 25, 29, 42, 52, 55, 94, 103, 117, 153, 170, 173, 174, 179, 195–197, 200, 209, 214, 217, 231, 232, 249, 251, 297, 318, 327, 331, 333, 334, 348
    constant elasticity of substitution (CES) production function, 71, 83, 93, 103, 107, 150, 171, 194, 211, 250, 251, 263, 318, 331, 348, 349
    constant returns to scale production function, 25, 71, 107, 150, 173, 197, 211, 232, 263, 297, 333
    neoclassical production function, 251, 276
    production function of effective output, 209, 297, 331, 352
    production function of potential output, 25, 28, 29, 52, 83, 94, 105, 106, 126, 127, 129–131, 133, 137, 149, 153, 173, 174, 179, 196, 197, 200, 209, 210, 214, 227, 229, 231, 249, 251, 262, 267, 282, 297, 317, 331, 333, 349, 351, 352

short-run production function, 24, 29, 31, 55, 116, 117, 209, 232, 251, 354
possibility frontier, 93, 211, 325, 327
process, 135, 173, 249, 250, 263, 316–319, 332
Productivity of labour function, 318
Profit, 31, 33, 71, 88, 93, 96, 107, 263, 265, 282, 295, 319, 327, 329, 338, 390
Project Link, ix, 22, 41, 86, 97, 101, 104, 105, 108, 116, 117, 128, 133, 134, 153, 174, 180, 191, 203, 205, 207, 208, 210, 213, 230, 241–244, 252, 269, 288
PROMETEIA models, 109, 114, 259, 267
Propensity
to consume, 12, 16, 17, 24, 309
to invest, 12
PROTEE model, 67, 70
Purchasing power parity (PPP), 246, 298, 388–392

Q

QEM-ECM model, 165, 175
QFM models, 38, 41
QPM models, 35, 40
Quantity adjustments, 41, 128, 131, 282, 289
Quarter, x, 306, 317, 323
Quasi-demand determined models, 228, 300, 379
Quasi-supply determined models, 379
QUEST model, 191, 196, 250, 251, 256, 261, 309, 310, 325, 329

R

Random errors, 6, 27, 32, 268, 334, 354
Rank of lags, 54, 129, 293, 339, 357
Rate, 10, 17–20, 23, 25–31, 33, 37, 51–53, 59, 71, 76, 82, 86–88, 92–97, 105–107, 109, 116, 126, 148–150, 153, 173–176, 178, 179, 192, 194–197, 200, 208–214, 226, 231–233, 235, 242, 243, 246, 248, 249, 251, 261–269, 282, 283, 292, 293, 296, 298, 299, 303–307, 309–311, 317, 320, 322–325, 327, 328, 330–333, 335, 338–342, 347, 351–355, 357–360, 362–364, 366, 368, 373–376, 378, 383–392
discount rate, 309
growth rate, 18, 96, 200, 265, 317, 333, 373
interest rate, 10, 20, 33, 51, 59, 87, 96, 109, 197, 200, 208–212, 226, 231, 232, 242, 246, 249, 251, 265–268, 292, 293, 296, 299, 303–305, 309–311, 322, 323, 376, 383–387, 392
nominal interest rate, 96, 200, 387
real interest rate, 59, 87, 96, 197, 208–210, 226, 303, 304, 309, 392
investment rate, 311
unemployment rate, 20, 29, 33, 52, 53, 71, 94, 97, 107, 116, 148, 194, 196, 197, 200, 263, 265, 266, 268, 292, 293, 296, 298, 299, 307, 311, 338–340, 342, 347, 354, 355
Rates of utilisation, 97, 352
rates of
capacity utilisation, 97
utilisation of fixed capital, 352
utilisation of labour, 97
utilisation of production factors, 352
Rational expectations, 5, 27, 31, 32, 47, 51, 53, 54, 88, 90, 93, 102, 103, 106, 110, 197, 198, 242, 248, 258, 259, 261–263, 269, 270, 276, 293, 294, 300, 306, 341, 396
RDX model, 34, 35, 37
Recession, 23, 152, 244
Reduced form, 242, 301, 375, 376, 386
Regression, 30, 136, 155, 178, 301
cointegration regression, 301
spurious regression, 301
Relationship, 30, 64, 96, 211, 295, 360, 373, 376, 387, 388
long-run relationship, 96
short-run relationship, 211
Representative agent, 282, 285, 300
Research and development (R&D)
cumulated outlays, 28, 153, 265
current outlays, 334, 335
domestic outlays, 297, 334, 335
expenditures, 23, 29, 297, 334, 335
outlays
on education, 79, 230, 324, 336
on innovation, 132, 153, 396
Reserve Bank of New Zealand models, 211
Residuals, 232
Restrictions, 5, 6, 27, 32, 56, 59, 97, 102, 126, 132, 133, 147, 155, 161, 170, 173, 208, 227, 228, 233, 290, 300, 301, 349, 368
Returns to scale, 25, 41, 71, 107, 150, 173, 197, 211, 232, 245, 263, 297, 318, 319, 331, 333
Revaluation, 172, 195, 214
RIMINI model, 91, 95

Risk
investment, 248, 299, 322, 326, 384, 388
premium, 298, 299, 322, 326, 384, 388
RWI models, 105, 106

S

SAFE model, 80, 82
SAPO model, 135, 143
Saving, 374
Scenario analyses, 25, 79, 153, 154, 208
long-run scenario, 79, 153, 208
short-run scenario, 79, 153, 208
Seasonal, 229
SECMON model, 80, 83
Sector, 5, 11, 17, 21, 23, 24, 26, 28, 29, 31–34, 36, 38–40, 42, 47, 49, 51, 52, 54, 56, 57, 63, 64, 66–71, 76, 79, 81–83, 86–88, 93, 94, 96, 102–106, 108–118, 123, 127, 129–131, 133, 137, 143, 149, 150, 154–156, 158, 159, 161, 165–168, 170–174, 176–178, 192, 193, 195–197, 204–206, 208–210, 212–217, 219, 222, 224–226, 228, 229, 231–234, 236, 246–251, 259–261, 265, 270, 276, 277, 282, 286–288, 298, 299, 305, 309, 315, 316, 331, 347, 348, 352, 353, 366, 371, 374, 376, 379, 380, 396
open, 42, 277
protected, 326, 340, 351
Short-run multiplier, 24, 175
Short-run production function, 29, 55, 196, 232, 251
Short-term forecasts, 18, 41, 42, 70, 78, 102, 109, 116, 141, 149, 151, 152, 154, 164, 178, 205, 212, 216, 262
Shortage models, 290
SIGAPE model, 223, 234
SIGMA model, 258, 262
Simulation, ix, 21, 22, 25–28, 32, 37, 39, 41, 49, 55–57, 65–68, 71, 77–81, 83, 86, 87, 89–93, 97, 103, 104, 106, 107, 110–118, 127–129, 133, 137–146, 148–151, 154, 156–160, 162–171, 178–181, 193, 198–200, 206, 210, 214, 216–224, 226, 231, 241–243, 245, 247, 252, 253, 256–259, 264, 267, 395
Simultaneous block, 20, 21, 171
Skeleton macromodel, 291
SLOPOL models, 166
Slutsky process, 48
SNEP model, 92, 96, 97
Social capital, 396
SOE model, 97
SOE-PL model, 154, 160
Solow residual, 333
SOVMOD model, 136, 145
Specification error, 5
Spurious correlation, 301
SR model, 164, 174
St. Louis F.R. Model, 21, 26, 32, 69
Stability, 3, 131, 180, 242, 252, 357, 364, 384
State
budget expenditures
current state budget, 251, 379
multiplier, 24, 380
debt, 35, 118, 209, 248, 309, 377, 379
Static simulation, 66, 160, 170
STEP model, 97
Stochastic, 6, 11, 16, 18–20, 22–26, 28, 30–32, 34–42, 55–58, 66–68, 71, 78–81, 86, 87, 89–92, 97, 101, 104, 107, 110–115, 118, 138–146, 148, 149, 152, 154–169, 173, 176, 192–199, 204, 207–226, 228, 231–236, 245, 247, 250, 252–263, 268, 269, 277, 282, 288, 289, 349, 392, 395
error, 5, 27, 295, 301, 306, 307, 324, 328
simulation, 26, 37, 41, 68, 71, 78, 81, 86, 91, 107, 110–115, 139, 141, 144–146, 148, 156–158, 163, 166, 168, 169, 199, 216, 218–221, 224, 231, 245
structure error, 5, 27, 32, 295, 301, 306, 307, 324, 328
trend, 32, 268
Stock
building, 10, 15, 17, 20, 64, 75
market, 54, 59, 69, 86, 87, 107, 171, 206, 263, 352, 383
of commodities, 10
Structural equations, 296, 300, 301
Structure, 4, 5, 16, 19, 22, 24–35, 41, 42, 47, 50, 54, 65, 67, 69, 70, 82, 83, 86–89, 92, 94, 97, 102, 105, 111, 112, 116, 117, 124, 126, 127, 132, 133, 140, 141, 146, 148, 150, 170, 175, 178, 181, 193, 195, 197, 200, 205, 206, 210, 215, 221, 222, 227, 230–232, 234, 235, 245, 247–250, 252, 254, 258, 262–264, 267–270, 276, 277, 281, 291, 293, 298, 300, 356, 363, 395
long-run structure, 24, 30, 31, 35, 67, 86, 150, 178, 195, 197, 200, 258, 262, 264

short-run structure, 24, 29–31, 41, 67, 69, 70, 89, 103, 116, 150, 200, 232, 248, 258, 262, 263, 300
Submodels, 4, 6, 22, 27, 31, 41, 48, 49, 56, 59, 82, 85, 104, 105, 127, 132, 140, 143, 147, 178, 180, 245, 275, 278, 288, 303, 315, 392
science sector submodel, 91, 337
Substitution, 17, 33, 48, 243, 246, 265, 308–310, 318, 322, 325, 331, 349
Supply, x, xi, 5, 10, 12, 20, 22, 25, 26, 28, 29, 31–33, 39, 42, 47, 51–54, 56, 65, 67–71, 76, 86, 97, 104, 106–109, 116–118, 124–148, 150, 152–155, 159–163, 165–167, 170–178, 181, 192–194, 196–199, 206, 208–215, 218, 219, 222, 223, 225–227, 229, 231–235, 242, 243, 245, 246, 248–251, 254, 256, 257, 259, 261, 262, 264–266, 268–270, 276, 282, 287–290, 292, 293, 295–300, 303, 307, 311, 312, 320, 330, 342, 347–351, 353, 354, 356–359, 361, 367, 368, 380, 384, 386–389
consumption multiplier, 170, 175
determined systems, 20, 31, 127, 129, 131, 132, 135, 136, 147, 148, 155, 170, 175, 181, 197, 206, 212, 225, 227, 231–233, 235, 243, 264, 266, 289, 297, 330, 342, 350, 356, 358, 367, 368, 380
functions, 32, 52, 128, 131, 141, 181, 295–297, 303, 311, 354, 358
gap, x, 173, 307, 308
multiplier, 170, 175, 368
production factors, 5, 12, 29, 54, 71, 86, 118, 125, 133, 150, 171, 177, 194, 196, 227, 229, 242, 246, 248, 262, 276, 277, 282, 288–290, 296, 299, 330, 331, 342, 347, 348, 380
shocks, 4, 298, 351, 357, 359
side, xi, 25, 31, 52, 53, 71, 86, 108, 118, 130, 171, 176
type accelerator, 129, 131, 132, 136, 141, 170, 175, 226, 368
Sustainability, 36–40, 305
Symmetric, 48
SYSIFO model, 105, 112
SZAMKI models, 128, 141

T

Targeting, 247, 283, 363
Taxes, 24, 208, 212, 231, 232, 249, 283, 355, 362, 363, 371, 373–378, 380
on corporate profits, 377, 380
on personal incomes, 372, 378
Technological gap, 365
Technological progress, 28, 29, 133, 276, 297, 333, 334, 354, 396
capital augmenting, 368, 375
disembodied technical progress, 334
embodied technical progress, 334
endogenous technological progress, 29, 297
exogenous technological progress, 29, 297, 333
Hicks type neutral technical progress, 12
labour augmenting, 41
neutral technical progress, 51, 377
technological progress effects, 297, 333, 334, 354
technological proximity, 335
Technology, 55, 71, 168, 178, 250, 286, 317, 325, 327, 331, 365
diffusion/transfer, 28, 335, 389
Temporary, 107, 307
Term structure, 31
Tests, 211, 232, 281
Theory, x, 3, 5, 6, 9–12, 24, 29, 31, 35, 47, 48, 55, 59, 63, 66, 77, 83, 102, 107, 123, 135, 200, 245, 248, 277, 278, 282, 295, 297, 298, 300, 308, 309, 325, 326, 334, 389, 396
endogenous growth theory, 29, 297
production theory, 77
Time series, 3, 18, 22, 28, 63, 148, 196, 232, 270, 275, 278, 295, 301, 331, 332, 334, 335
TINBERGEN model, x, 4, 10, 11, 75, 78, 214
Tobins Q-ratio, 86, 249, 296
Top-down approach, 102, 153, 293
Total factor productivity (TFP), 28, 29, 148, 153, 159, 170, 230, 297, 318, 319, 327, 329, 332–336, 396
Divisia formula TFP index, 331, 352, 391, 414
effective total factor productivity, 297
potential total factor productivity, 28, 297
total factor productivity decomposition, 334
TRACE model, 37, 41
Tracking simulation, 71, 77, 83, 86, 87, 93, 97, 103, 104, 106, 107, 116, 117, 128, 129, 137, 148, 149, 159, 162, 164–166, 168, 180, 193, 199, 206, 216–218, 220–222, 224, 226, 231, 243, 245, 247, 252, 253, 257–259, 395

Trade, 19, 23, 28, 34, 36, 53, 55, 63, 64, 71, 85, 87, 93, 96, 97, 101, 104, 107, 117, 118, 125–127, 130–132, 135, 137–139, 146, 149, 155–157, 159, 164, 165, 174, 175, 177, 179, 180, 192, 194, 195, 206–208, 210, 211, 213–215, 223, 225–227, 229, 231, 233–235, 241–247, 249, 250, 253, 254, 257, 258, 260, 263, 264, 266, 267, 269, 277, 282, 286–288, 292, 299, 347, 351, 352, 356, 357, 363, 364, 368, 388–392
Trade unions, 53, 71, 96
Trade-off, 23
Trend path, 11, 32, 33, 180, 265, 268, 297, 329, 332, 334, 338, 339, 341, 352, 354, 357, 358, 360, 384
TSUKUBA-FAIS model, 269
Turning point, 130, 132, 147, 148, 172, 244, 368

U
UIP exchange rate, 298, 389–391
UK models, 48–51, 55, 60, 262, 301, 305, 306, 329, 330, 332, 358, 392
UKR model, 136, 144
Uncertainty, 336
UNCTAD models, 203, 215, 219, 221, 222, 226, 230, 231, 236, 243
Unemployment, 12, 17, 20, 23, 24, 26, 29, 32, 33, 52, 53, 59, 70, 71, 76, 82, 86, 92–97, 102, 104, 107, 110, 116, 118, 148, 150, 152, 153, 166, 174–177, 194–197, 200, 210, 211, 233, 251, 263–268, 282, 289, 290, 292, 293, 296–299, 307, 311, 312, 330, 338–340, 342, 347, 348, 353–355
  level, 12, 53, 174, 251, 293, 353
  NAIRU, 53, 107, 116, 196, 197, 200, 282, 298, 354, 355
  NAWRU, 265, 282, 355
  rate, 17, 20, 23, 26, 29, 33, 52, 53, 59, 71, 76, 86, 93–97, 107, 116, 148, 150, 153, 175, 176, 194–197, 200, 210, 233, 263–266, 268, 292, 293, 296, 298, 299, 307, 311, 338–340, 342, 347, 353–355
Uniqueness, 49, 334, 391
US models, 4, 10, 11, 15, 16, 18, 21, 22, 25–28, 30–32, 42, 261, 262, 266, 268, 294, 304, 306, 309, 334, 338, 339, 352
User costs, 20, 24, 29, 31, 33, 59, 82, 88, 93, 153, 195, 197, 209, 248, 267, 286, 296, 299, 322, 323, 374–377

V
Value added, 21, 93, 96, 123, 130, 152, 153, 170, 197, 209, 211, 226, 229, 232–235, 246, 315, 331, 351, 372, 374–376
  tax, 251, 362, 377, 378
VAR models, 6, 103
Variables, 5, 9, 12, 16–20, 25, 28, 29, 32, 33, 35, 48, 51, 52, 55, 59, 60, 63, 70, 86, 93, 95, 96, 102, 104, 117, 127, 129–131, 138, 148, 155, 173, 176, 177, 200, 206–211, 225, 226, 232, 233, 245, 248, 250, 251, 263, 268, 291, 292, 295, 296, 298, 300, 301, 303–308, 310–312, 315, 318–320, 322, 324–329, 331–335, 337–339, 347, 349, 350, 352, 354, 355, 357–362, 364, 372, 373, 376, 379, 383, 384, 386, 388–392, 396
  cointegrated, 60
  cointegrated variables, 60
  deterministic variables, 268
  dummy variables, 232
  economic policy, 12, 59
  endogenous, 20, 29, 51, 60, 131, 155, 268, 291, 300
  integrated variables, 42
  lagged variables, 20, 59, 155, 268, 300, 311, 329, 379
  nonstationary variables, 5, 200, 295, 301, 306, 392
  stationary, 32, 268, 306
Vector, 32, 301
VINTAF model, 76
Vintage capital, 52, 55, 76–78, 80, 83, 118
VUSEiAR models, 125, 138, 139
VVS models, 124, 125, 127, 129, 138

W
W-1 models, 130–132, 147, 151
W-3 models, 131, 132
W-5 models, 133–135, 147, 151, 159, 180, 351
W-8 models, 151–154, 180
W-8D models, 153, 181
Wages, 4, 5, 11, 15, 16, 20, 22–24, 29, 31–33, 39, 41, 52–54, 57, 59, 64, 71, 86, 87, 89, 92–97, 103, 107, 108, 125–127, 129, 130, 132–134, 149, 150, 152, 153, 170, 172–176, 179, 180, 192, 195, 197, 198, 209–212,

215, 227, 229, 231, 233, 247, 250, 251, 262, 263, 266–268, 276, 277, 281, 282, 286, 287, 292–294, 296, 298, 299, 304, 311, 312, 316, 325, 328, 329, 336–338, 340–342, 353–356, 365, 366, 372, 373, 375, 376, 379, 380, 392
average wages, 95, 125, 127, 149, 153, 174–176, 180, 251, 266, 267, 282, 292, 293, 299, 316, 325, 336–338, 340–342, 353–356, 366, 372, 373, 375, 376, 379, 392
nominal wages, 53, 210, 233, 268, 341, 366
real wages, 16, 33, 53, 129, 150, 173, 176, 197, 209, 211, 212, 251, 263, 268, 293, 296, 311, 312, 328, 329, 340–342
wage
elasticities, 53
equations, 71, 93, 96, 116–118, 233, 246, 265, 296, 298, 341, 342
Wealth, 16, 24, 31–33, 35, 55, 57, 59, 64, 86, 88, 104, 109, 117, 150, 153, 176, 195, 196, 209–211, 226, 232, 245, 248, 249, 251, 263, 264, 267, 296, 299, 305, 307–310, 312, 371, 374, 385, 396
Wharton Annual and Industry model, 21, 22, 26–28, 40, 50, 136, 203, 235, 269, 311, 324, 326, 331, 352
Wharton quarterly model, 311
WK models, 151, 152, 158
Work experience, 337
World, economy, ix, x, 32, 47, 105, 128, 134, 155, 178, 179, 205, 208, 210, 213, 241–244, 250, 261–269, 276, 288, 356, 392
World, trade, 63, 96, 174, 179, 180, 207, 208, 210, 211, 213, 241, 242, 244, 245, 247, 249, 264, 266, 269, 287, 292, 356, 357, 363, 364
World Bank models, 6, 160, 235, 241, 247, 288, 306
World Scan model, 269